AQA

Biology for GCSE Combined Science: Trilogy
Third edition

Ann Fullick
Andrea Coates
Editor: Lawrie Ryan

Message from AQA

This textbook has been approved by AQA for use with our qualification. This means that we have checked that it broadly covers the specification and we are satisfied with the overall quality. Full details of our approval process can be found on our website.

We approve textbooks because we know how important it is for teachers and students to have the right resources to support their teaching and learning. However, the publisher is ultimately responsible for the editorial control and quality of this book.

Please note that when teaching the *AQA GCSE Biology* or *AQA GCSE Combined Science: Trilogy* course, you must refer to AQA's specification as your definitive source of information. While this book has been written to match the specification, it cannot provide complete coverage of every aspect of the course.

A wide range of other useful resources can be found on the relevant subject pages of our website: www.aqa.org.uk.

OXFORD
UNIVERSITY PRESS

OXFORD
UNIVERSITY PRESS

Great Clarendon Street, Oxford, OX2 6DP, United Kingdom

Oxford University Press is a department of the University of Oxford. It furthers the University's objective of excellence in research, scholarship, and education by publishing worldwide. Oxford is a registered trade mark of Oxford University Press in the UK and in certain other countries

British Library Cataloguing in Publication Data

Data available

978 0 19 835926 5

10 9 8 7 6 5 4 3 2 1

Paper used in the production of this book is a natural, recyclable product made from wood grown in sustainable forests.
The manufacturing process conforms to the environmental regulations of the country of origin.

Printed in Great Britain by Bell and Bain Ltd., Glasgow

Acknowledgements

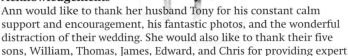

Ann would like to thank her husband Tony for his constant calm support and encouragement, his fantastic photos, and the wonderful distraction of their wedding. She would also like to thank their five sons, William, Thomas, James, Edward, and Chris for providing expert advice and a lot of fun throughout the project.

AQA examination questions are reproduced by permission of AQA.

Index compiled by INDEXING SPECIALISTS (UK) Ltd., Indexing house, 306A Portland Road, Hove, East Sussex, BN3 5LP, United Kingdom.

COVER: ETHAN DANIELS/SCIENCE PHOTO LIBRARY

p5(T): Gerd Guenther/Science Photo Library; p5(B): Adrian T Sumner/ Science Photo Library; p6: Jack Bostrack/Visuals Unlimited/Science Photo Library; p7(T): Steve Gschmeissner/Science Photo Library; p7(B): John Durham/Science Photo Library; p15: Dr R. Dourmashkin/Science Photo Library; p18(L): Michael Abbey/Science Photo Library; p18(R): Michael Abbey/Science Photo Library; p21: Anthony Short; p26: CNRI/Science Photo Library; p28(T): Deva Studio/Shutterstock; p28(B): William Perugini/ Shutterstock; p29: Biodisc/Visuals Unlimited/Science Photo Library; p30: Cordelia Molloy/Science Photo Library; p31: Anthony Short; p32: Julian Finney/Getty Images; p36(T): Eric Grave/Science Photo Library; p36(B): Steve Gschmeissner/Science Photo Library; p39: ISM/Science Photo Library; p40(R): Evgenia Sh./Shutterstock; p40(L): JPC-PROD/Shutterstock; p41: Africa Studio/ Shutterstock; p42: J.C. Revy, ISM/Science Photo Library; p44: Filip Fuxa/ Shutterstock; p48: Martyn F. Chillmaid/Science Photo Library; p49: Roblan/ Shutterstock; p53(R): St Bartholomew's Hospital/Science Photo Library; p53(L): National Cancer Institute/Science Photo Library; p57: Science Photo Library; p59: Raphael Gaillarde/Getty Images; p62: Kansak Buranapreecha/ Shutterstock; p64: Schankz/Shutterstock; p65(L): Anthony Short; p65(R): Steve Gschmeissner/Science Photo Library; p66: Welcomia/Shutterstock; p68: Anthony Short; p76: Nature's Geometry/Science Photo Library; p77: CDC/ Science Photo Library; p78: Mark Thomas/Science Photo Library; p79: CDC/ Science Photo Library; p80: Lowell Georgia/Science Photo Library; p81: Norm Thomas/Science Photo Library; p82: Sergiy Zavgorodny/Shutterstock; p83: Nigel Cattlin/Science Photo Library; p84(T): Anthony Short; p84(B): Anthony Short; p85(L): Smuay/Shutterstock; p85(R): Andy Crump, TDR, WHO/Science Photo Library; p86(T): Prof. P. Motta/Dept. of Anatomy/University "La Sapienza", Rome/Science Photo Library; p86(B): Steve Gschmeissner/Science Photo Library; p89: Anthony Short; p90: Valeriya Anufriyeva/Shutterstock; p92: Paul Whitehill/Science Photo Library; p93: CC Studio/Science Photo Library; p94: Jody Ann/Shutterstock; p95(L): Anthony Short; p95(R): St Mary's Hospital Medical School/Science Photo Library; p96: Sigrid Gombert/ Cultura/Science Photo Library; p103: Anthony Short; p112: Mllevphoto/ iStockphoto; p113(L): Jamie Farrant/iStockphoto; p113(R): Dr Keith Wheeler/Science Photo Library; p116: Lebendkulturen.de/Shutterstock; p117(T): Cordelia Molloy/Science Photo Library; p117(B): Cathy Keifer/ Shutterstock; p118: Lukasz Szwaj/Shutterstock; p119: Ngataringa/ iStockphoto; p123: Nancy Bauer/Shutterstock; p124: Microscape/Science Photo Library; p126: Liquorice Legs/Shutterstock; p128(L): SJ Travel Photo and Video/Shutterstock; p128(R): Panu Ruangjan/Shutterstock; p134(T): J. Helgason/Shutterstock; p134(B): Marek Velechovsky/Shutterstock; p135: Manzrussali/Shutterstock; p136(T): Daxiao Productions/Shutterstock; p136(B): Anthony Short; p138: Tony Wear/Shutterstock; p142: Rawpixel/ Shutterstock; p143: Sarah2/Shutterstock; p146: Image Point Fr/ Shutterstock; p147: Fotopool/Shutterstock; p152(T): Jose Luis Calvo/ Shutterstock; p152(B): Prof. P. Motta/Dept. of Anatomy/University "La Sapienza", Rome/Science Photo Library; p154: JPC-PROD/Shutterstock; p162: Crown Copyright/Health & Safety Laboratory Science Photo Library; p163: David Scharf/Science Photo Library; p165: Eye of Science/Science Photo Library; p167: Lawrence Berkeley National Laboratory/Science Photo Library; p168: Floris Slooff/Shutterstock; p172: Sovereign, ISM / Science Photo Library; p175: Public Health England/Science Photo Library; p177: Anthony Short; p178: Erik Lam/Shutterstock; p179: Diplomedia/ Shutterstock; p180(T): Brian A Jackson/Shutterstock; p180(BL): Anthony Short; p180(BR): 135pixels/Shutterstock; p182: Egyptian/Getty Images; p183(TL): Andyworks/iStockphoto; p183(TR): Anthony Short; p183(B): Henk Bentlage/Shutterstock; p185: Panda3800/Shutterstock; p186(L): Madlen/Shutterstock; p186(R): Zcw/Shutterstock; p190: Ria Novosti/Science Photo Library; p191: John Reader/Science Photo Library; p193: Mr. Suttipon Yakham/Shutterstock; p194: Vereshchagin Dmitry/Shutterstock; p195: Martin Bond/Science Photo Library; p198(T): Monkey Business Images/ Shutterstock; p198(CT): Nicholas Lee/Shutterstock; p198(CB): Anthony Short; p198(B): Toeytoey/Shutterstock; p199(L): Bildagentur Zoonar GmbH/ Shutterstock; p199(R): Rasmus Holmboe Dahl/Shutterstock; p200(T): Eye Of Science/Science Photo Library; p200(C): Steve Gschmeissner/Science Photo Library; p200(B): Jeep2499/Shutterstock; p201(T): Ehtesham/ Shutterstock; p201(B): Leungchopan/Shutterstock; p202(T): Vereshchagin Dmitry/Shutterstock; p202(B): Ria Novosti/Science Photo Library; p203: Dr. Morley Read/Shutterstock; p206(R): Ikordela/Shutterstock; p206(L): Jamesdavidphoto/Shutterstock; p207(L): Fotos593/Shutterstock; p207(R): Piotr Krzeslak/Shutterstock; p208(T): Dennis W. Donohue/Shutterstock; p208(B): Anthony Short; p209(T): Lobster20/Shutterstock; p209(B): Menno Schaefer/Shutterstock; p210: Martyn F. Chillmaid/Science Photo Library; p211: Anthony Short; p212(T): Leungchopan/Shutterstock; p212(B): Dirk Ercken/Shutterstock; p213(T): Anthony Short; p213(B): Hintau Aliaksei/ Shutterstock; p214: Anthony Short; p215(T): Solarseven/Shutterstock; p215(B): Smirnova Irina/Shutterstock; p216: Anthony Short; p217(T): Dante Fenolio/Science Photo Library; p217(B): B. Murton/Southampton Oceanography Centre/ Science Photo Library; p218(B): Louise Murray/ Science Photo Library; p218(T): Michael Pettigrew/Shutterstock; p219: Aleksandr Hunta/Shutterstock; p220: Francesco de marco/Shutterstock; p221(T): Anthony Short; p221(B): Alexander Mazurkevich/Shutterstock; p223: Bluestock/Dreamstime; p224(T): Corbis; p224(CT): Martin Fowler/ Shutterstock; p224(CB): Chris2766/Shutterstock; p224(B): Peter Louwers/ Shutterstock; p226(T): Anthony Short; p226(C): Anthony Short; p226(B): Anthony Short; p228: Anthony Short; p229: Gadag/Shutterstock; p231: Anthony Short; p232: NASA; p235: Michael Marten/Science Photo Library; p238(T): CNES, 2002 Distribution Spot Image/Science Photo Library; p238(B): Pichugin Dmitry/Shutterstock; p239: Dennis W. Donohue/ Shutterstock; p241: Mrlee1989/Shutterstock; p242(T): Dennis Jacobsen/ Shutterstock; p242(B): Aleksander Bolbot/Shutterstock; p251: Erni/ Shutterstock; p254(T): Roberto Piras/Shutterstock; p254(B): Wacomka/ Shutterstock; p255: Becris/Shutterstock; p257: Nomad_Soul/Shutterstock; p260: SpeedKingz/Shutterstock; p262: Richard Bowden/Shutterstock; p263(T): Ozgurdonmaz/iStockphoto; p263(B): Locrifa/Shutterstock; p266: Cavallini James/BSIP/Science Photo Library; p269: Bayanova Svetlana/ Shutterstock; p271: Annto/Shutterstock; p272: Kletr/Shutterstock; p273: Alexei Novikov/Shutterstock; p274: Hxdbzxy/Shutterstock; p275: Anthony Short; SO1: Power and Syred/Science Photo Library; SO2: Vovan/ Shutterstock; SO3: Semnic/Shutterstock; SO4: Merlin74/Shutterstock; SO5: Jack Hong/Shutterstock;

All artwork by Q2A media

Contents

Required Practicals

Practical work is a vital part of biology, helping to support and apply your scientific knowledge, and develop your investigative and practical skills. In this Biology part of your AQA Combined Science: Trilogy course, there are seven required practicals that you must carry out. Questions in your exams could draw on any of the knowledge and skills you have developed in carrying out these practicals.

A Required practical feature box has been included in this student book for each of your required practicals. Further support is available on Kerboodle.

Required practical	Topic
1 **Using a light microscope.** Use a light microscope to observe, draw, and label a selection of plant and animal cells and include a scale magnification.	B1.2
2 **Investigate the effect of a range of concentrations of salt or sugar solutions on the mass of plant tissue.** Investigate osmosis by measuring how the mass of plant tissue changes in a range of concentrations of salt or sugar solutions.	B1.8
3 **Use standard food tests to identify food groups.** Detect sugars, starch, and proteins in food using Benedict's test, the iodine test, and Biuret reagent.	B3.3
4 **Investigate the effect of pH on the rate of reaction of amylase enzyme.** Students should use a continuous sampling technique to determine the time taken to completely digest a starch solution at a range of pH values.	B3.6
5 **Investigate the effect of light intensity on the rate of photosynthesis** Use an aquatic plant to observe the effect light intensity has on the rate of photosynthesis.	B8.2
6 **Investigate the effect of a factor on human reaction time.** Plan and carry out an investigation, choosing appropriate ways to measure reaction time and considering the risks and ethics of the investigation.	B10.2
7 **Measure the population size of a common species in a habitat.** Use sampling techniques to investigate the effect of a factor on the distribution of this species.	B15.3

How to use this book

Learning objectives

- Learning objectives at the start of each spread tell you the content that you will cover.
- Any outcomes marked with the higher tier icon **H** are only relevant to those who are sitting the higher tier exams.

This book has been written by subject experts to match the new 2016 specifications. It is packed full of features to help you prepare for your course and achieve the very best you can.

Key words are highlighted in the text. You can look them up in the glossary at the back of the book if you are not sure what they mean.

Many diagrams are as important for your understanding as the text, so make sure you revise them carefully.

Synoptic link

Synoptic links show how the content of a topic links to other parts of the course. This will support you with the synoptic element of your assessment.

There are also links to the Maths skills for biology chapter, so you can develop your maths skills whilst you study.

Practical

Practicals are a great way for you to see science in action for yourself. These boxes may be a simple introduction or reminder, or they may be the basis for a practical in the classroom. They will help your understanding of the course.

Required practical

These practicals have important skills that you will need to be confident with for part of your assessment. Your teacher will give you additional information about tackling these practicals.

Study tip

Hints giving you advice on things you need to know and remember, and what to watch out for.

Anything in the Higher Tier spreads and boxes must be learnt by those sitting the higher tier exam. If you will be sitting foundation tier, you will not be assessed on this content.

Higher

Go further

Go further feature boxes encourage you to think about science you have learnt in a different context and introduce you to science beyond the specification. You do not need to learn any of the content in a Go further box.

Using maths

This feature highlights and explains the key maths skills you need. There are also clear step-by-step worked examples.

Summary questions

Each topic has summary questions. These questions give you the chance to test whether you have learnt and understood everything in the topic. The questions start off easier and get harder, so that you can stretch yourself.

The Literacy pen ✏ shows activities or questions that help you develop literacy skills.

Key points

Linking to the Learning objectives, the Key points boxes summarise what you should be able to do at the end of the topic. They can be used to help you with revision.

Any questions marked with the higher tier icon **H** are for students sitting the higher tier exams.

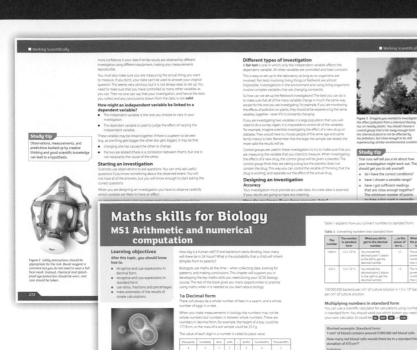

Working scientifically skills are an important part of your course. The working scientifically section describes and supports the development of some of the key skills you will need.

The Maths skills for biology chapter describes and supports the development of the important mathematical skills you will need for all aspects of your course. It also has questions so you can test your skills.

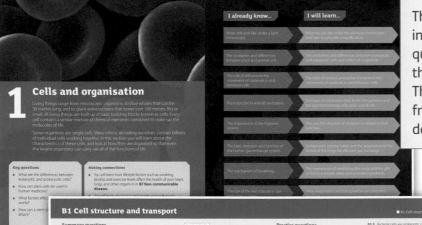

The section openers include an introduction to the section, some key questions the section will answer, and the required practicals in that section. They also introduce the key concepts from KS3 and tell you how they will be developed in that section.

At the end of every chapter there are summary questions and practice questions. The questions test your literacy, maths, and working scientifically skills, as well as your knowledge of the concepts in that chapter. The practice questions can also call on knowledge from any of the previous chapters to help support the synoptic element of your assessment.

There are also further practice questions at the end of the book to cover all of the content from your course.

Kerboodle

This book is also supported by Kerboodle, offering unrivalled digital support for building your practical, maths and literacy skills.

If your schools subscribes to Kerboodle, you will find a wealth of additional resources to help you with your studies and revision:

- animations, videos, and revision podcasts
- webquests
- maths and literacy skills activities and worksheets
- on your marks activities to help you achieve your best
- practicals and follow-up activities
- interactive quizzes that give question-by-question feedback
- self-assessment checklists

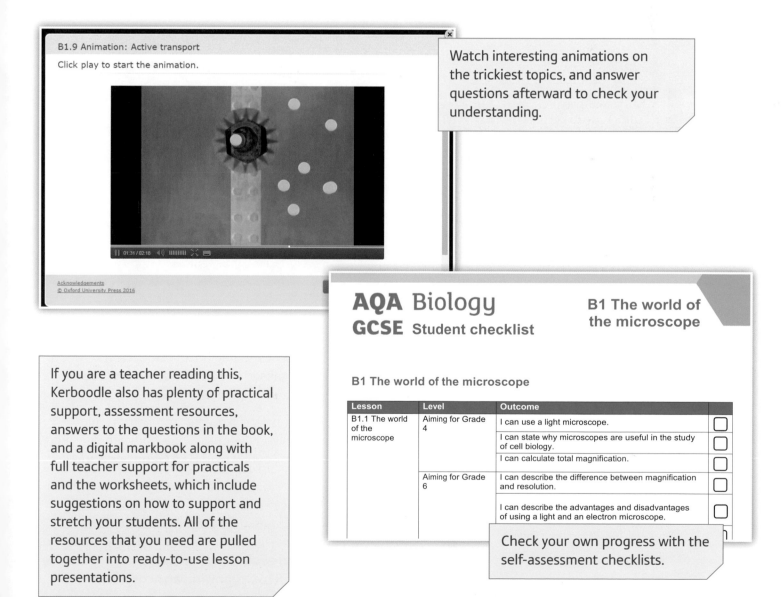

B1.9 Animation: Active transport

Click play to start the animation.

01:31 / 02:18

Acknowledgements
© Oxford University Press 2016

Watch interesting animations on the trickiest topics, and answer questions afterward to check your understanding.

If you are a teacher reading this, Kerboodle also has plenty of practical support, assessment resources, answers to the questions in the book, and a digital markbook along with full teacher support for practicals and the worksheets, which include suggestions on how to support and stretch your students. All of the resources that you need are pulled together into ready-to-use lesson presentations.

AQA Biology
GCSE Student checklist

B1 The world of the microscope

B1 The world of the microscope

Lesson	Level	Outcome	
B1.1 The world of the microscope	Aiming for Grade 4	I can use a light microscope.	☐
		I can state why microscopes are useful in the study of cell biology.	☐
		I can calculate total magnification.	☐
	Aiming for Grade 6	I can describe the difference between magnification and resolution.	☐
		I can describe the advantages and disadvantages of using a light and an electron microscope.	☐

Check your own progress with the self-assessment checklists.

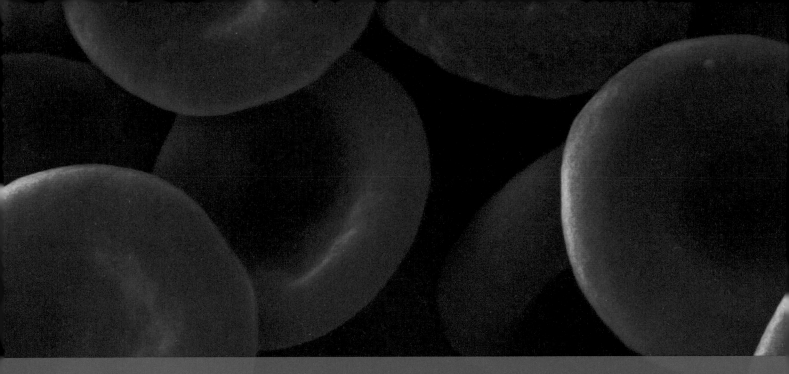

1 Cells and organisation

Living things range from microscopic organisms, to blue whales that can be 30 metres long, and to giant redwood trees that tower over 100 metres. Big or small, all living things are built up of basic building blocks known as cells. Every cell contains a similar mixture of chemical elements combined to make up the molecules of life.

Some organisms are single cells. Many others, including ourselves, contain billions of individual cells working together. In this section you will learn about the characteristics of these cells, and look at how they are organised so that even the largest organisms can carry out all of the functions of life.

Key questions

- What are the differences between eukaryotic and prokaryotic cells?

- How can stem cells be used in human medicine?

- What factors affect how an enzyme works?

- How can a stent prevent a heart attack?

Making connections

- You will learn how lifestyle factors such as smoking, alcohol, and exercise levels affect the health of your heart, lungs, and other organs in in **B7 Non-communicable diseases**

- You will learn about how eukaryotic and prokaryotic organisms have evolved over time, how they are classified, and how they are still evolving in **B14 Genetics and evolution**.

- You will find out much more about the role of bacteria in animal and plant diseases in **B5 Communicable diseases**, about their importance in genetic engineering and evolution in **B13 Variation and evolution**, and about their importance in decomposition in **B16 Organising an ecosystem**.

I already know...

What cells look like under a light microscope.

The similarities and differences between plant and animal cells.

The role of diffusion in the movement of materials in and between cells.

Reproduction in animals and plants.

The importance of the digestive system.

The basic structure and function of the human gas exchange system.

The mechanism of breathing.

The role of the leaf stomata in gas exchange in plants.

I will learn...

What we can see under the electron microscope – and how to calculate magnification.

The similarities and differences between prokaryotic and eukaryotic cells and orders of magnitude.

The roles of osmosis and active transport in the movement of materials in and between cells.

The type of cell division that forms the gametes and the way normal body cells grow and divide

The way the structure of enzymes is related to their function.

Surface area: volume ratios and the adaptations of the alveoli of the lungs for effective gas exchange.

The importance of ventilating the lungs and the gills of fish to maintain steep concentration gradients.

How evaporation and transpiration are controlled in plants.

Required Practicals

Practical		Topic
1	Looking at cells	B1.2
2	Investigating osmosis in plant cells	B1.8
3	Food tests	B3.3
4	The effect of pH on the rate of reaction of amylase	B3.6

1.1 The world of the microscope

Learning objectives

After this topic, you should know:

- how microscopy techniques have developed over time
- the differences in magnification and resolution between a light microscope and an electron microscope
- how to calculate the magnification, real size, and image size of a specimen.

Living things are all made up of cells, but most cells are so small you can only see them using a microscope. It is important to grasp the units used for such tiny specimens before you start to look at them.

Using units

1 kilometre (km) = 1000 metres (m)

1 m = 100 centimetres (cm)

1 cm = 10 millimetres (mm)

1 mm = 1000 micrometres (µm)

1 µm = 1000 nanometres (nm) – so a nanometre is 0.000 000 001 metres (or written in standard form as 1×10^{-9} m).

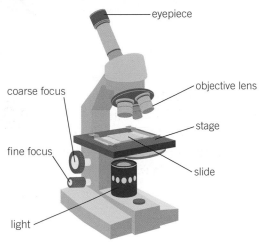

Figure 1 *A light microscope*

The first light microscopes were developed in the mid-17th century. Their development has continued ever since and they are still widely used to look at cells. Light microscopes use a beam of light to form an image of an object and the best can magnify around 2000 times (×2000), although school microscopes usually only magnify several hundred times. They are relatively cheap, can be used almost anywhere, and can magnify live specimens (Figures 1 and 2).

The invention of the electron microscope in the 1930s allowed biologists to see and understand more about the subcellular structures inside cells. These instruments use a beam of electrons to form an image and can magnify objects up to around 2 000 000 times. Transmission electron microscopes give 2D images with very high magnification and resolution. Scanning electron microscopes give dramatic 3D images but lower magnifications (Figure 3). Electron microscopes are large, very expensive, and have to be kept in special temperature, pressure, and humidity-controlled rooms.

Calculating magnification

You can calculate the magnification you are using with a light microscope very simply. You multiply the magnification of the eyepiece lens by the magnification of the objective lens. So if your eyepiece lens is ×4 and your objective lens is ×10, your overall magnification is:

$$4 \times 10 = \times 40$$

When you label drawings made using a microscope, make it clear that the magnification you give is the magnification at which you looked at the specimen (eg., as viewed at ×40).

Calculating the size of an object

You will want to calculate the size of objects under the microscope. There is a simple formula for this, based on the magnification triangle.

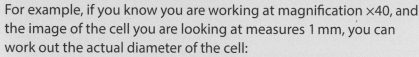

As long as you know or can measure two of the factors, you can find the third.

$$\text{magnification} = \frac{\text{size of image}}{\text{size of real object}}$$

For example, if you know you are working at magnification ×40, and the image of the cell you are looking at measures 1 mm, you can work out the actual diameter of the cell:

$$\text{size of real object} = \frac{\text{size of image}}{\text{magnification}}$$

so

$$= \frac{1}{40}\,\text{mm} = 0.025\,\text{mm or } 25\,\mu\text{m}$$

Your cell has a diameter of **25 μm**.

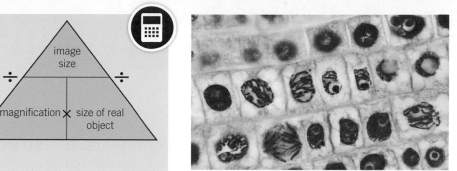

Figure 2 *Onion cells dividing as seen through a light microscope – magnification ×570*

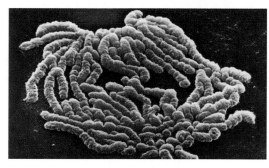

Figure 3 *Chromosomes during cell division seen with a scanning electron microscope – magnification ×4500*

Magnifying and resolving power

Microscopes are useful because they magnify things, making them look bigger. The height of an average person magnified by one of the best light microscopes would look about 3.5 km, and by an electron microscope about 3500 km. There is, however, a minimum distance between two objects when you can see them clearly as two separate things. If they are closer together than this, they appear as one object. Resolution is the ability to distinguish between two separate points and it is the **resolving power** of a microscope that affects how much detail it can show. A light microscope has a resolving power of about 200 nm, a scanning electron microscope of about 10 nm and a transmission electron microscope of about 0.2 nm – that is approximately the distance apart of two atoms in a solid substance!

1. State one advantage and one disadvantage of using:
   a a light microscope [2 marks] **b** an electron microscope. [2 marks]

2. **a** A student measured the diameter of a human capillary on a micrograph. The image measures 5 mm and the student knows the magnification is ×1000. How many micrometres is the diameter of the capillary? [3 marks]
   b A student is told the image of the cell has a diameter of 800 μm. The actual cell has a diameter of 20 μm. At what magnification has the cell been observed? [2 marks]

3. Evaluate the use of an electron microscope and a light microscope, giving one example where each type of microscope might be used. [6 marks]

Synoptic links

You can learn more about writing very small or very large numbers in standard form in Maths skills MS1b.

For more information on cell division look at Chapter B2.

Study tip

Make sure you can work out the magnification, the size of a cell, or the size of the image depending on the information you are given.

Key points

- Light microscopes magnify up to about ×2000, and have a resolving power of about 200 nm.
- Electron microscopes magnify up to about ×2 000 000, and have a resolving power of around 0.2 nm.
- $\text{magnification} = \dfrac{\text{size of image}}{\text{size of real object}}$

B1.2 Animal and plant cells

Learning objectives

After this topic, you should know:

- the main parts of animal cells
- the similarities and differences between plant and animal cells.

Synoptic link

You will find out more about classifying the living world in Topics B14.9 and B14.10

Go further

The ultrastructure of a cell – the details you can see under an electron microscope – includes structures such as the cytoskeleton, the Golgi apparatus, and the rough and smooth endoplasmic reticulum. They support and move the cell, modify and package proteins and lipids, and produce the chemicals that control the way your body works.

Study tip

Learn the parts of the cells shown on these diagrams, and their functions.

Synoptic link

For more information on photosynthesis, look at Topic B8.1.

The cells that make up your body are typical animal cells. All cells have some features in common. You can see these features clearly in animal cells.

Animal cells – structure and function

The structure and functions of the parts that make up a cell have been made clear by the electron microscope (Figure 1). You will learn more about how their structure relates to their functions as you study more about specific organ systems during your GCSE Biology course. An average animal cell is around 10–30 μm long (so it would take 100 000–300 000 cells to line up along the length of a metre ruler). Human beings are animals so human cells are just like most other animal cells, and you will see exactly the same structures inside them.

- The **nucleus** – controls all the activities of the cell and is surrounded by the nuclear membrane. It contains the genes on the chromosomes that carry the instructions for making the proteins needed to build new cells or new organisms. The average diameter is around 10 μm.

- The **cytoplasm** – a liquid gel in which the organelles are suspended and where most of the chemical reactions needed for life take place.

- The **cell membrane** – controls the passage of substances such as glucose and mineral ions into the cell. It also controls the movement of substances such as urea or hormones out of the cell.

- The **mitochondria** – structures in the cytoplasm where aerobic respiration takes place, releasing energy for the cell. They are very small: 1–2 μm in length and only 0.2–0.7 μm in diameter.

- The **ribosomes** – where protein synthesis takes place, making all the proteins needed in the cell.

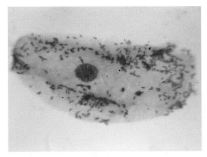 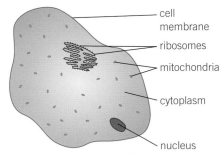

cell membrane
ribosomes
mitochondria
cytoplasm
nucleus

Figure 1 *Diagrams of cells are much easier to understand than the real thing seen under a microscope. This picture shows a simple animal cheek cell magnified ×1350 times under a light microscope. This is the way a model animal cell is drawn to show the main features common to most living cells*

Plant cells – structure and function

Plants are very different organisms from animals. They make their own food by photosynthesis. They do not move their whole bodies about from one place to another. Plant cells are often rather bigger than animal cells – they range from 10 to 100 μm in length.

Plant cells have all the features of a typical animal cell, but they also contain features that are needed for their very different functions (Figures 2 and 3). **Algae** are simple aquatic organisms. They also make their own food by photosynthesis and have many similar features to plant cells. For centuries they were classified as plants, but now they are classified as part of a different kingdom – the protista.

All plant and algal cells have a **cell wall** made of **cellulose** that strengthens the cell and gives it support.

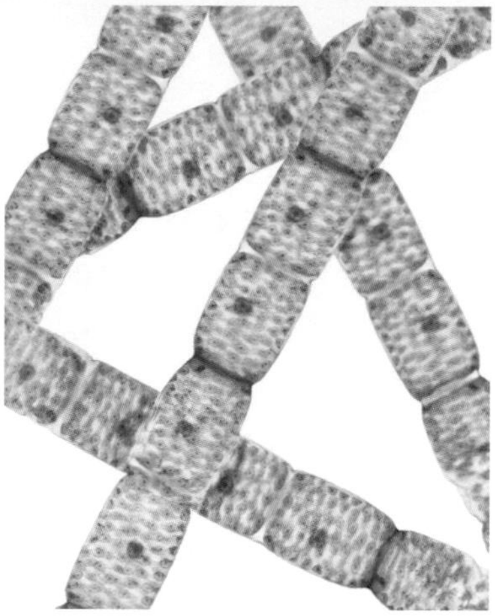

Many (but not all) plant cells also have these other features:

● **Chloroplasts** are found in all the green parts of a plant. They are green because they contain the green substance **chlorophyll**. Chlorophyll absorbs light so the plant can make food by photosynthesis. Each chloroplast is around 3–5 µm long. Root cells do not have chloroplasts because they are underground and do not photosynthesise.

● A **permanent vacuole** is a space in the cytoplasm filled with cell sap. This is important for keeping the cells rigid to support the plant.

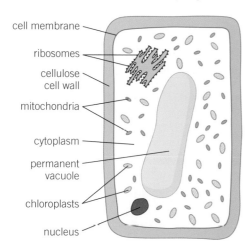

cell membrane
ribosomes
cellulose cell wall
mitochondria
cytoplasm
permanent vacuole
chloroplasts
nucleus

Figure 3 *A plant cell has many features in common with an animal cell, as well as other features that are unique to plants*

Figure 2 *Algal cells contain a nucleus and chloroplasts so that they can photosynthesise*

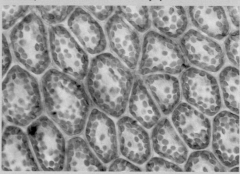

Figure 4 *Some of the common features of plant cells show up well under the light microscope. Here, the features are magnified ×40*

Study tip

Remember that not all plant cells have chloroplasts.

Do not confuse chloroplasts and chlorophyll.

1 a List the main structures you would expect to find in a human cell. [5 marks]
  b State the three extra features that may be found in plant cells but not animal cells. [3 marks]
  c Describe the main functions of these three extra structures. [3 marks]

2 Suggest why the nucleus and the mitochondria are so important in all cells. [4 marks]

3 Chloroplasts are found in many plant cells but not all of them. Suggest two types of plant cells that are unlikely to have chloroplasts and in each case explain why they have none. [4 marks]

Key points

● Animal cell features common to all cells – a nucleus, cytoplasm, cell membrane, mitochondria, and ribosomes.
● Plant and algal cells contain all the structures seen in animal cells as well as a cellulose cell wall.
● Many plant cells also contain chloroplasts and a permanent vacuole filled with sap.

B1.3 Eukaryotic and prokaryotic cells

Learning objectives

After this topic, you should know:

- the similarities and differences between eukaryotic cells and prokaryotic cells
- how bacteria compare to animal and plant cells
- the size and scale of cells including order of magnitude calculations.

Synoptic link

You will learn more about bacteria that cause disease in Topic B5.7, and about bacteria that are important in the environment in Topic B16.2 and Topic B17.3.

Go further

The plasmids found in bacteria are used extensively in genetic engineering to carry new genes into the genetic material of other organisms, ranging from bananas to sheep.

Eukaryotic cells

Animal and plant cells are examples of **eukaryotic cells**. Eukaryotic cells all have a cell membrane, cytoplasm, and genetic material that is enclosed in a nucleus.

The genetic material is a chemical called DNA and this forms structures called chromosomes that are contained within the nucleus. All animals (including human beings), plants, fungi, and protista are eukaryotes.

Prokaryotes

Bacteria are single-celled living organisms. They are examples of prokaryotes. At 0.2–2.0 μm in length prokaryotes are 1–2 orders of magnitude smaller than eukaryotes. You could fit hundreds of thousands of bacteria on to the full stop at the end of this sentence, so you cannot see individual bacteria without a powerful microscope. When you culture bacteria on an agar plate, you grow many millions of bacteria. This enables you to see the bacterial colony with your naked eye.

Bacteria have cytoplasm and a cell membrane surrounded by a cell wall, but the cell wall does not contain the cellulose you see in plant cells. In prokaryotic cells the genetic material is not enclosed in a nucleus. The bacterial chromosome is a single DNA loop found free in the cytoplasm.

Prokaryotic cells may also contain extra small rings of DNA called plasmids. Plasmids code for very specific features such as antibiotic resistance.

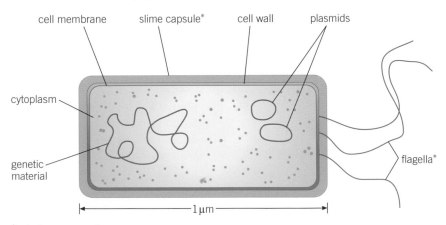

*not always present

Figure 1 *Bacteria come in a variety of shapes, but they all have the same basic structure*

Some bacteria have a protective slime capsule around the outside of the cell wall. Some types of bacterium have at least one flagellum (plural: flagella), that is, a long protein strand that lashes about. These bacteria use their flagella to move themselves around.

Many bacteria have little or no effect on other organisms and many are very useful.

Some bacteria are harmful. Bacteria can cause diseases in humans and other animals and also in plants. They can also decompose and destroy stored food.

Relative sizes

In cell biology it is easy to forget just how small everything is – and how much bigger some cells are than others. It is also important to remember just how large the organisms built up from individual cells can be. Figure 2 shows you some relative sizes.

Orders of magnitude

Orders of magnitude are used to make approximate comparisons between numbers or objects. If one number is about 10 times bigger than another, it is an order of magnitude bigger. You show orders of magnitude using powers of 10. If one cell or organelle is 10 times bigger than another, it is an order of magnitude bigger or 10^1. If it is approximately 100 times bigger it is two orders of magnitude bigger or 10^2.

If you have two numbers to compare, as a rule of thumb you can work out orders of magnitude as follows:

If the bigger number divided by the smaller number is less than 10, then they are the same order of magnitude.

If the bigger number divided by the smaller number is around 10, then it is 10^1 or an order of magnitude bigger.

If the bigger number divided by the smaller number is around 100, then it is two orders of magnitude or 10^2 bigger.

Example:
A small animal cell has a length of around 10 µm. A large plant cell has a length of around 100 µm.

$$\frac{100}{10} = 10$$

So, a large plant cell is an order of magnitude or 10^1 bigger than a small animal cell.

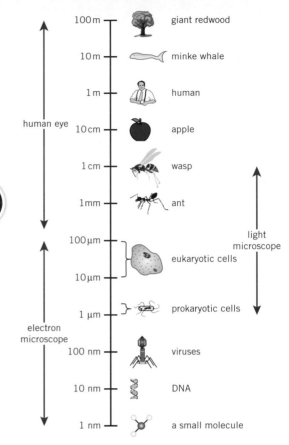

Figure 2 *The relative sizes of different cells and whole organisms and how they can be seen*

Be clear about the similarities and differences between animal, plant, and bacterial cells and between eukaryotic cells and prokaryotic cells.

1 **a** Describe the difference between the genetic material in a prokaryotic cell and the genetic material in the eukaryotic cell. [2 marks]
  b i Describe what flagella are. [1 mark]
    ii State one use of flagella in a prokaryote. [1 mark]
2 A cell nucleus has an average length of 6 µm. Calculate the order of magnitude comparison between the nucleus of a cell and:
  a a small animal cell [2 marks] **b** a large plant cell. [2 marks]
3 Describe the similarities and differences between the features found in prokaryotic and eukaryotic plant and animal cells. [6 marks]

Key points

- Eukaryotic cells all have a cell membrane, cytoplasm, and genetic material enclosed in a nucleus.
- Prokaryotic cells consist of cytoplasm and a cell membrane surrounded by a cell wall. The genetic material is not in a distinct nucleus. It forms a single DNA loop. Prokaryotes may contain one or more extra small rings of DNA called plasmids.
- Bacteria are all prokaryotes.

B1.4 Specialisation in animal cells

Learning objectives

After this topic, you should know:

- how cells differentiate to form specialised cells
- animal cells may be specialised to carry out a particular function
- how the structure of different types of animal cells relates to their function.

Synoptic links

You can find out much more about the organisation of specialised cells into tissues, organs and organ systems in Topic B3.1 and Topic B3.2.

Synoptic link

You can find out more about specialised nerve cells in Chapter B10.

Observing specialised cells

Try looking at different specialised cells under a microscope.

When you look at a specialised cell, there are two useful questions you can ask yourself:

- How is this cell different in structure from a generalised cell?
- How does the difference in structure help the cell to carry out its function?

Although the smallest living organisms are only single cells, they can carry out all of the functions of life. Most organisms are bigger and are made up of lots of cells. Some of these cells become specialised to carry out particular jobs.

As an organism develops, cells differentiate to form different types of specialised cells. Most types of animal cells differentiate at an early stage of development, whereas many types of plant cells retain the ability to differentiate throughout life. As a cell differentiates, it gets different sub-cellular structures that enable it to carry out a particular function. It has become a specialised cell. Some specialised cells, such as egg and sperm cells, work individually. Others are adapted to work as part of a tissue, an organ, or a whole organism.

Nerve cells

Nerve cells are specialised to carry electrical impulses around the body of an animal (Figure 1). They provide a rapid communication system between the different parts of the body. They have several adaptations including:

- Lots of dendrites to make connections to other nerve cells.
- An axon that carries the nerve impulse from one place to another. They can be very long – the axon of a nerve cell in a blue whale can be up to 25 m long! The longest axon in your body runs from the base of your spine to your big toe.
- The nerve endings or synapses are adapted to pass the impulses to another cell or between a nerve cell and a muscle in the body using special transmitter chemicals. They contain lots of mitochondria to provide the energy needed to make the transmitter chemicals.

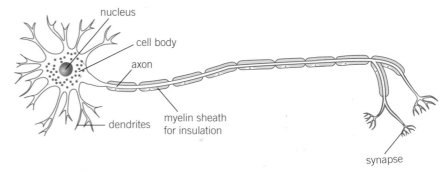

Figure 1 *A nerve cell is specialised to carry electrical impulses from one part of the body to another*

Muscle cells

Muscle cells are specialised cells that can contract and relax. Striated (striped) muscle cells work together in tissues called muscles (Figure 2). Muscles contract and relax in pairs to move the bones of the skeleton, so vertebrates can move on land and in water, and in some cases fly. Smooth muscle cells form one of the layers of tissue in your digestive system and they contract to squeeze the food through your gut.

Striated muscle cells have three main adaptations:

- They contain special proteins that slide over each other making the fibres contract.
- They contain many mitochondria to transfer the energy needed for the chemical reactions that take place as the cells contract and relax.
- They can store glycogen, a chemical that can be broken down and used in cellular respiration by the mitochondria to transfer the energy needed for the fibres to contract.

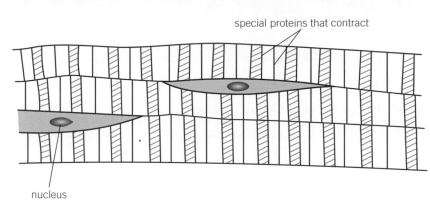

Figure 2 *A striated muscle cell is specialised to contract and relax*

Sperm cells

Sperm cells are usually released a long way from the egg they are going to fertilise. They contain the genetic information from the male parent. Depending on the type of animal, sperm cells need to move through water or the female reproductive system to reach an egg. Then they have to break into the egg.

Sperm cells have several adaptations to make all this possible (Figure 3):

- A long tail whips from side to side to help move the sperm through water or the female reproductive system.
- The middle section is full of mitochondria, which transfer the energy needed for the tail to work.
- The acrosome stores digestive enzymes for breaking down the outer layers of the egg.
- A large nucleus contains the genetic information to be passed on.

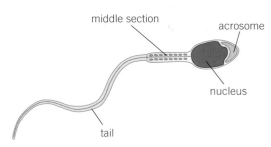

Figure 3 *A sperm cell*

1 State one adaptation for each of the following specialised animal cells. Describe how this adaptation helps the cell carry out its function:
   a nerve cell [2 marks]
   b muscle cell [2 marks]
   c sperm cell [2 marks]

2 Cone cells are specialised nerve cells in the eye. They contain a chemical that changes in coloured light. As a result of the change, an impulse is sent along another nerve cell to the brain. Cone cells usually contain many mitochondria. Suggest why this is an important adaptation. [4 marks]

3 Describe the features you would look for to decide on the function of an unknown specialised animal cell. [6 marks]

Key points

- As an organism develops, cells differentiate to form different types of cells.
- As an animal cell differentiates to form a specialised cell it acquires different sub-cellular structures to enable it to carry out a certain function.
- Examples of specialised animal cells are nerve cells, muscle cells, and sperm cells.
- Animal cells may be specialised to function within a tissue, an organ, organ systems, or whole organisms.

B1.5 Specialisation in plant cells

Learning objectives

After this topic, you should know:

- how plant cells may be specialised to carry out a particular function
- how the structure of different types of plant cells relates to their function.

Animals are not the only organisms to have cells specialised for a particular function within a tissue or an organ. Plants also have very specialised cells with clear adaptations for the job they carry out. Here are four examples.

Root hair cells

You find root hair cells close to the tips of growing roots. Plants need to take in lots of water (and dissolved mineral ions). The root hair cells help them to take up water and mineral ions more efficiently. Root hair cells are always relatively close to the xylem tissue. The xylem tissue carries water and mineral ions up into the rest of the plant. Mineral ions are moved into the root hair cell by active transport (Topic B1.9).

Root hair cells (Figure 1) have three main adaptations:

- They greatly increase the surface area available for water to move into the cell.
- They have a large permanent vacuole that speeds up the movement of water by osmosis from the soil across the root hair cell.
- They have many mitochondria that transfer the energy needed for the active transport of mineral ions into the root hair cells.

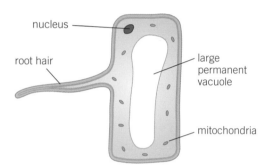

Figure 1 *A root hair cell*

Photosynthetic cells

One of the ways plants differ from animals is that plants can make their own food by photosynthesis. There are lots of plant cells that can carry out photosynthesis – and lots that cannot. Photosynthetic cells (Figure 2) usually have a number of adaptations including:

- They contain specialised green structures called chloroplasts containing chlorophyll that trap the light needed for photosynthesis.
- They are usually positioned in continuous layers in the leaves and outer layers of the stem of a plant so they absorb as much light as possible.
- They have a large permanent vacuole that helps keep the cell rigid as a result of osmosis (Topic B1.8). When lots of these rigid cells are arranged together to form photosynthetic tissue they help support the stem. They also keep the leaf spread out so it can capture as much light as possible.

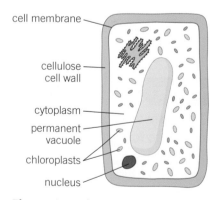

Figure 2 *A photosynthetic plant cell*

Xylem cells

Xylem is the transport tissue in plants that carries water and mineral ions from the roots to the highest leaves and shoots. The xylem is also important in supporting the plant. The xylem is made up of xylem cells (Figure 3) that are adapted to their functions in two main ways:

- The xylem cells are alive when they are first formed but a special chemical called lignin builds up in spirals in the cell walls. The cells die and form long hollow tubes that allow water and mineral ions

Synoptic link

You will learn much more about photosynthesis in Chapter B8.

to move easily through them, from one end of the plant to the other.

- The spirals and rings of lignin in the xylem cells make them very strong and help them withstand the pressure of water moving up the plant. They also help support the plant stem.

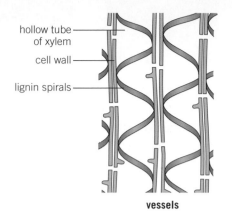

Figure 3 *The adaptations of xylem cells*

Synoptic link

You will learn more about the movement of water up the xylem and the process of transpiration in Chapter B4.

Phloem cells

Phloem is the specialised transport tissue that carries the food made by photosynthesis around the body of the plant. It is made up of phloem cells that form tubes rather like xylem cells, but phloem cells do not become lignified and die. The dissolved food can move up and down the phloem tubes to where it is needed. The adaptations of the phloem cells (Figure 4) include:

- The cell walls between the cells break down to form special sieve plates. These allow water carrying dissolved food to move freely up and down the tubes to where it is needed.

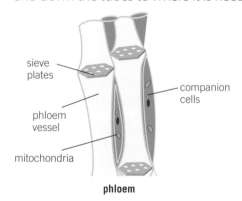

Figure 4 *The adaptations of phloem cells*

- Phloem cells lose a lot of their internal structures but they are supported by companion cells that help to keep them alive. The mitochondria of the companion cells transfer the energy needed to move dissolved food up and down the plant in phloem.

1 State one adaptation for each of the following specialised plant cells. Describe how this adaptation helps the cell carry out its function:
   a root hair cell [2 marks]
   b xylem cell [2 marks]
   c phloem cell [2 marks]
   d photosynthetic cell [2 marks]

2 Suggest why a cell within the trunk of a tree cannot carry out photosynthesis. [2 marks]

3 Describe the features you would look for to decide on the function of an unknown specialised plant cell. 🖊 [6 marks]

Key points

- Plant cells may be specialised to carry out a particular function.
- Examples of specialised plant cells are root hair cells, photosynthetic cells, xylem cells, and phloem cells
- Plant cells may be specialised to function within tissues, organs, organ systems, or whole organisms.

B1.6 Diffusion

Learning objectives

After this topic, you should know:

- how diffusion takes place and why it is important in living organisms
- what affects the rate of diffusion.

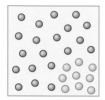

At the moment when the blue particles are added to the red particles they are not mixed at all

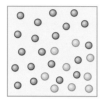

As the particles move randomly, the blue ones begin to mix with the red ones

As the particles move and spread out, they bump into each other. This helps them to keep spreading randomly

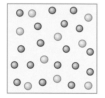

Eventually, the particles are completely mixed and diffusion is complete, although they do continue to move randomly

Figure 1 *The random movement of particles results in substances spreading out, or diffusing, from an area of higher concentration to an area of lower concentration*

Study tip

Particles move randomly, but the net movement is from a region of high concentration to a region of low concentration.

Your cells need to take in substances such as glucose and oxygen for respiration. They also need to get rid of waste products, and chemicals that are needed elsewhere in your body. Dissolved substances and gases can move into and out of your cells across the cell membrane. One of the main ways in which they move is by **diffusion**.

Diffusion

Diffusion is the spreading out of the particles of a gas, or of any substance in solution (a solute). This results in the net movement (overall movement) of particles. The net movement is from an area of higher concentration to an area of lower concentration of the particle. It takes place because of the random movement of the particles (molecules or ions). The motion of the particles causes them to bump into each other, and this moves them all around.

Imagine a room containing a group of boys on one side and a group of girls on the other. If everyone closes their eyes and moves around briskly but randomly, they will bump into each other. They will scatter until the room contains a mixture of boys and girls. This gives you a good model of diffusion (see Figure 1).

Rates of diffusion

If there is a big difference in concentration between two areas, diffusion will take place quickly. Many particles will move randomly towards the area of low concentration. Only relatively few will move randomly in the other direction.

However, if there is only a small difference in concentration between two areas, the net movement by diffusion will be quite slow. The number of particles moving into the area of lower concentration by random movement will only be slightly more than the number of particles that are leaving the area.

net movement = particles moving in − particles moving out

In general, the greater the difference in concentration, the faster the rate of diffusion. This difference between two areas of concentration is called the concentration gradient. The bigger the difference, the steeper the concentration gradient and the faster the rate of diffusion. In other words, diffusion occurs *down* a concentration gradient.

Temperature also affects the rate of diffusion. An increase in temperature means the particles in a gas or a solution move around more quickly. When this happens, diffusion takes place more rapidly as the random movement of the particles speeds up.

Diffusion in living organisms

Dissolved substances move into and out of your cells by diffusion across the cell membrane. These include simple sugars, such as glucose, gases such as oxygen and carbon dioxide, and waste products such as urea from the breakdown of amino acids in your liver. The urea passes from the liver cells into the blood plasma and is excreted by the kidneys.

The oxygen you need for respiration passes from the air in your lungs into your red blood cells through the cell membranes by diffusion. The oxygen moves down a concentration gradient from a region of high oxygen concentration to a region of low oxygen concentration.

Oxygen then also moves by diffusion down a concentration gradient from the blood cells into the cells of the body where it is needed. Carbon dioxide moves out from the body cells into the red blood cells and then into the air in the lungs by diffusion down a concentration gradient in a similar way. The diffusion of oxygen and carbon dioxide in opposite directions in the lungs is known as gas exchange.

Individual cells may be adapted to make diffusion easier and more rapid. The most common adaptation is to increase the surface area of the cell membrane (Figure 2). By folding up the membrane of a cell, or the tissue lining an organ, the area over which diffusion can take place is greatly increased. Therefore the rate of diffusion is also greatly increased, so that much more of a substance moves in a given time.

Synoptic link

You will learn more about gas exchange in Topic B4.5.

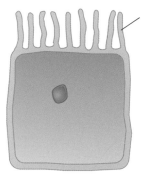

Folds in the cell membrane form microvilli, which increase the surface area of the cell

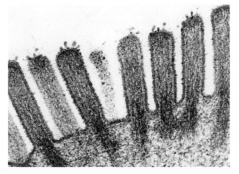

Figure 2 *An increase in the surface area of a cell membrane means diffusion can take place more quickly. This is an intestinal cell*

1 Define the process of diffusion in terms of the particles involved. [2 marks]

2 a Explain why diffusion takes place faster when there is an increase in temperature. [3 marks]
 b Explain why so many cells have folded membranes along at least one surface. [2 marks]

3 Describe the process of diffusion occurring in each of the following statements. Include any adaptations that are involved. 🖉
 a Digested food products move from your gut into the bloodstream. [3 marks]
 b Carbon dioxide moves from the blood in the capillaries of your lungs to the air in the lungs. [3 marks]
 c Male moths can track down a mate from up to 3 miles away because of the special chemicals produced by the female. [3 marks]

Key points

- Diffusion is the spreading out of particles of any substance, in solution or a gas, resulting in a net movement from an area of higher concentration to an area of lower concentration, down a concentration gradient.
- The rate of diffusion is affected by the difference in concentrations, the temperature, and the available surface area.
- Dissolved substances such as glucose and urea and gases such as oxygen and carbon dioxide move in and out of cells by diffusion.

B1.7 Osmosis

Learning objectives

After this topic, you should know:

● how osmosis differs from diffusion
● why osmosis is so important in animal cells.

Study tip

Remember, any particles can diffuse from an area of high concentration to an area of low concentration, provided they are **soluble** and **small** enough to pass through the membrane.

Osmosis in organisms refers only to the diffusion of *water* molecules through the partially permeable cell membrane.

Investigating osmosis

You can make model cells using bags made of partially permeable membrane (see Figure 1). You can find out what happens to them if the concentrations of the solutions inside or outside the 'cell' change.

Diffusion takes place when particles can spread out freely from a higher to a lower concentration. However, the solutions inside cells are separated from those outside by the cell membrane. This membrane does not let all types of particles through. Membranes that only let some types of particles through are called **partially permeable membranes**.

How osmosis differs from diffusion

Partially permeable cell membranes let water move across them. Remember:

● A **dilute** solution of sugar contains a *high* concentration of water (the solvent). It has a *low* concentration of sugar (the solute).

● A **concentrated** sugar solution contains a relatively *low* concentration of water and a *high* concentration of sugar.

The cytoplasm of a cell is made up of chemicals dissolved in water inside a partially permeable cell membrane. The cytoplasm contains a fairly concentrated solution of salts and sugars. Water moves from a dilute solution (with a high concentration of water molecules) to a concentrated solution (with fewer water molecules in a given volume) across the membrane of the cell.

This special type of diffusion, where only water moves across a partially permeable membrane from a dilute solution to a concentrated solution is called **osmosis**.

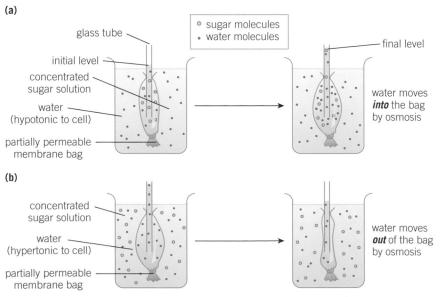

Figure 1 *A model of osmosis in a cell. In (a) the model cell is in a hypotonic solution. In (b) the model cell is in a hypertonic solution*

The concentration of solutes inside your body cells needs to stay at the same level for them to work properly. However, the concentration of the solutions outside your cells may be very different to the concentration inside them. This concentration gradient can cause water to move into or out of the cells by osmosis (see Figure 1).

● If the concentration of solutes in the solution outside the cell is **the same** as the internal concentration, the solution is **isotonic** to the cell.

● If the concentration of solutes in the solution outside the cell is **higher** than the internal concentration, the solution is **hypertonic** to the cell.

● If the concentration of solutes in the solution outside the cell is **lower** than the internal concentration, the solution is **hypotonic** to the cell.

Osmosis in animals

If a cell uses up water in its chemical reactions, the cytoplasm becomes more concentrated. The surrounding fluid becomes hypotonic to the cell and more water immediately moves in by osmosis.

If the cytoplasm becomes too dilute because more water is made in chemical reactions, the surrounding fluid becomes hypertonic to the cell and water leaves the cell by osmosis. Osmosis restores the balance in both cases.

However, osmosis can also cause big problems. If the solution outside the cell becomes much more dilute (hypotonic) than the cell contents, water will move in by osmosis. The cell will swell and may burst. If the solution outside the cell becomes much more concentrated (hypertonic) than the cell contents, water will move out of the cell by osmosis. The cytoplasm will become too concentrated and the cell will shrivel up and can no longer survive. Once you understand the effect osmosis can have on cells, the importance of maintaining constant internal conditions becomes clear.

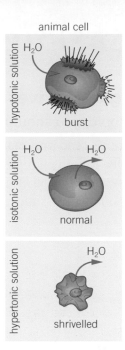

Figure 2 *Osmosis can have a dramatic effect on animal cells*

> ### Study tip
>
> When writing about osmosis, be careful to specify whether it is the concentration of water or of solutes that you are referring to.
>
> Make sure you understand exactly what is meant by the terms isotonic, hypertonic, and hypotonic.

1 a State the difference between osmosis and diffusion. [2 marks]

   b Explain how osmosis helps to maintain the cytoplasm of plant and body cells at a specific concentration. [2 marks]

2 a Define the following terms:
      i isotonic solution [1 mark]
      ii hypotonic solution [1 mark]
      iii hypertonic solution. [1 mark]

   b Explain why it is so important for the cells of the human body that the solute concentration of the fluid surrounding the cells is kept as constant as possible. [4 marks]

3 Animals that live in fresh water have a constant problem with their water balance. The single-celled organism called *Amoeba* has a special vacuole in its cell. The vacuole fills with water and then moves to the outside of the cell and bursts. A new vacuole starts forming straight away. Explain in terms of osmosis why the *Amoeba* needs one of these vacuoles. ⦸ [4 marks]

> ### Key points
>
> ● Osmosis is a special case of diffusion. It is the movement of water from a dilute to a more concentrated solute solution through a partially permeable membrane that allows water to pass through.
> ● Differences in the concentrations of solutions inside and outside a cell cause water to move into or out of the cell by osmosis.
> ● Animal cells can be damaged if the concentration outside the cell changes dramatically.

B1.8 Osmosis in plants

Learning objectives

After this topic, you should know:

- why osmosis is so important in plant cells
- how to investigate the effect of osmosis on plant tissues.

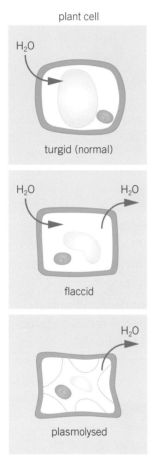

plant cell

H_2O

turgid (normal)

H_2O H_2O

flaccid

H_2O

plasmolysed

Figure 1 *Osmosis in plant cells*

As you have seen, osmosis is the movement of water through a partially permeable membrane down a water concentration gradient. The effect of osmosis on animal cells can be dramatic. Animals have many complex ways of controlling the concentrations of their body solutions to prevent cell damage as a result of osmosis. In plants, osmosis is key to their whole way of life.

Osmosis in plants

Plants rely on osmosis to support their stems and leaves. Water moves into plant cells by osmosis. This causes the vacuole to swell, which presses the cytoplasm against the plant cell wall. The pressure builds up until no more water can physically enter the cell – this pressure is known as **turgor**. Turgor pressure makes the cells hard and rigid, which in turn keeps the leaves and stems of the plant rigid and firm.

Plants need the fluid surrounding the cells to always be hypotonic to the cytoplasm, with a lower concentration of solutes and a higher concentration of water than the plant cells themselves. This keeps water moving by osmosis in the right direction and the cells are turgid. If the solution surrounding the plant cells is hypertonic to (more concentrated than) the cell contents, water will leave the cells by osmosis. The cells will no longer be firm and swollen – they become flaccid (soft) as there is no pressure on the cell walls. At this point, the plant wilts as turgor no longer supports the plant tissues.

If more water is lost by osmosis, the vacuole and cytoplasm shrink, and eventually the cell membrane pulls away from the cell wall. This is **plasmolysis**. Plasmolysis is usually only seen in laboratory experiments. Plasmolysed cells die quickly unless the osmotic balance is restored.

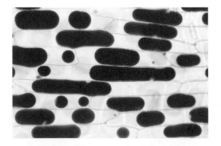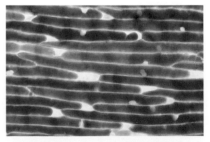

Figure 2 *Micrographs of red onion cells in hypertonic and hypotonic solutions show the effect of osmosis on the contents of the cell*

Investigating osmosis in plant cells

Plant tissue reacts so strongly to the concentration of the external solution that you can use it as an osmometer – a way of measuring osmosis. There are lots of ways you can investigate the effect of osmosis on plant tissue, each with advantages and disadvantages.

The basis of many experiments is to put plant tissue into different concentrations of salt solutions or sugar solutions. You can even use squash to give you the sugar solution. If plant tissue is placed in a hypotonic solution, water will move in to the cells by osmosis. If it is placed in a hypertonic solution, water will move out by osmosis. These changes can be measured by the effect they have on the tissue sample.

- Suggest why salt and sugar are used in osmosis experiments. How could you decide which gives the clearest results?
- Potato is commonly used as the experimental plant tissue. It can be cut into cylinders, rectangular 'chips' or smaller discs. Suggest why potato is so often used as a test plant tissue.
- Sweet potato and beetroot are other common sources of plant tissue for osmosis experiments – suggest possible advantages and disadvantages of using them. How could you determine which is the best experimental plant tissue?

Measuring changes in mass is a widely used method for investigating the uptake or loss of water from plant tissues by osmosis. You must take care not to include any liquid left on the outside of the plant tissue in your measurements as this can have a big effect on your results.

- Discuss the possible advantages and disadvantages of cylinders, chips, or discs for assessing the effect of osmosis on plant tissue. How would you determine which method is the most effective?
- Investigate the effect of surface area on osmosis.
- Explain how you think the surface area of the plant tissue samples might affect osmosis.
- Plan an investigation to see if your ideas are right.
- Show your plan to your teacher and then carry out your investigation.

Safety: Take care when using cutting instruments.

Go further

Scientists have discovered ways of measuring the turgor pressure inside individual cells using very tiny probes. The pressures inside the root or leaf cell of a plant are far higher than human blood pressure, or even the pressure in a car tyre.

1 Define the term osmosis. [1 mark]
2 Students carried out an investigation into the effects of osmosis on plant tissues, placing three sets of beetroot cylinders in three different sugar solutions for 30 mins. One set had gained mass, another lost mass and the third set did not change. One student thought the last experiment hadn't worked. Another disagreed. Explain the results in terms of osmosis in plant cells. [6 marks]
3 Suggest and explain why osmosis is so important in the structural support systems of plants. [6 marks]

Key points

- Osmosis is important to maintain turgor in plant cells.
- There are a variety of practical investigations that can be used to show the effect of osmosis on plant tissues.

B1.9 Active transport

Learning objectives

After this topic, you should know:

- how active transport works
- the importance of active transport in cells.

People with cystic fibrosis have thick, sticky mucus in their lungs, gut and reproductive systems. This causes many different health problems and it happens because an active transport system in their mucus-producing cells does not work properly. Sometimes diffusion and osmosis are not enough.

All cells need to move substances in and out. Water often moves across the cell boundaries by osmosis. Dissolved substances also need to move in and out of cells. There are two main ways in which this happens:

- Substances move by diffusion, down a concentration gradient. This must be in the right direction to be useful to the cells.
- Sometimes the substances needed by a cell have to be moved against a concentration gradient, across a partially permeable membrane. This needs a special process called **active transport**.

Moving substances by active transport

Active transport allows cells to move substances from an area of low concentration to an area of high concentration. This movement is *against* the concentration gradient. As a result, cells can absorb ions from very dilute solutions. It also enables cells to move substances, such as sugars and ions, from one place to another through the cell membrane.

Energy is needed for the active transport system to carry a molecule across the membrane and then return to its original position. This energy is produced during cell respiration. Scientists have shown in a number of different cells that the rate of respiration and the rate of active transport are closely linked (Figure 1).

In other words, if a cell respires and releases a lot of energy, it can carry out lots of active transport. Examples include root hair cells in plants and the cells lining your gut. Cells involved in a lot of active transport usually have many mitochondria to release the energy they need.

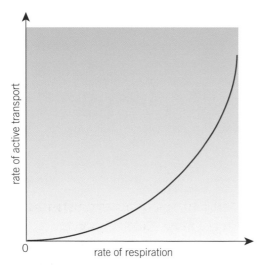

Figure 1 *The rate of active transport depends on the rate of respiration*

The importance of active transport

Active transport is widely used in cells. There are some situations where it is particularly important. For example, mineral ions in the soil, such as nitrate ions, are usually found in very dilute solutions. These solutions are more dilute than the solution within the plant root hair cells. By using active transport, plants can absorb these mineral ions, even though it is against a concentration gradient (see Figure 2).

Sugar, such as glucose, is always actively absorbed out of your gut and kidney tubules into your blood. This is often done against a large concentration gradient.

For example, glucose is needed for cell respiration so it is important to get as much as possible out of the gut. The concentration of glucose in your blood is kept steady, so sometimes it is higher than the concentration of glucose in your gut. When this happens, active transport is used to move the glucose from your gut into your blood against the concentration gradient.

Synoptic links

You can find out more about cystic fibrosis in Topic B12.6, and about the absorption of glucose in the gut in Topic B3.6.

Study tip

Do not refer to movement *along* a concentration gradient. Always refer to movement as *down* a concentration gradient (from higher to lower) for diffusion or osmosis and *against* a concentration gradient (from lower to higher) for active transport.

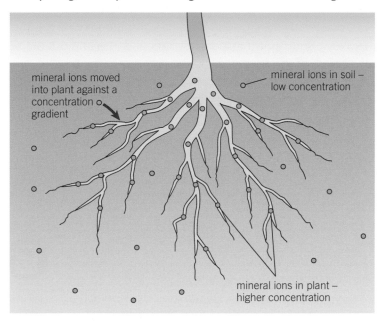

mineral ions moved into plant against a concentration gradient

mineral ions in soil – low concentration

mineral ions in plant – higher concentration

Figure 2 *Plants use active transport to move mineral ions from the soil into the roots against a concentration gradient*

Figure 3 *Some crocodiles have special salt glands in their tongues. These remove excess salt from the body against the concentration gradient by active transport. That's why members of the crocodile species Crocodylus porosus can live in estuaries and even the sea*

1 Describe how active transport works in a cell. [4 marks]

2 a Describe how active transport differs from diffusion and osmosis. [3 marks]

b Explain why cells that carry out a lot of active transport also usually have many mitochondria. [2 marks]

3 Explain fully why active transport is so important to: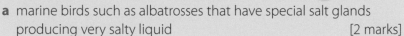

a marine birds such as albatrosses that have special salt glands producing very salty liquid [2 marks]

b plants. [3 marks]

Key points

- Active transport moves substances from a more dilute solution to a more concentrated solution (against a concentration gradient).
- Active transport uses energy released from food in respiration to provide the energy required.
- Active transport allows plant root hairs to absorb mineral ions required for healthy growth from very dilute solutions in the soil against a concentration gradient.
- Active transport enables sugar molecules used for cell respiration to be absorbed from lower concentrations in the gut into the blood where the concentration of sugar is higher.

B1.10 Exchanging materials

Learning objectives

After this topic, you should know:

- how the surface area to volume ratio varies depending on the size of an organism
- why large multicellular organisms need special systems for exchanging materials with the environment.

Synoptic links

You will use the idea of surface area to volume ratio when you study the adaptations of animals and plants for living in a variety of different habitats in Topics B15.7 and B15.8.

Synoptic links

You will find out much more about gas exchange in the lungs in Topic B4.5, and about the adaptations of the small intestine in Topic B3.2.

You can find out more about the transpiration stream in Topic B4.8.

For many single-celled organisms, diffusion, osmosis, and active transport are all that is needed to exchange materials with their environment because they have a relatively large surface area compared to the volume of the cell. This allows sufficient transport of molecules into and out of the cell to meet the needs of the organism.

Surface area to volume ratio

The surface area to volume ratio is very important in biology. It makes a big difference to the way animals can exchange substances with the environment. Surface area to volume ratio is also important when you consider how energy is transferred by living organisms, and how water evaporates from the surfaces of plants and animals.

Surface area to volume ratio

The ratio of surface area to volume falls as objects get bigger. You can see this clearly in Figure 1. In a small object, the surface area to volume (SA:V) ratio is relatively large. This means that the diffusion distances are short and that simple diffusion is sufficient for the exchange of materials.

As organisms get bigger, the surface area to volume ratio falls. As the distances between the centre of the organism and the surface get bigger, simple diffusion is no longer enough to exchange materials between the cells and the environment.

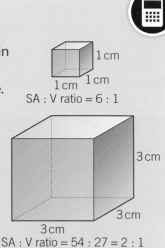

1 cm
1 cm
1 cm
SA : V ratio = 6 : 1

3 cm
3 cm
3 cm
SA : V ratio = 54 : 27 = 2 : 1

Figure 1 *Relationship of surface area to volume*

Getting bigger

As living organisms get bigger and more complex, their surface area to volume ratio gets smaller. This makes it increasingly difficult to exchange materials quickly enough with the outside world:

- Gases and food molecules can no longer reach every cell inside the organism by simple diffusion
- Metabolic waste cannot be removed fast enough to avoid poisoning the cells.

In many larger organisms, there are special surfaces where the exchange of materials takes place. These surfaces are adapted to be as effective as possible. You can find them in humans, in other animals, and in plants.

Adaptations for exchanging materials

There are various adaptations to make the process of exchange more efficient. The effectiveness of an exchange surface can be increased by:

● having a large surface area over which exchange can take place

● having a thin membrane or being thin to provide a short diffusion path

● in animals, having an efficient blood supply moves the diffusing substances away from the exchange surfaces and maintains a steep concentration (diffusion) gradient

● in animals, being **ventilated** makes gas exchange more efficient by maintaining steep concentration gradients.

Sacs for gas exchange

Muscular opening pumps water in and out

Finger-like folds on lining provide a large surface area and blood supply for gas exchange.

Figure 2 *Fitzroy river turtles can get oxygen from the water through a specialised excretory opening*

Examples of adaptations

Different organisms have very different adaptations for the exchange of materials. For example, the Australian Fitzroy river turtle (locally known as the bum-breathing turtle) can 'breathe' underwater (Figure 2). Inside the rear opening are two large sacs lined with finger-like folds which provide a large surface area and rich blood supply for gas exchange. The muscular opening pumps water in and out, ventilating the folds and maintaining a steep concentration gradient for gas exchange.

The human surface area to volume ratio is so low that the cells inside your body cannot possibly get the food and oxygen they need, or get rid of the waste they produce, by simple diffusion. Air is moved into and out of your lungs when you breathe, ventilating the millions of tiny air sacs called **alveoli**. The alveoli have an enormous surface area and a very rich blood supply, for effective gas exchange. The villi of the small intestine also provide a large surface area, short diffusion paths and a rich blood supply to make exchange of materials more effective.

Fish need to exchange oxygen and carbon dioxide between their blood and the water in which they swim. This happens across the gills, which are made up of stacks of thin filaments, each with a rich blood supply. Fish need a constant flow of water over their gills to maintain the concentration gradients needed for gas exchange. They get this by pumping water over the gills using a flap that covers the gills called the operculum.

Plant roots have a large surface area, made even bigger by the root hair cells, to make the uptake of water and mineral ions more efficient. Water constantly moves away from the roots in the transpiration stream, maintaining a steep concentration gradient in the cells.

Plant leaves are also modified to make gas and solute exchange as effective as possible. Flat, thin leaves, the presence of air spaces in the leaf tissues, and the **stomata** all help to provide a big surface area and maintain a steep concentration gradient for the diffusion of substances such as water, mineral ions, and carbon dioxide.

Key points

● Single-celled organisms have a relatively large surface area to volume ratio so all necessary exchanges with the environment take place over this surface.

● In multicellular organisms, many organs are specialised with effective exchange surfaces.

● Exchange surfaces usually have a large surface area and thin walls, which give short diffusion distances. In animals, exchange surfaces will have an efficient blood supply or, for gaseous exchange, be ventilated.

1 Describe two adaptations of an effective exchange surface. [2 marks]

2 Compare the gas exchange system of fish with that of the Australian Fitzroy river turtle shown in Figure 2. [5 marks]

3 a Explain how the surface area to volume ratio of an organism affects the way it exchanges materials with the environment. [2 marks]

 b Summarise the adaptations you would expect to see in effective exchange surfaces and explain the importance of each adaptation. [6 marks]

B1 Cell structure and transport

Summary questions

1 Summarise the importance of microscopes in the development of our understanding of cell biology. [6 marks]

2 a Name the structures labelled A–F in the bacterial cell in Figure 1. [6 marks]

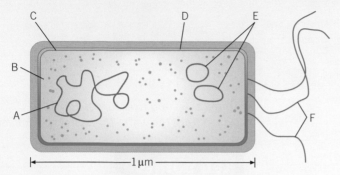

Figure 1

b Draw and label a typical eukaryotic cell to show the main characteristics and indicate the expected size range. [6 marks]

c Explain the similarities and differences between a bacterial cell and a plant cell. [6 marks]

d Some people think that structures found in plant and animal cells such as chloroplasts and mitochondria may have originally been free-living bacteria. Discuss this possibility using the relative sizes of prokaryotic cells, eukaryotic cells, and eukaryotic organelles to support the argument. [5 marks]

3 a State one similarity and one difference between diffusion and osmosis. [2 marks]

b State one similarity and one difference between diffusion and active transport. [2 marks]

c In an experiment to investigate osmosis, two Visking tubing bags were set up, with sugar solution inside the bags and water outside the bags. Bag A was kept at 20 °C and bag B was kept at 30 °C (Figure 2).

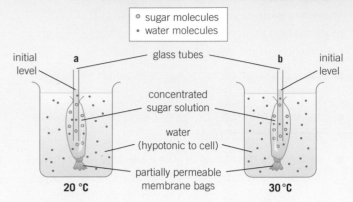

Figure 2

Describe what you would expect to happen and explain it in terms of osmosis and particle movements. [5 marks]

d Evaluate the use of these model cells in practical investigations to demonstrate the importance of osmosis in:
  i animal cells [2 marks]
  ii plant cells. [2 marks]

4 *Amoeba* is a single-celled animal that lives in ponds. It obtains oxygen for cell respiration from the water by diffusion across the cell membrane. Sticklebacks are small fish that live in the same habitat. They have a complicated structure of feathery gills to obtain oxygen. Water is pushed over the gills by muscular action.

a Explain why *Amoeba* can obtain sufficient oxygen for respiration by simple diffusion across its outer surface but the stickleback requires a special structure. [4 marks]

b Explain how the gills and the circulating blood will increase the diffusion of oxygen into the cells of the stickleback. [5 marks]

5 Exchanging materials with the outside world by diffusion is vital for most living organisms.

a Give **two** different adaptations that are found in living organisms to make this more efficient. [2 marks]

b For each adaptation in part **a**, explain how it makes the exchange process more efficient and give at least one example of where this adaptation is seen. [6 marks]

Practice questions

01   **Figure 1** shows an animal cell.

Figure 1

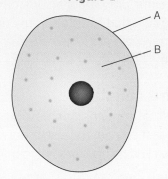

01.1   What is structure **A**?
Choose the correct answer from the following options.

chloroplast   chromosome   nucleus   cell membrane

[1 mark]

01.2   What is structure **B**?
Choose the correct answer from the following options.

cell membrane   cytoplasm   ribosome   vacuole

[1 mark]

01.3   How can you tell that the cell in **Figure 1** is an animal cell and not a plant cell?
Give **one** reason.   [1 mark]

02   **Figure 2** shows a drawing of a bacterial cell.

Figure 2

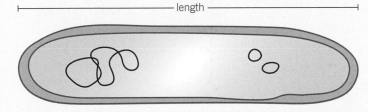

scale: |———|
0.5 micrometres

02.1   Use the scale to determine the length of the bacterial cell in micrometres.   [1 mark]
02.2   A different bacterial cell has a real size of 6.4 micrometres.
Use the equation to work out the image size of this cell when magnified 600 times.
Give your answer in millimetres.

$$\text{magnification} = \frac{\text{size of image}}{\text{size of real object}}$$

[3 marks]

02.3   Bacterial cells are prokaryotic cells.
Describe **two** differences between a prokaryotic cell and a eukaryotic cell.   [2 marks]

03   Match each cell structure to its description.

Cell structure	Description
Chloroplast	Controls what enters and leaves a cell
Mitochondrion	Ring of DNA
	Site of aerobic respiration
Plasmid	Site of photosynthesis
	Site of protein synthesis
Ribosome	Strengthens the cell

04.1   Describe the process by which water enters cells.   [3 marks]
04.2   **Figure 3** shows some equipment that can be used to determine the concentration of salt solution inside potato cells.

Figure 3

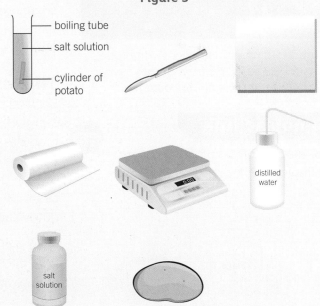

Describe how the equipment could be used to determine the concentration of salt solution inside potato cells.
You should include:

• variables that you would need to keep the same
• measurements you would need to take
• how you would conclude what the concentration of salt solution inside the potato cells is.   [6 marks]

Learning objectives

After this topic, you should know:

- the role of the chromosomes in cells
- the importance of the cell cycle
- how cells divide by mitosis.

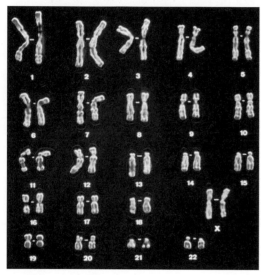

Figure 1 *This special image, a karyotype, shows the 23 pairs of chromosomes from a body cell of a female human being*

Synoptic link

For more information on genes look at Topics B12.4, B12.5 and B12.6 and about DNA in Topic B12.3.

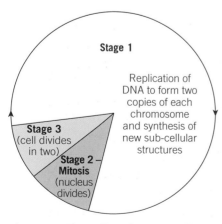

Stage 1

Replication of DNA to form two copies of each chromosome and synthesis of new sub-cellular structures

Stage 3 (cell divides in two)

Stage 2 – Mitosis (nucleus divides)

Figure 2 *The cell cycle. In rapidly growing tissue, stage 1 may only be a few hours, but in adult animals it can last for years*

New cells are needed for an organism, or part of an organism, to grow. They are also needed to replace cells that become worn out and to repair damaged tissue. However, the new cells must have the same genetic information as the originals so they can do the same job.

The information in the cells

Each of your cells has a nucleus that contains chromosomes. Chromosomes carry the genes that contain the instructions for making both new cells and all the tissues and organs needed to make an entire new you.

A gene is a small packet of information that controls a characteristic, or part of a characteristic, of your body. It is a section of DNA, the unique molecule that makes up your chromosomes.

Most of your characteristics are the result of many different genes rather than a single gene. The genes are grouped together on chromosomes. A chromosome may carry several hundred or even thousands of genes.

You have 46 chromosomes in the nucleus of your body cells. They are arranged in 23 pairs (see Figure 1). In each pair of chromosomes, one chromosome is inherited from the father and one from the mother. As such, sex cells (gametes) only have one chromosome from each pair, so only have 23 chromosomes in total.

The cell cycle and mitosis

Body cells divide in a series of stages known as the **cell cycle** (Figure 2). Cell division in the cell cycle involves a process called **mitosis** and it produces two identical cells. As a result, all your normal body cells have the same chromosomes and so the same genetic information. Cell division by mitosis produces the additional cells needed for growth and development in multicellular organisms, and for the replacement of worn out or damaged cells.

In asexual reproduction, the cells of the offspring are produced by mitosis from the cells of their parent. This is why they contain exactly the same genes as their parent with little or no genetic variation.

The cell cycle

The length of the cycle varies considerably. It can take less than 24 hours, or it can take several years, depending on the cells involved and the stage of life of the organism. The cell cycle is short as a baby develops before it is born, when new cells are being made all the time. It remains fairly rapid during childhood, but the cell cycle slows down once puberty is over and the body is adult. However, even in adults, there are regions where there is continued growth or a regular replacement of cells. They include the hair follicles, the skin, the blood, and the lining of the digestive system.

The cell cycle in normal, healthy cells follows a regular pattern (Figure 2):

● Stage 1: this is the longest stage in the cell cycle. The cells grow bigger, increase their mass, and carry out normal cell activities. Most importantly they replicate their DNA to form two copies of each chromosome ready for cell division. They also increase the number of sub-cellular structures such as mitochondria, ribosomes and chloroplasts ready for the cell to divide.

● Stage 2 – Mitosis: in this process one set of chromosomes is pulled to each end of the dividing cell and the nucleus divides.

● Stage 3: this is the stage during which the cytoplasm and the cell membranes also divide to form two identical daughter cells.

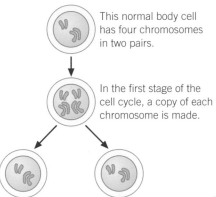

This normal body cell has four chromosomes in two pairs.

In the first stage of the cell cycle, a copy of each chromosome is made.

The cell divides in two to form two daughter cells, each with a nucleus containing four chromosomes identical to the ones in the original parent cell.

Figure 3 *Two identical cells are formed by mitotic division in the cell cycle. This cell is shown with only two pairs of chromosomes rather than 23*

In some parts of an animal or plant, mitotic cell division carries on rapidly all the time. For example, you constantly lose cells from the skin's surface and make new cells to replace them. In fact, about 300 million of your body cells die every minute, so cell division by mitosis is very important. In a child, mitotic divisions produce new cells faster than the old ones die. As an adult, cell death and mitosis keep more or less in balance. When you get very old, mitosis slows down and you show the typical signs of ageing.

1 Define the terms:
  a chromosome [1 mark]
  b gene [1 mark]
  c DNA. [1 mark]
2 Describe what happens during the three stages of the cell cycle. [6 marks]
3 a Explain why cell division by mitosis is so important in the body. [2 marks]
  b Suggest why it is important for the chromosome number to stay the same when the cells divide to make other normal body cells. [3 marks]

Observing cell division
View a special preparation of a growing root tip under a microscope. When cells divide, the membrane round the nucleus disappears and the chromosomes take up stains, making them relatively easy to see. You should be able to see the chromosomes dividing to form two identical nuclei.

● Describe your observations of mitosis.

Study tip

Remember – cells produced when the nucleus divides by mitosis are genetically identical.

Key points

● In body cells, chromosomes are found in pairs.
● Body cells divide in a series of stages called the cell cycle.
● During the cell cycle the genetic material is doubled. It then divides into two identical nuclei in a process called mitosis.
● Before a cell can divide it needs to grow, replicate the DNA to form two copies of each chromosome and increase the number of sub-cellular structures. In mitosis one set of chromosomes is pulled to each end of the cell and the nucleus divides. Finally the cytoplasm and cell membranes divide to form two identical cells
● Mitotic cell division is important in the growth, repair, and development of multicellular organisms.

B2.2 Growth and differentiation

Learning objectives

After this topic, you should know:
- how cell differentiation varies in animals and plants
- the production and use of plant clones.

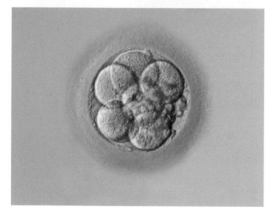

Figure 1 *This early embryo has only 8 cells – a lot of mitosis is needed before it becomes a teenager with around 3.7×10^{13} cells!*

At the moment of conception, a potential new human being is just one cell. By the time you are an adult, scientists have estimated that your body will contain around 37.2 trillion (3.72×10^{13}) cells – although estimates vary from 15 to 100 trillion! Almost all of these cells are the result of mitosis. The growth that takes place is amazing. Growth is a permanent increase in size as a result of cell division or cell enlargement (Figure 1).

The cells of your body, or any complex multicellular organism, are not all the same. They are not the same as the original cell either. This is because, as cells divide, grow and develop, they also begin to **differentiate**.

Differentiation in animal cells

In the early development of animal and plant embryos, the cells are unspecialised. Each one of them (known as a **stem cell**) can become any type of cell that is needed.

In animals, many types of cells become specialised very early in life. By the time a human baby is born, most of its cells are specialised to carry out a particular job, such as nerve cells, skin cells, or muscle cells. They have differentiated. Some of their genes have been switched on and others have been switched off. As a result, different types of specialised cells have different sub-cellular structures to carry out specific functions.

Most specialised cells can divide by mitosis, but they can only form the same sort of cell. Muscle cells divide to produce more muscle cells, for example. Some differentiated cells, such as red blood cells and skin cells, cannot divide at all and so **adult stem cells** replace dead or damaged cells. Nerve cells do not divide once they have differentiated and they are not replaced by stem cells. As a result, when nerve cells are damaged they are not usually replaced.

In a mature animal, little or no growth takes place. Cell division is almost entirely restricted to repair and replacement of damaged cells, and each differentiated cell type divides only to make more of the same cells.

Differentiation in plant cells

In contrast to animal cells, most plant cells are able to differentiate all through their lives. Undifferentiated cells are formed at active regions of the stems and roots, known as the meristems (Figure 2). In these areas, mitosis takes place almost continuously. The cells then elongate and grow before they finally differentiate.

Plants keep growing all through their lives at these 'growing points'. The plant cells produced do not differentiate until they are in their final position in the plant. Even then, the differentiation is not permanent. You can move a plant cell from one part of a plant to another. There it can redifferentiate and become a completely different type of cell. You cannot do that with animal cells – once a muscle cell, always a muscle cell.

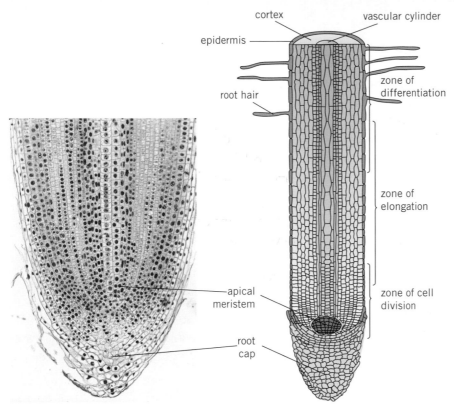

Figure 2 *The main zones of division, elongation, and differentiation in a plant root*

Cloning plants

Producing identical offspring is known as **cloning**. Huge numbers of identical plant clones can be produced from a tiny piece of leaf tissue. This is because, in the right conditions, a plant cell will become unspecialised and undergo mitosis many times. Each of these undifferentiated cells will produce more cells by mitosis. Given different conditions, these will then differentiate to form tissues such as xylem, phloem, photosynthetic cells, and root hair cells that are needed to form a tiny new plant. The new plant will be identical to the original parent.

It is difficult to clone animals because, as you have seen, most animal cells differentiate permanently early in embryo development. The cells cannot change back. As a result, artificial animal clones can only be made by cloning embryos in some way, although adult cells can be used to make an embryo.

1 **a** Define differentiation [1 mark]
   b Describe why differentiation is important in living organisms. [2 marks]

2 Explain how differentiation differs in animal and plant cells. [4 marks]

3 Calculate by what order of magnitude an adult human is bigger than the original fertilised ovum (Topic B1.3). [4 marks]

4 Discuss how the difference in differentiation patterns affects our ability to clone plants and animals. [6 marks]

Synoptic links

You learnt about some specialised cells that result from differentiation in Topic B1.4 and Topic B1.5.

You will learn more about the results of differentiation in Topic B3.1.

Study tip

Cells produced by mitosis are genetically identical to the parent cell.

Key points

- In plant cells, mitosis takes place throughout life in the meristems found in the shoot and root tips.
- Many types of plants cells retain the ability to differentiate throughout life.
- Most types of animal cell differentiate at an early stage of development.

B2.3 Stem cells

Learning objectives

After this topic, you should know:

- how stem cells are different from other body cells
- the functions of stem cells in embryos, in adult animals, and in plants
- how treatment with stem cells may be used to treat people with different medical conditions.

Synoptic links

You will learn more about the spinal nerves in Topic B10.2.

You will learn more about insulin and the control of blood glucose levels in Topic B11.2 and Topic B11.3.

Figure 1 *This is what the world looks like to someone with macular degeneration. The light-sensitive cells in the middle of their retina stop working. Soon stem cell therapy might be able to restore the lost vision*

The function of stem cells

An egg and sperm cell fuse to form a **zygote**, a single new cell. That cell divides and becomes a hollow ball of cells – the embryo. The inner cells of this ball are the **embryonic stem cells** that differentiate to form all of the specialised cells of your body. Even when you are an adult, some of your stem cells remain. An adult stem cell is an undifferentiated cell of an organism that can give rise to many more cells of the same type. Certain other types of cell can also arise from stem cells by differentiation. Your bone marrow is a good source of **adult stem cells**. Scientists now think there may be a tiny number of stem cells in most of the different tissues in your body including your blood, brain, muscle, and liver.

Many of your differentiated cells can divide to replace themselves. However, some tissues cannot do this and stem cells can stay in these tissues for years, only needed if the cells are injured or affected by disease. Then they start dividing to replace the different types of damaged cell.

Using stem cells

Many people suffer and even die because parts of their body stop working properly. For example, spinal injuries can cause paralysis, because the spinal nerves cannot repair themselves. People with type 1 diabetes have to inject themselves with insulin every day because specialised cells in their pancreas do not work. Millions of people would benefit if we could replace damaged or diseased body parts.

In 1998, there was a breakthrough. Two scientists managed to culture human embryonic stem cells, capable of forming other types of cell. Scientists hope that the embryonic stem cells can be encouraged to grow into almost any different type of cell needed in the body. Already scientists have used nerve cells grown from embryonic stem cells to restore some movement to the legs of paralysed rats. In 2010, the first trials testing the safety of injecting nerve cells grown from embryonic stem cells into the spinal cords of paralysed human patients were carried out. The scientists and doctors hope it will not be long before they can use stem cells to help people who have been paralysed to walk again.

In 2014, doctors transplanted embryonic stem cells into the eyes of people going blind as a result of macular degeneration (Figure 1). It was a small study to check the safety of the technique but all of the patients found they could see better. Larger trials are now taking place. Scientists are also using different types of stem cells to try and grow cells that are sensitive to blood sugar levels and produce the hormone insulin to help treat people with diabetes.

We might also be able to grow whole new organs from embryonic stem cells. These could then be used in transplant surgery (Topic B2.4). Conditions from infertility to dementia could eventually be treated using stem cells.

Stem cells in plants

The stem cells from plant meristems can be used to make clones of the mature parent plant very quickly and economically. This is important as it gives us a way of producing large numbers of rare plants reliably and safely. We may be able to save some rare plants from extinction in this way. Plant cloning also gives us a way of producing large populations of identical plants for research. This is important as scientists can change variables and observe the effects on genetically identical individuals.

Cloning large numbers of identical plants from the stem cells in plant meristems is also widely used in horticulture producing large numbers of plants such as orchids for sale (Figure 2). In agriculture it is used to produce large numbers of identical crop plants with special features, such as disease resistance. For example, every banana you eat is produced by a cloned plant.

Figure 2 *Cloning exotic plants, like this orchid, from plant stem cells makes them relatively cheap and available for everyone to enjoy*

1 **a** Identify the differences between a stem cell and a normal body cell.
    [4 marks]
   b Give three sources of stem cells. [3 marks]
2 Identify the advantages of using stem cells to treat diseases. [4 marks]
3 **a** Explain why the ability to clone large numbers of individual plants is such an advantage in plant research [3 marks]
   b Suggest how it may enable us to save rare species of plants from extinction. ⓘ [3 marks]

Key points

- Embryonic stem cells (from human embryos) and adult stem cells (from adult bone marrow) can be cloned and made to differentiate into many different types of cell.
- Treatment with stem cells may be able to help conditions such as paralysis and diabetes.
- Stem cells from plant meristems are used to produce new plant clones quickly and economically for research, horticulture, and agriculture.

B2.4 Stem cell dilemmas

Learning objectives

After this topic, you should know:

- the process of therapeutic cloning
- some of the potential benefits, risks, and social and ethical issues of the use of stem cells in medical research and treatments.

As you saw in Topic B2.3, there are many potential benefits in using stem cells in human medicine and they are gradually being used to treat real patients. However, the technology is still very new so there are still practical risks as well as social and ethical issues raised by the use of stem cells in both medical research and in treatments.

Problems with embryonic stem cells

Many embryonic stem cells come from aborted embryos. Others come from spare embryos from fertility treatment, donated because they will not otherwise be used. Some people question the use of a potential human being as a source of cells, even to cure others. Some people feel that, as the embryo cannot give permission, using it is a violation of its human rights. The religious beliefs of others mean they cannot accept any interference with the process of human reproduction.

In addition, progress in developing therapies using embryonic stem cells has been relatively slow, difficult, expensive, and hard to control. However, it is easy to forget that scientists have only been working with them for around 20 years. The signals that control cell differentiation are still not completely understood. Not surprisingly it is proving difficult to persuade embryonic stem cells to differentiate into the type of cells needed to treat patients.

Embryonic stem cells divide and grow rapidly. This is partly why they are potentially so useful but there is some concern that embryonic stem cells might cause cancer if they are used to treat people. This has sometimes been a problem when they have been used to treat mice and in early human treatments for autoimmune diseases.

There is a risk that adult stem cells might be infected with viruses, and so could transfer the infections to patients. If stem cells from an adult are used to treat another unrelated person, they may trigger an immune response. The patient may need to take immunosuppressant drugs to stop their body rejecting the new cells. Scientists hope embryonic stem cells will solve this problem. The body of a mother does not reject the embryo, so they hope that embryonic stem cells will not be rejected by the patient.

Some people feel that a great deal of money and time is being wasted on stem cell research that would be better spent on research into other areas of medicine. Yet in spite of all these concerns, there is a lot of investment into stem cell research as many scientists and doctors are convinced stem cells have the potential to benefit many people.

Figure 1 *Dream Alliance won the Welsh Grand National after revolutionary stem cell treatment on a badly damaged tendon. Doctors hope people will soon have the same benefits*

The future of stem cell research

Scientists have found embryonic stem cells in the umbilical cord blood of newborn babies and even in the amniotic fluid that surrounds the fetus as it grows. Using these instead of cells from spare embryos may help to overcome some of the ethical concerns about their use.

Scientists are also finding ways of growing adult stem cells, although so far they have only managed to develop them into a limited range of cell types. Adult stem cells avoid the controversial use of embryonic tissue. They have been used successfully to treat some forms of heart disease and to grow some new organs such as tracheas (windpipes).

The area of stem cell research known as **therapeutic cloning** (Figure 2) has much potential but is proving very difficult. It involves using cells from an adult to produce a cloned early embryo of themselves. This would provide a source of perfectly matched embryonic stem cells. In theory, these could then be used for medical treatments such as growing new organs for the original donor. The new organs would not be rejected by the body because they have been made from the body's own cells and have the same genes.

Scientists have discovered stem cells in some of the tubes that connect the liver and the pancreas to the small intestine. They have managed to make these cells turn into the special insulin-producing cells in the pancreas that are so important for controlling blood sugar. These are the cells that are missing or destroyed in people with type 1 diabetes. Scientists have transplanted these modified stem cells into diabetic mice, which worked to control the blood sugar levels. The next stage is to work towards the same success in humans.

At the moment, after years of relatively slow progress, hopes are high again that stem cells will change the future of medicine. Currently, in the UK, stem cell research is being carried out into potential therapies to treat:

- spinal cord after injuries
- diabetes
- heart after damage in a heart attack
- eyesight in the blind
- damaged bone and cartilage.

It is not known how many of these hopes will be fulfilled – only time will tell.

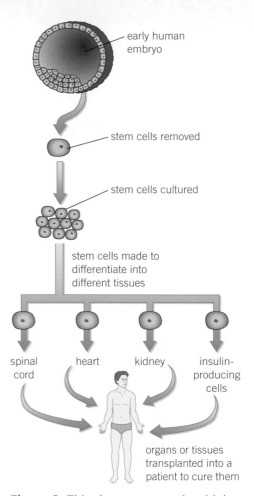

Figure 2 *This shows one way in which scientists hope embryonic stem cells might be formed into adult cells and used as human treatments in the future*

1 Describe three areas of medical research where the use of stem cells could provide valuable medical treatments. [3 marks]

2 Summarise the main arguments for and against the use of embryonic stem cells in medical research. [4 marks]

3 Explain how scientists are hoping to overcome the ethical objections to using embryonic stem cells in their research. 🖉 [5 marks]

Key points

- Treatment with stem cells, from embryos or adult cell cloning, may be able to help with conditions such as diabetes.
- In therapeutic cloning, an embryo is produced with the same genes as the patient so the stem cells produced are not rejected and may be used for medical treatment.
- The use of stem cells has some potential risks and some people have ethical or religious objections.

B2 Cell division

Summary questions

1. a What is the cell cycle? [1 mark]
   b Explain how and why you would expect the length of the cell cycle to vary:
   i between an early embryo in the first days after fertilisation and a 5-year-old child [4 marks]
   ii between a 13-year-old student and a 70-year-old adult. [5 marks]

2. a State what mitosis is and explain its role in the cell cycle. [3 marks]
   b Explain, using diagrams, the stages of the cell cycle. [5 marks]
   c The cell cycle is very important during the development of a baby from a fertilised egg. It is also important all through life. Explain why. [5 marks]
   d The rate of the cell cycle can vary greatly.
   i Which stage of the cell cycle is variable? [1 mark]
   ii Discuss when the cell cycle is likely to be very rapid in an human being, and when it is likely to be relatively slow. [5 marks]

3. a State what stem cells are. [2 marks]
   b It is hoped that many different medical problems may be cured using stem cells. Explain how this might work. [4 marks]
   c There are some ethical issues associated with the use of embryonic stem cells. Explain the arguments
   i for and [4 marks]
   ii against their use. [4 marks]

4. Plants have stem cells just as animals do.
   a Where would you expect to find stem cells in a plant? [2 marks]
   b Describe how plant stem cells differ from animal stem cells. [2 marks]
   c Suggest two examples of the use of plant stem cells to produce plant clones. For each explain the advantages of using cloning from stem cells over normal plant reproduction. [6 marks]

5. In 2014, a paper in the journal *Cell Transplantation* reported on the case of a Polish man whose spinal cord had been severed in a stabbing, causing paralysis. He had been given a novel treatment – cells from the olfactory lobe of his brain (the area that analyses smells) were grown in culture and then injected around the site of the injury. Thin strips of nerves were attached between the two ends of the spinal cord to give a framework for recovery. He was given intensive physiotherapy. Over a period of months the patient began to recover some control of his legs. He also regained some control of his bladder, bowel, and sexual function. He can now walk with a frame.
   a The cells he was given were not stem cells – they were cells that encourage the growth of nerve cells – but the research was supported and funded by groups involved in stem cell research. Suggest why stem cells might also be involved in this type of therapy in the future. [2 marks]
   b The sight of a paralysed man walking, with difficulty, and the apparent regeneration of some of his spinal cord, caused a lot of excitement in the media. The scientists involved were very cautious but excited. Other scientists were very reluctant to see it as major progress. Evaluate this research from the information given here and indicate some of the problems still to be overcome before there is a useful cure for spinal injuries. [6 marks]

6. The racehorse Dream Alliance (Topic B2.4, Figure 1) won the 2009 Welsh Grand National after stem cell treatment for a badly injured tendon in his leg.
   a Tendons do not heal easily. Explain how stem cells might help overcome such an injury. [3 marks]
   b Horses and other animals have been having successful stem cell treatment for tendon injuries for around 10 years. People also suffer from tendon injuries but human trials are still in the relatively early stages. Suggest reasons why human treatments are so far behind those used by vets. [4 marks]

Practice questions

01.1 Which of the following structures is the **smallest**?

| cell | chromosome | gene | nucleus |

[1 mark]

01.2 Name the molecule that makes up chromosomes.
[1 mark]

02 A body cell of a cat has 18 pairs of chromosomes that control characteristics, plus one pair of chromosomes that determines the sex of the cat. Body cells divide by a process called mitosis.

02.1 State how many new daughter cells are formed when a cell divides by mitosis. [1 mark]

02.2 State how many chromosomes each new cat cell will contain after mitosis. [1 mark]

02.3 Why is mitosis important to organisms? Give **two** reasons. [2 marks]

03 **Figure 1** shows a simplified diagram of the cell cycle.

Figure 1

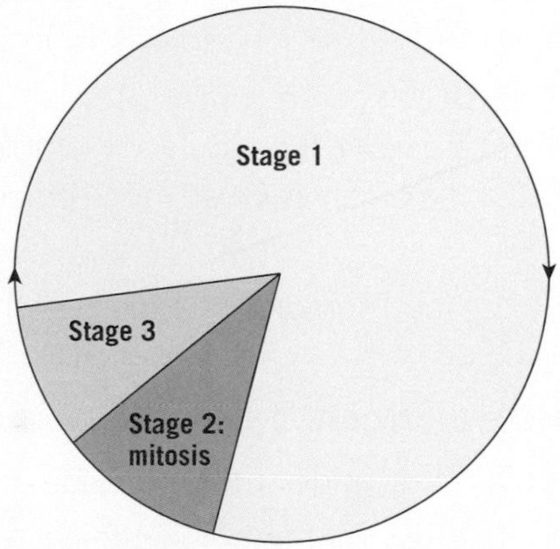

03.1 **H** The length of the cell cycle varies greatly between different types of cell.
In which of the following cells would the cycle be the shortest?

| an adult brain cell | an embryonic cell |
| a teenager's kidney cell | an unfertilised egg cell |

[1 mark]

03.2 Suggest a reason for your answer to **03.1**.
[1 mark]

03.3 A student prepares a microscope slide with cells from a plant shoot and counts the number of cells he can see in the field of view. He calculates that 17% of the cells are in Stage 3 of the cell cycle. If 3 cells are in Stage 3 of the cell cycle, calculate the total number of cells that the student can see.
[2 marks]

03.4 **H** Look at **Figure 1**.
Describe the changes that occur during each Stage of the cell cycle. [6 marks]

04 A scientist is studying a rare plant she found in the wild. She finds that when she cuts off a leaf, a new leaf will grow out of the stem.

04.1 Explain how a new leaf can grow even though there is no trace left of the original leaf. [2 marks]

04.2 Unlike the plant, most animals **cannot** grow back parts of their body after they have been cut off. Explain why this is. [2 marks]

04.3 The scientist wants more specimens of this rare plant for testing, but is unable to get hold of any others. Suggest one way in which the scientist could obtain more specimens. [1 mark]

05 Stem cells are undifferentiated cells. They divide to produce many identical cells from which other cells can develop by differentiation.

05.1 Name **one** part of the adult human body in which stem cells are relatively common. [1 mark]

05.2 Stem cells are used to treat some human diseases such as leukaemia. In the future, they may be used to help conditions such as diabetes and paralysis, but further research is required.
Therapeutic cloning is a technique in which embryonic stem cells are produced with the same DNA as the patient.
The nucleus is removed from a donated egg cell and replaced with the nucleus of a body cell from the patient. The cell is stimulated to divide and form a cloned embryo of the patient.
Stem cells are removed from the embryo and the embryo dies. These cells are cultured to form many embryonic stem cells, which can be stimulated to differentiate and form the required cells.
Explain the benefits and issues of using therapeutic cloning in medicine. [6 marks]

3.1 Tissues and organs

Learning objectives

After this topic, you should know:

- how specialised cells become organised into tissues
- how several different tissues work together to form an organ.

As you have seen, cells are the basic building blocks of all living organisms. Unicellular and simple multicellular organisms carry out all the exchanges they need across their cell membranes. Large multicellular organisms may contain billions of cells and they have to overcome the problems linked to their size. They have evolved different ways of exchanging materials. During the development of a multicellular organism, cells **differentiate**, becoming specialised to carry out particular jobs. However, the adaptations of multicellular organisms go beyond specialised cells. Similar specialised cells are often found grouped together to form a tissue.

Tissues

A **tissue** is a group of cells with similar structure and function working together. For example, muscular tissue can contract to bring about movement (Figure 1). Glandular tissue contains secretory cells that can produce and release substances such as enzymes and hormones. Epithelial tissue covers the outside of your body as well as your internal organs.

Organs

Organs are collections of tissues. Each organ contains several tissues, all working together to perform a specific function. For example, the stomach, as shown in Figure 3, is an organ involved in the digestion of food. It contains:

- muscular tissue, to churn the food and digestive juices of the stomach together
- glandular tissue, to produce the digestive juices that break down food
- epithelial tissue, which covers the inside and the outside of the organ.

The pancreas is an organ that has two important functions. It makes hormones to control blood sugar, as well as some of the enzymes that digest food. It contains two very different types of tissue, which produce these different secretions (Figure 2).

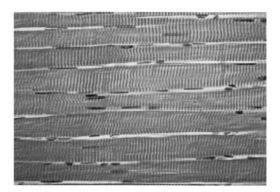

Figure 1 *Muscle tissue contracts to move your skeleton around*

Synoptic links

For more information on specialised cells, look back at Topic B1.4 and Topic B1.5.

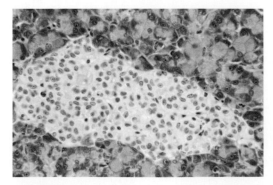

Figure 2 *The pancreas showing the tissue that makes hormones (stained yellow) and the tissue that makes enzymes (stained red)*

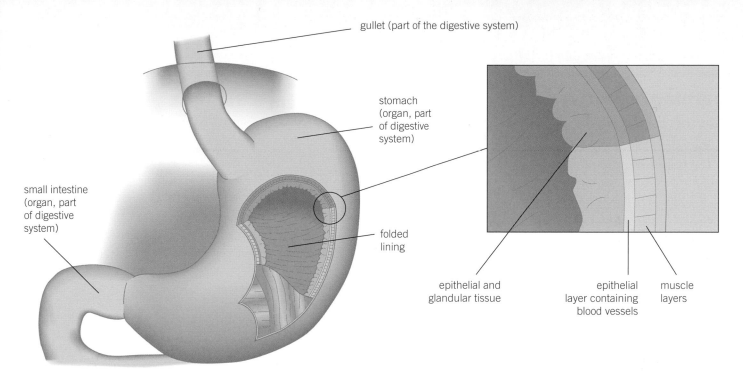

gullet (part of the digestive system)

stomach (organ, part of digestive system)

small intestine (organ, part of digestive system)

folded lining

epithelial and glandular tissue

epithelial layer containing blood vessels

muscle layers

Figure 3 *The stomach contains several different tissues, each with a different function in the organ*

Organ systems

A whole multicellular organism is made up of a number of **organ systems** working together. Organ systems are groups of organs that all work together to perform specific functions. The way in which one organ functions often depends on other organs in the system. Organ systems work together to form organisms. Organ systems in the human body include the digestive system, the circulatory system, and the gas exchange system. All of these systems have adaptations in some of their organs that make them effective as exchange surfaces. These adaptations include features to increase the surface area of part of an organ system, a rich blood supply to areas where exchange takes place, areas with short diffusion distances for exchange, and mechanisms to increase the concentration gradients by ventilating surfaces or moving materials on.

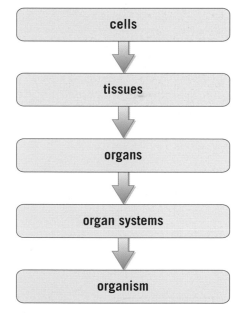

cells

↓

tissues

↓

organs

↓

organ systems

↓

organism

Figure 4 *Larger multicellular organisms have many levels of organisation*

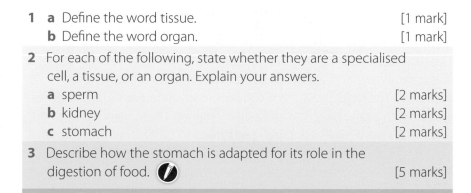

1　a　Define the word tissue.　　[1 mark]
　　b　Define the word organ.　　[1 mark]
2　For each of the following, state whether they are a specialised cell, a tissue, or an organ. Explain your answers.
　　a　sperm　　[2 marks]
　　b　kidney　　[2 marks]
　　c　stomach　　[2 marks]
3　Describe how the stomach is adapted for its role in the digestion of food. 🖉　　[5 marks]

Key points

- A tissue is a group of cells with similar structure and function.
- Organs are collections of tissues performing specific functions.
- Organs are organised into organ systems, which work together to form organisms.

B3.2 The human digestive system

Learning objectives

After this topic, you should know:

● the position of the main organs of the human digestive system.

Study tip

Learn the names of the parts of the digestive system. Make sure you know the difference between the larger, lobed liver and the thinner leaf-like pancreas.

The digestive system

Your digestive system is between 6 and 9 m long – 9 million times or 6 orders of magnitude longer than an average human cell! The digestive system of humans and other mammals exchanges substances with the environment. The food you take in and eat is made up of large insoluble molecules. Your body cannot absorb and use these molecules. They need to be broken down or digested to form smaller, soluble molecules that can then be absorbed and used by your cells. This process of digestion takes place in your **digestive system**, one of the major organ systems of the body.

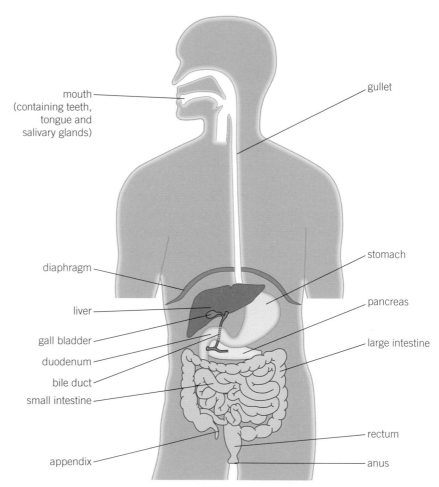

mouth (containing teeth, tongue and salivary glands)

gullet

diaphragm

stomach

liver

pancreas

gall bladder

duodenum

large intestine

bile duct

small intestine

appendix

rectum

anus

Figure 1 *The main organs of the human digestive system*

The digestive system is a muscular tube that squeezes your food through it. It starts at one end with your mouth, and finishes at the other with your anus. The digestive system contains many different organs. There are glands such as the pancreas and salivary glands. These glands make and release digestive juices containing **enzymes** to break down your food.

The stomach and the small intestine are the main organs where food is digested. Enzymes break down the large insoluble food molecules into smaller, soluble ones. Your small intestine is also where the soluble food molecules are absorbed into your blood. Once there, they get transported in the bloodstream around your body. The small intestine is adapted to have a very large surface area as it is covered in villi. It also has a good blood supply and short diffusion distances to the blood vessels. This greatly increases diffusion and active transport from the small intestine to the blood.

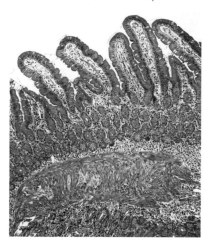

Figure 2 *The large surface area of the villi of the small intestine helps make it possible to absorb the digested food molecules from the gut into the blood*

The muscular walls of the small intestine squeeze the undigested food onwards into your large intestine. This is where water is absorbed from the undigested food into your blood. The material left forms the faeces. Faeces are stored and then pass out of your body through the rectum and anus back into the environment.

Other organs associated with the digestive system include the liver. The liver is a large organ that carries out many different functions in your body. The function of the liver that is most closely linked to the digestive system is the production of bile, which helps in the digestion of lipids.

1 Match each of the following organs to its correct function.

A	Liver	1	Breaking down large insoluble molecules into smaller soluble molecules and absorption
B	Stomach	2	Absorbing water from undigested food
C	Small intestine	3	Producing bile
D	Large intestine	4	Breaking down large insoluble molecules into smaller soluble molecules

[4 marks]

2 Explain the difference between organs and organ systems, giving two examples. [4 marks]

3 Using the human digestive system as an example, explain how the organs in an organ system rely on each other to function properly. 🖊 [6 marks]

Synoptic links

You can remind yourself about the adaptations of the villi in the small intestine as an exchange surface in Topic B1.10.

Key points

- Organ systems are groups of organs that perform specific functions in the body.
- The digestive system in a mammal is an organ system where several organs work together to digest and absorb food.

B3.3 The chemistry of food

Carbohydrates, lipids, and proteins are the main compounds that make up the structure of a cell. They are vital components in the balanced diet of any organism that cannot make its own food. Carbohydrates, lipids, and proteins are all large molecules that are often made up by smaller molecules joined together as part of the cell metabolism.

Carbohydrates

Carbohydrates provide us with the fuel that makes all of the other reactions of life possible. They contain the chemical elements carbon, hydrogen, and oxygen.

All carbohydrates are made up of units of sugars.

- Some carbohydrates contain only one sugar unit. The best known of these single sugars is glucose, $C_6H_{12}O_6$. Other carbohydrates are made up of two sugar units joined together, for example sucrose, the compound we call 'sugar' in everyday life. These small carbohydrate units are referred to as **simple sugars**.

- Complex carbohydrates such as starch and cellulose are made up of long chains of simple sugar units bonded together (Figure 1).

Carbohydrate-rich foods include bread, potatoes, rice, and pasta. Most of the carbohydrates you eat will be broken down to glucose used in cellular respiration to provide energy for metabolic reactions in your cells. The carbohydrate cellulose is an important support material in plants.

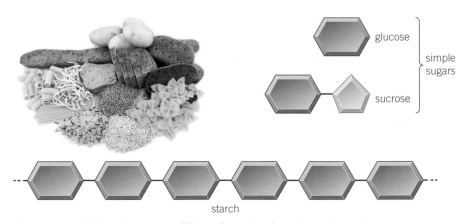

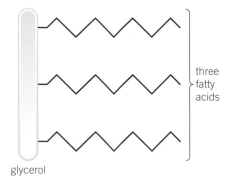

glycerol

three fatty acids

Figure 1 *Carbohydrates are all based on simple sugar units*

Lipids

Lipids are fats (solids) and oils (liquids). They are the most efficient energy store in your body and an important source of energy in your diet. Combined with other molecules, lipids are very important in your cell membranes, as hormones, and in your nervous system. Like carbohydrates, lipids are made up of carbon, hydrogen, and oxygen. All lipids are insoluble in water.

Figure 2 *Lipids are made of three molecules of fatty acids joined to a molecule of glycerol*

Lipids are made up of three molecules of **fatty acids** joined to a molecule of **glycerol** (Figure 2). The glycerol is always the same, but the fatty acids vary. Lipid-rich food includes all the oils, such as olive oil and corn oil, as well as butter, margarine, cheese, and cream. The different combination of fatty acids affects whether the lipid will be a liquid oil or a solid fat.

Proteins

Proteins are used for building up the cells and tissues of your body, as well as the basis of all your enzymes. Between 15 and 16% of your body mass is protein. Protein is found in tissues ranging from your hair and nails to the muscles that move you around and the enzymes that control your body chemistry. Proteins are made up of the elements carbon, hydrogen, oxygen, and nitrogen. Protein-rich foods include meat, fish, pulses, and cheese.

A protein molecule is made up of long chains of small units called **amino acids** (Figure 3). There are around 20 different amino acids, and they are joined together into long chains by special bonds. Different arrangements of the various amino acids give you different proteins.

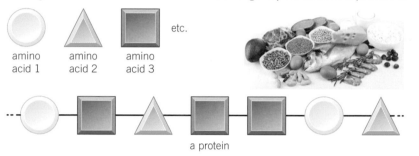

Figure 3 *Amino acids are the building blocks of proteins. They can join in an almost endless variety of ways to produce different proteins*

The long chains of amino acids that make up a protein are folded, coiled, and twisted to make specific 3D shapes. It is these specific shapes that enable other molecules to fit into the protein. The bonds that hold the proteins in these 3D shapes are very sensitive to temperature and pH, and can easily be broken. If this happens, the shape of the protein is lost and it may not function any more in your cells. The protein is **denatured**.

Proteins carry out many different functions in your body. They act as:

- structural components of tissues such as muscles and tendons
- hormones such as insulin
- antibodies, which destroy pathogens and are part of the immune system
- enzymes, which act as catalysts.

1 **a** State what a protein is. [1 mark]
   b Describe how proteins are used in the body. [4 marks]

2 Describe the main similarities and differences between the three main groups of chemicals (carbohydrates, proteins, and lipids) in the body. 🖊 [6 marks]

3 Describe how you would test a food sample to see if it contained:
   a starch [2 marks]
   b lipids. [2 marks]

4 Explain why lipids can be either fats or oils. [3 marks]

5 Explain how simple sugars are related to complex carbohydrates. 🖊 [3 marks]

Key points

- Carbohydrates are made up of units of sugar.
- Simple sugars are carbohydrates that contain only one or two sugar units – they turn blue Benedict's solution brick red on heating.
- Complex carbohydrates contain long chains of simple sugar units bonded together. Starch turns yellow-red iodine solution blue-black.
- Lipids consist of three molecules of fatty acids bonded to a molecule of glycerol. The ethanol test indicates the presence of lipids in solutions.
- Protein molecules are made up of long chains of amino acids. Biuret reagent turns from blue to purple in the presence of proteins.

B3.4 Catalysts and enzymes

Learning objectives

After this topic, you should know:

- what a catalyst is
- how enzymes work as biological catalysts
- what the metabolism of the body involves.

In everyday life, you control the rate of chemical reactions all the time. You increase the temperature of your oven to speed up chemical reactions when you cook, and you cool food down in the fridge to slow down the reactions that cause food to go off.

Sometimes people use special chemicals known as **catalysts** to speed up reactions. A catalyst speeds up a chemical reaction, but it is not used up in the reaction. You can use a catalyst over and over again.

Enzymes – biological catalysts

In your body, the rate of chemical reactions is controlled by enzymes. These are special biological catalysts that speed up reactions. Each enzyme interacts with a particular substrate (reactant).

Enzymes are large protein molecules. The shape of an enzyme is vital for the enzyme to function. The long chains of amino acids are folded to produce a molecule with an **active site** that has a unique shape so it can bind to a specific substrate molecule.

How do enzymes work?

The lock and key theory is a simple model of how enzymes work. The substrate of the reaction to be catalysed fits into the active site of the enzyme. You can think of it like a lock and key. Once it is in place, the enzyme and the substrate bind together. The reaction then takes place rapidly and the products are released from the surface of the enzyme (Figure 2). Remember that enzymes can join small molecules together as well as break up large ones. There are other, more complex models of how enzymes work but they are all based on the lock and key theory.

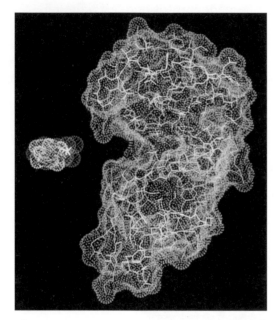

Figure 1 *Enzymes are made up of chains of amino acids folded together to make large complex molecules, as you can see in this computer-generated image*

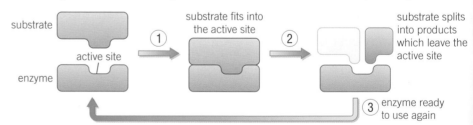

substrate — substrate fits into the active site — substrate splits into products which leave the active site

active site

enzyme

① ② ③ enzyme ready to use again

Figure 2 *Enzymes act as catalysts using the 'lock and key' mechanism shown here*

Metabolic reactions

Enzymes do not change a reaction in any way, they just make it happen faster. Enzymes control the **metabolism** – that is the sum of all the reactions in a cell or in the body. Different enzymes catalyse (speed up) specific types of metabolic reactions:

- Building large molecules from lots of smaller ones. This includes building starch, glycogen or cellulose from glucose; lipids from fatty acids; or proteins from amino acids. Plant cells also combine carbon dioxide and water to make glucose, and use glucose and nitrate ions to make amino acids.

Study tip

Remember that it is the shape of the active site of the enzyme that allows it to bind with the substrate.

- Changing one molecule into another. This includes changing one simple sugar into another, such as glucose to fructose, and converting one amino acid into another.

- Breaking down large molecules into smaller ones. This includes breaking down carbohydrates, lipids, and proteins into their constituent molecules during digestion; breaking down glucose in cellular respiration; and breaking down excess amino acids to form urea, and other molecules that can be used in respiration.

Each of your cells can have a hundred or more chemical reactions going on within it at any one time. Each of the different types of reaction is controlled by a different specific enzyme. Enzymes deliver the control that makes it possible for your cell chemistry to work without one reaction interfering with another.

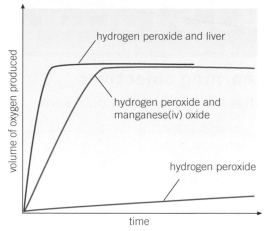

Figure 3 *The decomposition of hydrogen peroxide to oxygen and water happens much faster using a catalyst. The reaction takes place faster when catalysed by enzymes (found in liver) than when catalysed by manganese(IV) oxide*

Breaking down hydrogen peroxide

You can investigate the impact of both an inorganic catalyst and an enzyme on the breakdown of dilute hydrogen peroxide solution into oxygen and water using:

a manganese(IV) oxide (an inorganic catalyst) and

b raw liver or potato (which contain the enzyme catalase).

Hydrogen peroxide is a poisonous compound that is often a waste product of reactions in cells. It breaks down slowly itself but it is important that it gets broken down into harmless oxygen and water quickly, before it causes any damage.

You can determine the rate of the reaction by measuring the volume of oxygen produced over time. A simple way to do a quick comparison between the inorganic catalyst and the enzyme is to add a drop of washing-up liquid to the hydrogen peroxide. Add the inorganic catalyst or the enzyme (the liver or potato) and measure how quickly the foam produced by the bubbles of gas rises up the test tube!

- Describe your observations and interpret the graph (Figure 3).

Safety: Wear eye protection. 20 vol hydrogen peroxide – irritant. Manganese(IV) oxide – harmful.

1 Define each of the following terms:
  a a catalyst [1 mark]
  b an enzyme [1 mark]
  c the active site of an enzyme. [1 mark]
2 **a** State what enzymes are made of. [1 mark]
  b Explain in detail how enzymes act to speed up reactions in your body. [5 marks]

3 **a** Give three clear examples of the type of reactions that are catalysed by enzymes. [3 marks]
  b Explain how enzymes are important in the metabolism of a cell or organism. [6 marks]

Key points

- Catalysts increase the rate of chemical reactions without changing chemically themselves.
- Enzymes are biological catalysts and catalyse specific reactions in living organisms due to the shape of their active site. This is the lock and key theory of enzyme action.
- Enzymes are proteins. The amino acid chains are folded to form the active site, which matches the shape of a specific substrate molecule.
- The substrate binds to the active site and the reaction is catalysed by the enzyme.
- Metabolism is the sum of all the reactions in a cell or the body.

B3.5 Factors affecting enzyme action

Learning objectives

After this topic, you should know:

- how temperature and pH affect enzyme action
- different enzymes work fastest at different temperatures and pH values.

Go further

Extremophiles are organisms that live in the most extreme environments. Many extremophiles are prokaryotes, and their enzymes have chemical adaptations enabling them to function in extremes of saltiness, high temperature, and cold. We use enzymes from extremophile bacteria that live in hot springs in the polymerase chain reaction (PCR). PCR is a key process in DNA fingerprinting and genome analysis.

Figure 2 *In an extreme environment, such as the hot springs found in Iceland, it is amazing that any enzymes function at all*

A container of milk left at the back of your fridge for a week or two will be disgusting. The milk will go off as enzymes in bacteria break down the protein structure. Leave your milk in the sun for a day and the same thing happens – but much faster. Temperature affects the rate at which chemical reactions take place, even when they are controlled by biological catalysts.

Biological reactions are affected by the same factors as any other chemical reactions. These factors include concentration, temperature, and surface area. However, in living organisms, an increase in temperature only increases the rate of reaction up to a certain point.

The effect of temperature on enzyme action

The reactions that take place in cells happen at relatively low temperatures. As with other reactions, the rate of enzyme-controlled reactions increases as the temperature increases.

However, for most organisms this is only true up to temperatures of about 40 °C. After this, the protein structure of the enzyme is affected by the high temperature. The long amino acid chains begin to unravel, and as a result, the shape of the active site changes. The substrate will no longer fit in the active site. The enzyme is said to have been **denatured**. It can no longer act as a catalyst, so the rate of the reaction drops dramatically. Most human enzymes work best at 37 °C, which is human body temperature.

Without enzymes, none of the reactions in your body would happen fast enough to keep you alive. This is why it is dangerous if your temperature goes too high when you are ill. Once your body temperature reaches about 41 °C, your enzymes start to be denatured, which will result in death.

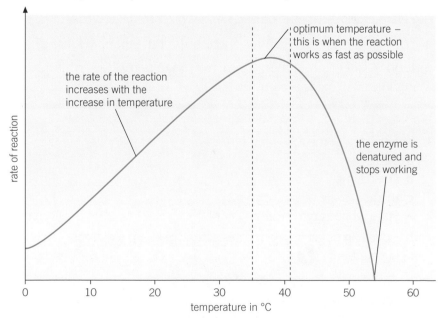

Figure 1 *The rate of an enzyme-controlled reaction increases as the temperature rises – but only until the protein structure of the enzyme breaks down*

Not all enzymes work best at around 40 °C. Bacteria living in hot springs survive at temperatures up to 80 °C and higher (Figure 2). On the other hand, some bacteria that live in the very cold, deep seas have enzymes that work effectively at 0 °C and below.

Effect of pH on enzyme action

The shape of the active site of an enzyme comes from forces between the different parts of the protein molecule. These forces hold the folded chains in place. A change in pH affects these forces. That's why it changes the shape of the molecule. As a result, the specific shape of the active site is lost, so the enzyme no longer acts as a catalyst. Different enzymes work best at different pH levels. A change in pH can stop them working completely. You will learn more about digestive enzymes and pH ranges in Topic B3.6.

Plotting graphs

Drawing graphs

When you investigate the effect of different conditions on the rate of enzyme controlled reactions you will often need to plot a graph of your results.

- Choose your scale carefully – look at the size of your graph paper and the range of your results before deciding on your scale.

- Label your *x* and *y* axes carefully.

- Make sure you show the units on your labelled axes.

- Plot each point as accurately as possible.

- Draw the line of best fit through your points. Don't worry if all your points don't fit on the line. Practice drawing a line of best fit a few times, but make sure your line of best fit is only ever one clear line.

- You can use a graph to calculate the rate of your enzyme controlled reaction. Plot a line graph of some change in the reaction mixture over time. The rate of reaction at any given time is found by calculating the gradient of the tangent drawn at that point on the line.

You can find more about drawing graphs in the Maths skills Topics MS4c and MS4d.

1 Describe the effect of temperature on an enzyme-controlled reaction. Use Figure 1 to help you. [3 marks]

2 Explain the effect of temperature and pH on enzyme action. [4 marks]

3 When you get an infectious disease you may 'get a temperature'. This is a way your body defends you as many microorganisms cannot reproduce at high temperatures. However, people always try to bring the temperature of an ill person down. Explain why this may be the case. [4 marks]

Study tip

Enzymes aren't killed (they are molecules, not living things themselves) – so make sure that you use the term denatured.

Key points

- Enzyme activity is affected by temperature and pH.
- High temperatures denature the enzyme, changing the shape of the active site.
- pH can affect the shape of the active site of an enzyme and make it work very efficiently or stop it working.

B3.6 How the digestive system works

Learning objectives

After this topic, you should know:

- how the food you eat is digested in your body
- the roles played by the different digestive enzymes.

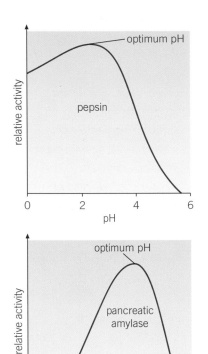

Figure 1 *These two digestive enzymes need very different pH levels to work at their maximum rate. Pepsin is found in the stomach, along with hydrochloric acid, while pancreatic amylase is in the first part of the small intestine along with alkaline bile*

Synoptic links

For more information on moving substances in and out of cells, see Topic B1.6, Topic B1.7, Topic B1.8, and Topic B1.9, and on adaptations for effective absorption, see Topic B1.10.

The food you take in and eat is made up of large insoluble molecules, including starch (a carbohydrate), proteins, and fats. Your body cannot absorb and use these molecules, so they need to be broken down or digested to form smaller, soluble molecules. These can then be absorbed in your small intestine and used by your cells. It is this chemical breakdown of your food that is controlled by the digestive enzymes in your digestive system.

Digestive enzymes

Most of your enzymes work *inside* the cells of your body, controlling the rate of the chemical reactions. Your digestive enzymes are different. They work *outside* your cells. They are produced by specialised cells in glands (such as your salivary glands and your pancreas), and in the lining of your digestive system. The enzymes then pass out of these cells into the digestive system itself, where they come into contact with food molecules.

Your digestive system is a hollow, muscular tube that squeezes your food. It helps to break up your food into small pieces that have a large surface area for your enzymes to work on. It mixes your food with your digestive juices so that the enzymes come into contact with as much of the food as possible. The muscles of the digestive system move your food along from one area to the next. Different areas of the digestive system have different pH levels which allow the enzymes in that region to work as efficiently as possible. For example, the mouth and small intestine are slightly alkaline, while the stomach has a low, acidic pH value.

Digesting carbohydrates

Enzymes that break down carbohydrates are called carbohydrases. Starch is one of the most common carbohydrates that you eat. It is broken down into sugars in your mouth and small intestine. This reaction is catalysed by an enzyme called **amylase**.

Amylase is produced in your salivary glands, so the digestion of starch starts in your mouth. Amylase is also made in the pancreas. No digestion takes place inside the pancreas. All the enzymes made there flow into your small intestine, where most of the starch you eat is digested.

Digesting proteins

The breakdown of protein foods such as meat, fish, and cheese into amino acids is catalysed by protease enzymes. Proteases are produced by your stomach, your pancreas, and your small intestine. The breakdown of proteins into **amino acids** takes place in your stomach and small intestine.

Digesting fats

The lipids (fats and oils) that you eat are broken down into **fatty acids** and **glycerol** in the small intestine. The reaction is catalysed by **lipase** enzymes, which are made in your pancreas and your small intestine. Again, the enzymes made in the pancreas are passed into the small intestine.

Once your food molecules have been completely digested into soluble glucose, amino acids, fatty acids, and glycerol, they leave your small intestine. They pass into your bloodstream to be carried around the body to the cells that need them.

Discovering the roles of the different areas of the digestive system hasn't been easy. For example, when Alexis St Martin suffered a terrible gunshot wound in 1822, Dr William Beaumont managed to save his life. However, Alexis was left with a hole (or fistula) from his stomach to the outside world. Dr Beaumont then used this hole to find out what happened in Alexis's stomach as he digested food!

The effect of pH on the rate of reaction of amylase

Investigating the effect of different pH on the rate of reaction of amylase helps show why the varying pH of the digestive system is so important.

Steps in this investigation include:

- Placing several different starch solutions of a known volume and concentration in a water bath not higher than 37 °C
- Adding a buffer solution at a different pH to each starch solution
- Setting up spotting tiles for each test solution with a drop of iodine in each well
- Mixing the same volume and concentration of amylase into each tube
- Starting a stop-clock as soon as the enzyme is added.
- Taking samples every 30 seconds using a pipette and adding each sample to an iodine-filled well.
- Observing and recording results that can be displayed graphically to compare the effect of pH on the rate of an amylase-catalysed reaction.

1 Explain why amylase, starch, and iodine are used in this investigation.
2 Explain why is it important that the concentration and volume of all the test starch solutions and the enzyme added are known and are the same.
3 Explain why all the test solutions are kept in a water bath at the same temperature and why that temperature must be controlled below 37 °C.
4 State the purpose of the spotting tiles with iodine in the wells.
5 The pipettes used to take samples must be rinsed out with clean water between each sample. Suggest a reason for this.
6 If all the starch is broken down before the first sampling, or if no starch was broken down after an hour, it would be hard to get useful results. Suggest reasons for both of these situations and ways of overcoming the difficulties.

Key points

- Digestion involves the breakdown of large insoluble molecules into soluble substances that can be absorbed into the blood across the wall of the small intestine.
- Digestive enzymes are produced by specialised cells in glands and in the lining of the digestive system.
- Carbohydrases such as amylase catalyse the breakdown of carbohydrates to simple sugars.
- Proteases catalyse the breakdown of proteins to amino acids.
- Lipases catalyse the breakdown of lipids to fatty acids and glycerol.

1 Three types of enzymes found in the body are called amylase, protease, and lipase.
   a State where each enzyme is made in the body.   [3 marks]
   b State which reaction each enzyme catalyses.   [3 marks]
   c State where each reaction works in the digestive system.   [3 marks]

2 Look at Figure 1.
   a At which pH does pepsin work best?   [1 mark]
   b At which pH does pancreatic amylase work best?   [1 mark]
   c What happens to the activity of the enzymes as pH increases?   [2 marks]
   d Explain why this change in activity happens.   [4 marks]

3 Explain the importance of the digestion of food in terms of the molecules involved and the role of enzymes in the gut. ✍   [6 marks]

B3.7 Making digestion efficient

Learning objectives

After this topic, you should know:

- the roles of hydrochloric acid and bile in making digestion more efficient.

Breaking down protein

You can see the effect of acid on pepsin (the protease found in the stomach), quite simply. Set up three test tubes: one containing pepsin, one containing hydrochloric acid, and one containing a mixture of the two. Keep them at body temperature in a water bath. Add a similar-sized chunk of meat to all three of them. Set up a webcam and watch for a few hours to see what happens.

- What conclusions can you make?

Figure 1 *These test tubes show clearly the importance of protein-digesting enzymes and hydrochloric acid in your stomach. Meat was added to each tube at the same time*

Safety: Wear eye protection.

Your digestive system produces many enzymes that speed up the breakdown of the food you eat. As your body is kept at a fairly steady 37 °C, your enzymes have an optimum temperature that allows them to work as fast as possible.

Keeping the pH in your digestive system at optimum levels isn't that easy, because different enzymes work best at different pH levels. For example, the protease enzyme found in your stomach works best in acidic conditions, while the proteases made in your pancreas need alkaline conditions to work at their best. So, your body makes a variety of different chemicals that help to keep conditions ideal for your enzymes all the way through your digestive system.

Changing pH in the digestive system

You have around 35 million glands in the lining of your stomach. These secrete pepsin, a protease enzyme, to digest the protein you eat. Pepsin works best in an acidic pH. Your stomach also produces a relatively concentrated solution of hydrochloric acid from the same glands. In fact, your stomach produces around 3 litres of hydrochloric acid a day! This acid allows your stomach protease enzymes to work very effectively. It also kills most of the bacteria that you take in with your food.

Your stomach also produces a thick layer of mucus. This coats your stomach walls and protects them from being digested by the acid and the enzymes. If someone develops a stomach ulcer, the protecting mucus is lost and acid production may increase. The lining of the stomach is then attacked by the acid and the protein-digesting enzymes, which can be very painful.

After eating a meal, a few hours later – depending on the size and type of the meal – your food leaves your stomach. It moves on into your small intestine. Some of the enzymes that catalyse digestion in your small intestine are made in your pancreas. Some are also made in the small intestine itself. They all work best in an alkaline environment.

The acidic liquid coming from your stomach needs to become an alkaline mixture in your small intestine. So this can happen, your liver makes a green-yellow alkaline liquid called **bile**. Bile is stored in your gall bladder until it is needed.

As food comes into the small intestine from the stomach, bile is squirted onto it through the bile duct. The bile neutralises the acid that was added to the food in the stomach. This provides the alkaline conditions necessary for the enzymes in the small intestine to work most effectively.

Altering the surface area

It is very important for the enzymes of the digestive system to have the largest possible surface area of food to work on. This is not a problem with carbohydrates and proteins. However, the fats that you eat do not mix with all the watery liquids in your digestive system. They stay as large globules (like oil in water) that make it difficult for the lipase enzymes to act.

This is the second important function of the bile – it emulsifies the fats in your food. This means bile physically breaks up large drops of fat into smaller droplets. This provides a much bigger surface area of fats for the lipase enzymes to act upon. The larger surface area helps the lipase chemically break down the fats more quickly into fatty acids and glycerol.

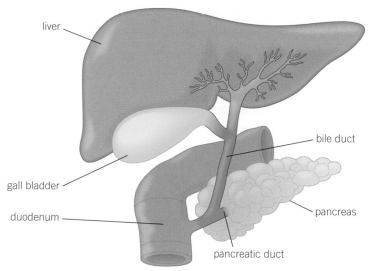

liver

bile duct

gall bladder

duodenum

pancreas

pancreatic duct

Figure 2 *Bile drains down small bile ducts in the liver. Most of it is stored in the gall bladder until it is needed*

Sometimes gall stones form and they can block the gall bladder and bile ducts. The stones can range from a few millimetres to several centimetres in diameter and can cause terrible pain. They can also stop bile being released onto the food and reduce the efficiency of digestion.

Figure 3 *Gall stones can be very large and can cause extreme pain*

> ### Study tip
>
> Understand that:
>
> ● Hydrochloric acid gives the stomach a low pH suitable for the protease secreted there to work efficiently.
>
> ● Alkaline bile neutralises the acid and gives a high pH for the enzymes from the pancreas and small intestine to work well.
>
> ● Bile is *not* an enzyme as it does *not* break down fat molecules. Instead it emulsifies the fat into tiny droplets, which increases the surface area for lipase to increase the rate of digestion.

1 Look at Figure 1 in Topic B3.6
   a State the conditions needed for the protease enzyme pepsin from the stomach to work best. [1 mark]
   b Describe how your body creates the right pH in the stomach for this enzyme. [2 marks]
   c State in what conditions the proteases in the small intestine work best. [1 mark]
   d Describe how your body creates the right pH in the small intestine for these enzymes. [2 marks]
2 a Describe how bile results in a large surface area for lipase to work. [2 marks]
   b Explain why this is important. [3 marks]
3 Describe the passage of a meal containing bread through your digestive system. Your description should include everything you have learnt about digestion in Chapter B3. 🖉 [6 marks]

> ### Key points
>
> ● The protease enzymes of the stomach work best in acid conditions. The stomach produces hydrochloric acid, which maintains a low pH.
>
> ● The enzymes made in the pancreas and the small intestine work best in alkaline conditions.
>
> ● Bile produced by the liver, stored in the gall bladder, and released through the bile duct neutralises acid and emulsifies fats.

B3 Organisation and the digestive system

Summary questions

1

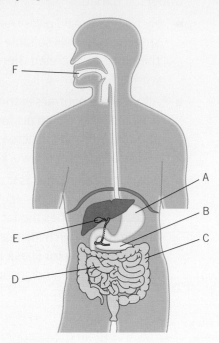

Figure 1

a What is an organ system? [1 mark]

b Name the parts of the human digestive system labelled A–F in Figure 1. [6 marks]

c Select an example of an individual tissue that you would find in an organ of the digestive system and explain how it is specialised for its role. [4 marks]

2 a Describe the difference between a simple sugar and a complex carbohydrate. [2 marks]

b Explain why carbohydrates are so important in the body. [2 marks]

c Explain carefully how you would test for a simple sugar such as glucose. [4 marks]

3 The results in these tables come from a student who was investigating the breakdown of hydrogen peroxide using manganese(IV) oxide and grated raw potato.

Table 1 *Using manganese(IV) oxide*

Temperature in °C	Time taken in s
20	106
30	51
40	26
50	12

Table 2 *Using raw grated potato*

Temperature in °C	Time taken in s
20	114
30	96
40	80
50	120
60	no reaction

a Draw a graph of the results using manganese(IV) oxide. [4 marks]

b State what these results tell you about the effect of temperature on a catalysed reaction. Explain your observation. [3 marks]

c Draw a graph of the results when raw grated potato was added to the hydrogen peroxide. [4 marks]

d What is the name of the enzyme found in living cells that catalyses the breakdown of hydrogen peroxide? [1 mark]

e What does this graph tell you about the effect of temperature on an enzyme-catalysed reaction? [2 marks]

f Explain the difference between the reactions catalysed by an enzyme and by manganese(IV) oxide. [4 marks]

g How could you change the second investigation to find the temperature at which the enzyme works best? [1 mark]

4 Students added samples of two protease enzymes, A and B, to test tubes containing solutions at a range of pH values. After 20 minutes they tested the protease activity in each tube. The table shows their results.

pH of solution in test tube	2	4	6	8	10	12
Activity of enzyme A	0	0	12	32	24	8
Activity of enzyme B	26	20	6	0	0	0

a Name two variables that the students should have controlled in this investigation. [2 marks]

b State one way the students could have improved the quality of the data they collected. [1 marks]

c What conclusions can the students make from these results about the enzymes A and B? [4 marks]

d The students are told that the two enzymes are pepsin from the stomach and trypsin from the pancreas. Suggest which letter represents which enzyme. Give reasons for your answer. [6 marks]

Practice questions

01 Use the correct words from the box to complete each sentence.

| a cell an organ an organism an organ system a tissue |

The basic building block of living organisms is called ....

A group of cells with similar structure and function is called ....

The brain is an example of .... [3 marks]

02 **Figure 1** shows some organs of the human body.

Figure 1

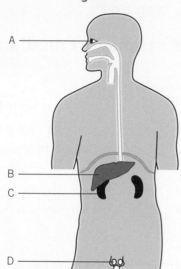

02.1 Name organs **A**, **B**, **C,** and **D**. [4 marks]

02.2 Which organ is part of the nervous system? [1 mark]

03 The digestive system is an example of an organ system.

03.1 What is an organ system? [1 mark]

Figure 2 shows a diagram of the digestive system.

Figure 2

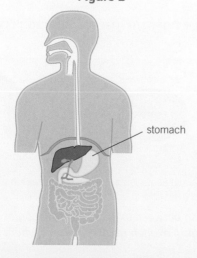

stomach

03.2 Give the two main functions of the digestive system. [2 marks]

03.3 Protein digestion begins in the stomach. Explain how the stomach is adapted to digest protein. [3 marks]

04 Amylase is an enzyme that breaks down starch into sugar molecules.

A student investigated the effect of pH on the activity of amylase. The activity of amylase can be measured by how quickly starch is digested. The students used the following method.

- Mix amylase solution and starch suspension in a boiling tube.
- Put the boiling tube in a water bath at 37 °C.
- Remove a drop of the mixture from the tube every 30 seconds and test it for the presence of starch.
- Repeat the investigation at different pH values.

04.1 One control variable was the temperature. Explain why it was important to use the same temperature for each test. [2 marks]

04.2 Describe the test for the presence of starch and state what result you would see if the test is positive. [2 marks]

04.3 What was the dependent variable in this investigation? [1 mark]

Table 1 shows the results of the investigation.

Table 1

pH	Time when no starch was detected in minutes
5.0	7.0
5.5	4.5
6.0	3.0
6.5	2.0
7.0	1.5
7.5	1.5
8.0	3.0

04.4 Plot the results on graph paper. Choose suitable scales, label both axes, and draw a line of best fit. [4 marks]

04.5 What is the optimum pH for this enzyme's activity? [1 mark]

04.6 Suggest **two** reasons why this conclusion may not be valid. [2 marks]

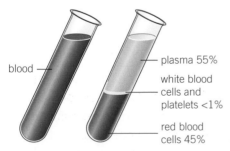

Figure 1 *The main components of blood. The red colour of your blood comes from the red blood cells*

Multicellular organisms with a small surface area to volume ratio often have specialised transport systems. The human circulatory system consists of the blood, the blood vessels, and the heart.

The components of the blood

Your blood is a unique tissue, based on a liquid called **plasma**. Plasma carries **red blood cells**, **white blood cells**, and **platelets** suspended in it. It also carries many dissolved substances around your body. The average person has between 4.7 and 5 litres of blood.

The blood plasma as a transport medium

Your blood plasma is a yellow liquid. The plasma transports all of your blood cells and some other substances around your body.

- Waste carbon dioxide produced by the cells is carried to the lungs.
- **Urea** formed in your liver from the breakdown of excess proteins is carried to your kidneys where it is removed from your blood to form urine.
- The small, soluble products of digestion pass into the plasma from your small intestine and are transported to the individual cells.

Red blood cells

There are more red blood cells than any other type of blood cell in your body – about 5 million in each cubic millimetre of blood. These cells pick up oxygen from the air in your lungs and carry it to the cells where it is needed. Red blood cells have adaptations that make them very efficient at their job:

- They are biconcave discs. Being concave (pushed in) on both sides, gives them an increased surface area to volume ratio for diffusion.
- They are packed with a red pigment called **haemoglobin** that binds to oxygen.
- They have no nucleus, making more space for haemoglobin.

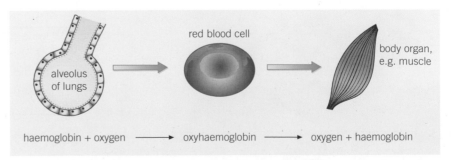

Figure 2 *The reversible reaction between oxygen and haemoglobin makes life as we know it possible by carrying oxygen to all the places where it is needed*

White blood cells

White blood cells are much bigger than red blood cells and there are fewer of them. They have a nucleus and form part of the body's defence system against harmful microorganisms. Some white blood cells (lymphocytes) form antibodies against microorganisms. Some form antitoxins against poisons made by microorganisms. Yet others (phagocytes) engulf and digest invading bacteria and viruses.

Platelets

Platelets are small fragments of cells. They have no nucleus. They are very important in helping the blood to clot at the site of a wound. Blood clotting is a series of enzyme-controlled reactions that result in converting fibrinogen into fibrin. This produces a network of protein fibres that capture lots of red blood cells and more platelets to form a jelly-like clot that stops you bleeding to death. The clot dries and hardens to form a scab. This protects the new skin as it grows and stops bacteria entering the body through the wound.

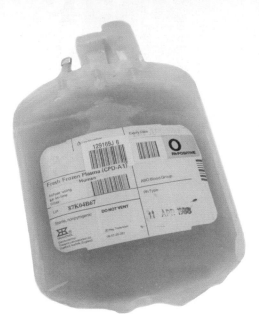

Figure 3 *Blood plasma is a yellow liquid that transports everything you need – and need to get rid of – around your body*

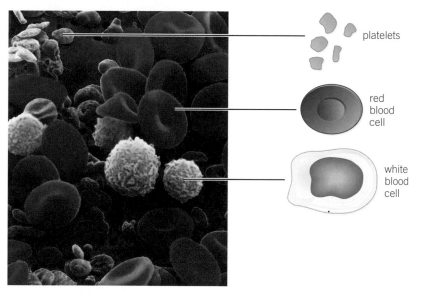

platelets

red blood cell

white blood cell

Figure 4 *Red blood cells, white blood cells, and platelets are suspended in the blood plasma*

1 State three functions of the blood. [3 marks]

2 **a** Explain why it is not accurate to describe the blood as a red liquid. [2 marks]

   b What actually makes the blood red? [1 mark]

   c Identify three important functions of blood plasma. [3 marks]

3 Discuss the main ways in which the blood helps you to avoid infection. Include a description of the parts of the blood involved. 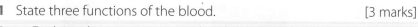 [6 marks]

Key points

- The blood, blood vessels, and heart make up the human circulatory system which transports substances to and from the body cells.
- Plasma has blood cells suspended in it and transports proteins and other chemicals around the body.
- Your red blood cells contain haemoglobin that binds to oxygen to transport it from the lungs to the tissues.
- White blood cells help to protect the body against infection.
- Platelets are cell fragments that start the clotting process at wound sites.

B4.2 The blood vessels

Learning objectives

After this topic, you should know:

- how the blood flows round the body
- that there are different types of blood vessels
- why valves are important
- the importance of a double circulatory system.

The substances transported in the blood need to reach the individual cells. Every cell in your body is within 0.05 mm of a capillary – the tiniest blood vessels in your circulatory system.

The blood vessels

Blood is carried around your body in three main types of blood vessels, each adapted for a different function.

- Your **arteries** carry blood away from your heart to the organs of your body. This blood is usually bright-red oxygenated blood. The arteries stretch as the blood is forced through them and go back into shape afterwards. You can feel this as a pulse where the arteries run close to the skin's surface (e.g., at your wrist). Arteries have thick walls containing muscle and elastic fibres. As the blood in the arteries is under pressure, it is very dangerous if an artery is cut, because the blood will spurt out rapidly every time the heart beats.

- The **veins** carry blood away from the organs towards your heart. This blood is usually low in oxygen and therefore a deep purple-red colour. Veins do not have a pulse. They have much thinner walls than arteries and often have valves to prevent the backflow of blood. The valves open as the blood flows through them towards the heart, but if the blood starts to flow backwards the valves close and prevent a backflow of blood. The blood is squeezed back towards the heart by the action of the skeletal muscles (Figure 2).

- Throughout the body, **capillaries** form a huge network of tiny vessels linking the arteries and the veins. Capillaries are narrow with very thin walls. This enables substances, such as oxygen and glucose, to diffuse easily out of your blood and into your cells. The substances produced by your cells, such as carbon dioxide, pass easily into the blood through the walls of the capillaries.

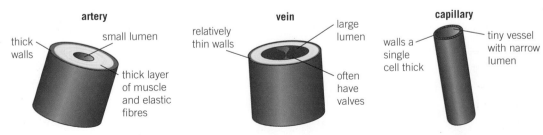

Figure 1 *The three main types of blood vessels*

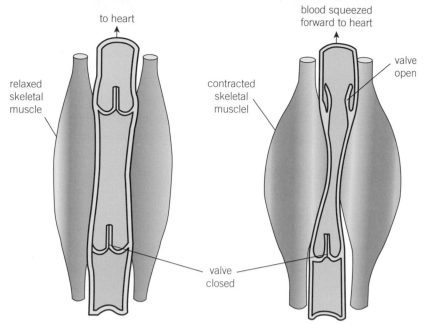

Figure 2 *How the valves and the muscles between them ensure that blood is moved from the body towards the heart*

In your circulatory system, arteries carry blood away from your heart to the organs of the body. Blood returns to your heart in the veins. The two are linked by the capillary network.

Double circulation

In humans and other mammals the blood vessels are arranged into a **double circulatory system**.

- One transport system carries blood from your heart to your lungs and back again. This allows oxygen and carbon dioxide to be exchanged with the air in the lungs.

- The other transport system carries blood from your heart to all other organs of your body and back again.

A double circulation like this is vital in warm-blooded, active animals such as humans. It makes our circulatory system very efficient. Fully oxygenated blood returns to the heart from the lungs. This blood can then be sent off to different parts of the body at high pressure, so more areas of your body can receive fully oxygenated blood quickly.

1 State the function of each of the following blood vessels. Describe how the structure of the blood vessel relates to its function.
   a arteries [3 marks]
   b veins [3 marks]
   c capillaries. [2 marks]

2 **a** Describe how the heart, arteries, veins, and capillaries are linked together in the circulatory system. [2 marks]
   b Describe what happens in capillaries in a cell. [2 marks]

3 Fish have a single circulation system. The blood goes from the heart, through the gills, around the body, and back to the heart. Describe the disadvantages of a single circulation system like this for an active land mammal such as a human being. ⊘ [4 marks]

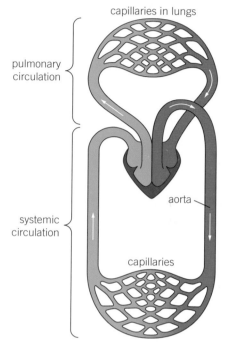

Figure 3 *The two separate circulation systems supply the lungs and the rest of the body*

Study tip

Remember:

Arteries carry blood away from the heart, and veins carry blood back to the heart – this applies to the circulation system of the lungs as well!

Key points

- Blood flows around the body in the blood vessels. The main types of blood vessels are arteries, veins, and capillaries.
- Substances diffuse in and out of the blood in the capillaries.
- The valves prevent backflow, ensuring that blood flows in the right direction.
- Human beings have a double circulatory system.

B4.3 The heart

Learning objectives

After this topic, you should know:

- the structure and functions of the heart
- ways of solving problems with the blood supply to the heart and problems with valves.

Your heart is the organ that pumps blood around your body. It is made up of two pumps (for the double circulation) that beat together about 70 times each minute. The walls of your heart are almost entirely muscle. This muscle is supplied with oxygen by the **coronary arteries**.

The heart as a pump

The structure of the human heart is perfectly adapted for pumping blood to your lungs and your body. The two sides of the heart fill and empty at the same time, giving a strong, coordinated heartbeat. Blood enters the top chambers of your heart, which are called the **atria**. The blood coming into the right atrium from the **vena cava** is deoxygenated blood from your body. The blood coming into the left atrium in the **pulmonary vein** is oxygenated blood from your lungs. The atria contract together and force blood down into the **ventricles**. Valves close to stop the blood flowing backwards out of the heart.

- The ventricles contract and force blood out of the heart.
- The right ventricle forces deoxygenated blood to the lungs in the **pulmonary artery**.
- The left ventricle pumps oxygenated blood around the body in a big artery called the **aorta**.

As the blood is pumped into the pulmonary artery and the aorta, valves close to make sure the blood flows in the right direction. The noise of the heartbeat you hear through a stethoscope is the sound of the valves of the heart closing to prevent the blood flowing backwards.

The muscle wall of the left ventricle is noticeably thicker than the wall of the right ventricle. This allows the left ventricle to develop the pressure needed to force the blood through the arterial system all over your body. The blood leaving the right ventricle moves through the pulmonary arteries to your lungs, where high pressure would damage the delicate capillary network where gas exchange takes place.

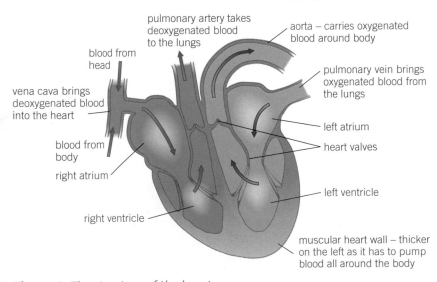

pulmonary artery takes deoxygenated blood to the lungs

blood from head

vena cava brings deoxygenated blood into the heart

blood from body

right atrium

right ventricle

aorta – carries oxygenated blood around body

pulmonary vein brings oxygenated blood from the lungs

left atrium

heart valves

left ventricle

muscular heart wall – thicker on the left as it has to pump blood all around the body

Figure 1 *The structure of the heart*

Problems with blood flow through the heart

In coronary heart disease the coronary arteries that supply blood to the heart muscle become narrow. A common cause is a buildup of fatty material on the lining of the vessels. If the blood flow through the coronary arteries is reduced, the supply of oxygen to the heart muscle is also reduced. This can cause pain, a heart attack, and even death.

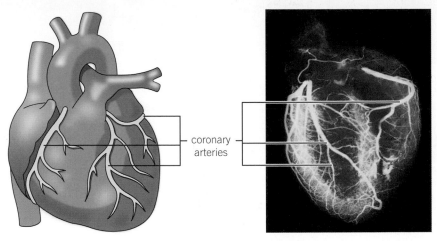

Figure 2 *The muscles of the heart work hard so they need a good supply of oxygen and glucose. This is supplied by the blood in the coronary arteries*

Doctors often solve the problem of coronary heart disease with a **stent**. A stent is a metal mesh that is placed in the artery. A tiny balloon is inflated to open up the blood vessel and the stent at the same time. The balloon is deflated and removed but the stent remains in place, holding the blood vessel open. As soon as this is done, the blood in the coronary artery flows freely. Doctors can put a stent in place without a general anaesthetic.

Stents can be used to open up a blocked artery almost anywhere in the body. Many stents also release drugs to prevent the blood clotting, although some studies suggest that the benefits do not justify the additional expense.

Doctors can also carry out bypass surgery, replacing the narrow or blocked coronary arteries with bits of veins from other parts of the body. This works for badly blocked arteries where stents cannot help. The surgery is expensive and involves the risk associated with a general anaesthetic.

Increasingly doctors prescribe **statins** to anyone at risk from cardiovascular disease. They reduce blood cholesterol levels and this slows down the rate at which fatty material is deposited in the coronary arteries.

1   Draw a flow chart to show how blood passes through the heart.
    [4 marks]

2   Explain the importance of the following in making the heart an effective pump in the circulatory system of the body:
    a heart valves   [2 marks]
    b coronary arteries   [2 marks]
    c the thickened muscular wall of the left ventricle.   [3 marks]
3   Blood in the arteries is usually bright red because it is full of oxygen. This is not true of the blood in the pulmonary arteries. Explain this observation.   [3 marks]
4   **a** Describe what a stent is.   [2 marks]
    b Construct a table to show the advantages and disadvantages of using a stent to improve the blood flow through the coronary arteries compared with bypass surgery.   [4 marks]

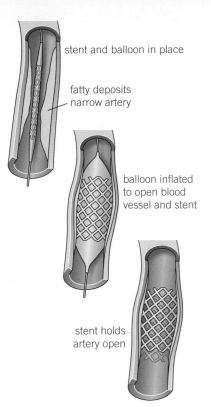

stent and balloon in place

fatty deposits narrow artery

balloon inflated to open blood vessel and stent

stent holds artery open

Figure 3 *A stent being positioned in an artery*

Study tip

Remember:
- the heart has *four* chambers
- ventricles pump blood *out* of the heart
- blood comes from the veins into the atria, through valves to the ventricles, and then out via arteries.

Key points

- The heart is an organ that pumps blood around the body.
- Heart valves keep the blood flowing in the right direction.
- Stents can be used to keep narrowed or blocked arteries open.
- Statins reduce cholesterol levels in the blood, reducing the risk of coronary heart disease.

B4.4 Helping the heart

Learning objectives

After this topic, you should know:

- how the heart keeps its natural rhythm
- how artificial pacemakers work
- what artificial hearts can do.

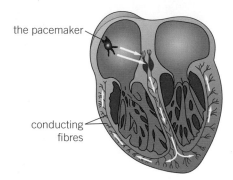

Figure 1 *The pacemaker region controls the basic rhythm of your heart*

Go further

The natural pacemaker regions of the heart are very complex. Doctors and scientists are developing ever more complex pacemakers to try and mimic the natural responses of the heart as closely as possible.

The heart can be affected by a number of problems. Doctors, scientists, and engineers have worked out some amazing ways to help solve them.

Leaky valves

Heart valves have to withstand a lot of pressure. Over time they may start to leak or become stiff and not open fully, making the heart less efficient. People affected may become breathless and without treatment, will eventually die.

Doctors can operate and replace faulty heart valves. Mechanical valves are made of materials such as titanium and polymers. They last a very long time. However, with a mechanical valve you have to take medicine for the rest of your life to prevent your blood from clotting around it. Biological valves are based on valves taken from animals such as pigs or cattle, or even human donors. These work extremely well and the patient does not need any medication. However, they only last about 12–15 years.

Artificial pacemakers

The resting rhythm of a healthy heart is around 70 beats a minute. It is controlled by a group of cells found in the right atrium of your heart that acts as your natural pacemaker (Figure 1). If the natural pacemaker stops working properly, this can cause serious problems. If the heart beats too slowly, the person affected will not get enough oxygen. If the heart beats too fast, it cannot pump blood properly.

Problems with the rhythm of the heart can often be solved using an artificial pacemaker. This is an electrical device used to correct irregularities in the heart rate, which is implanted into your chest. Artificial pacemakers only weigh between 20 and 50 g, and they are attached to your heart by two wires. The artificial pacemaker sends strong, regular electrical signals to your heart that stimulate it to beat properly. Modern pacemakers are often very sensitive to what your body needs and only work when the natural rhythm goes wrong. Some even stimulate the heart to beat faster when you exercise.

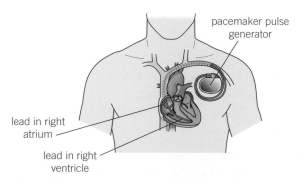

Figure 2 *An artificial pacemaker is positioned under the skin of the chest with wires running to the heart itself*

If you have a pacemaker fitted, you will need regular medical check-ups throughout your life. However, most people feel that this is a small price to pay for the increase in the quality and length of life that a pacemaker brings.

Artificial hearts

An artificial pacemaker may keep the heart beating steadily, but sometimes it is not enough to restore a person's health. When the heart fails completely, a donor heart or heart and lungs can be transplanted. When people need a heart transplant, they have to wait for a donor heart that is a tissue match. As a result of this wait, many people die before they get a chance to have a transplant.

Scientists have developed temporary hearts that can support your natural heart until it can be replaced. Although replacing your heart permanently with a machine is still a long way off, by 2015 almost 1500 people worldwide had been fitted with a completely artificial heart. These artificial hearts need a lot of machinery to keep them working. Most patients have to stay in hospital until they have their transplant.

In the past few years artificial hearts have improved considerably although there is always a risk of the blood clotting in an artificial heart, which can lead to death. However, this new technology gives people a chance to live a relatively normal life while they wait for a heart transplant. In 2011, 40-year-old Matthew Green became the first UK patient to leave hospital and go home with a completely artificial heart carried in a backpack. This kept him alive for two years until he had a heart transplant and no longer needed this life-saving machine.

Artificial hearts can also be used to give a diseased heart a rest, so that it can recover. Patients have a part or whole artificial heart implanted that removes the strain of keeping the blood circulating for a few weeks or months. However, the resources needed to develop artificial hearts and the cost of each one means they are not yet widely used in patients.

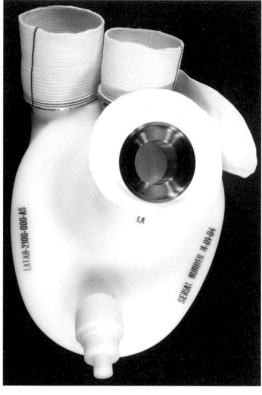

Figure 3 *This amazing artificial heart uses air pressure to pump blood around the body*

1 Describe what a natural pacemaker is. [2 marks]

2 Describe how an artificial pacemaker works. [3 marks]

3 **a** Explain how a leaky heart valve can cause health issues. [4 marks]
   b Give one advantage and one disadvantage of:
      i a biological replacement heart valve [2 marks]
      ii a mechanical replacement heart valve. [2 marks]

4 Evaluate some of the scientific and social arguments for and against the continued development of artificial hearts. [6 marks]

Key points

- Damaged heart valves can be replaced using biological or mechanical valves.
- The resting heart rate is controlled by a group of cells in the right atrium that form a natural pacemaker.
- Artificial pacemakers are electrical devices used to correct irregularities in the heart rhythm.
- Artificial hearts are occasionally used to keep patients alive while they wait for a transplant, or for their heart to rest as an aid to recovery.

B4.5 Breathing and gas exchange

Learning objectives

After this topic, you should know:

- the structure of the human gas exchange system
- how gases are exchanged in the alveoli of the lungs.

3
atmospheric air at higher pressure than chest – so air is drawn into the lungs

2
increased volume means **lower pressure** in the chest

1
as ribs move up and out and diaphragm flattens, the **volume** of the chest **increases**

breathing in

3
pressure in chest higher than outside – so air is forced out of the lungs

2
decreased volume means **increased pressure** in the chest

1
as ribs fall and diaphragm moves up, the **volume** of the chest **gets smaller**

breathing out

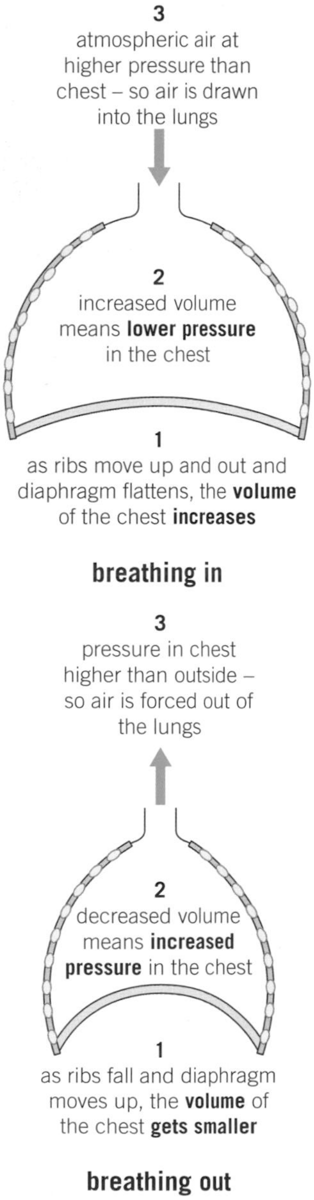

Figure 2 *Ventilation of the lungs*

For a gas exchange system to work efficiently, you need a large difference in concentrations of the gas on different sides of the exchange membrane (a steep concentration gradient). Many large animals, including humans, move air in and out of their lungs regularly. By changing the composition of the air in the lungs, they maintain a steep concentration gradient for both oxygen diffusing into the blood and carbon dioxide diffusing out of the blood. This is known as ventilating the lungs or breathing. It takes place in a specially adapted gas exchange system.

The gas exchange system

Your lungs are found in your chest (or thorax) and are protected by your ribcage. They are separated from the digestive organs beneath (in your abdomen) by the diaphragm. The diaphragm is a strong sheet of muscle. The job of your ventilation system is to move air in and out of your lungs, which provide an efficient surface for gas exchange in the alveoli (Figure 1). Ventilating the lungs is brought about by the contraction and relaxation of the intercostal muscles between the ribs and the diaphragm, changing the pressure inside the chest cavity so air is forced in or out of the lungs as a result of differences in pressure.

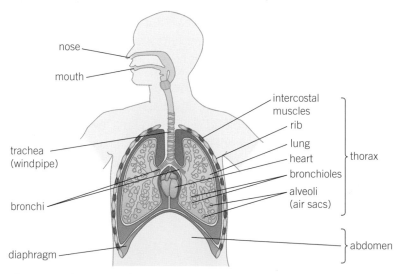

nose
mouth
intercostal muscles
rib
trachea (windpipe)
lung
heart
thorax
bronchioles
alveoli (air sacs)
bronchi
diaphragm
abdomen

Figure 1 *The gas exchange system supplies your body with vital oxygen and removes waste carbon dioxide*

When you breathe in, oxygen-rich air moves into your lungs. This maintains a steep concentration gradient with the blood. As a result, oxygen continually diffuses into your bloodstream through the gas exchange surfaces of your alveoli. Breathing out removes carbon dioxide-rich air from the lungs. This maintains a concentration gradient so carbon dioxide can continually diffuse out of the bloodstream into the air in the lungs.

Table 1 *The composition of inhaled and exhaled air (~ means approximately)*

Atmospheric gas	% of air breathed in	% of air breathed out
nitrogen	~80	~80
oxygen	~20	~16
carbon dioxide	0.04	~4

Adaptations of the alveoli

Your lungs are specially adapted to make gas exchange more efficient. They are made up of clusters of alveoli that provide a very large surface area. This is important for achieving the most effective diffusion of oxygen and carbon dioxide. The alveoli also have a rich supply of blood **capillaries**. This maintains a concentration gradient in both directions. The blood coming to the lungs is always relatively low in oxygen and high in carbon dioxide compared to the inhaled air.

As a result, gas exchange takes place down the steepest concentration gradients possible. This makes the exchange rapid and effective. The layer of cells between the air in the lungs and the blood in the capillaries is also very thin (only one cell wide). This allows diffusion to take place over the shortest possible distance. If all of the alveoli in your lungs were spread out flat, they would have a surface area equivalent to 10–15 table tennis tables.

Synoptic links

You can find out more about diffusion and concentration gradients in Topic B1.6 and about exchange surfaces in Topic B1.10.

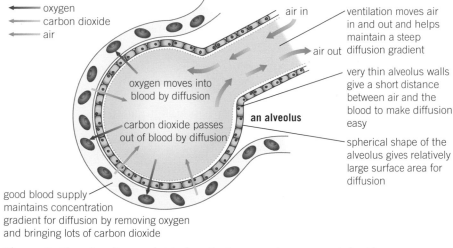

oxygen
carbon dioxide
air

air in
air out

oxygen moves into blood by diffusion

carbon dioxide passes out of blood by diffusion

an alveolus

ventilation moves air in and out and helps maintain a steep diffusion gradient

very thin alveolus walls give a short distance between air and the blood to make diffusion easy

spherical shape of the alveolus gives relatively large surface area for diffusion

good blood supply maintains concentration gradient for diffusion by removing oxygen and bringing lots of carbon dioxide

Figure 3 *The alveoli are adapted so that gas exchange can take place as efficiently as possible in the lungs*

1 Describe how air is moved in to and out of your lungs. [3 marks]

2 a Describe what is meant by the term gaseous exchange [1 mark]
b Explain why it is so important in your body. [2 marks]

3 a Draw a bar chart to show the difference in composition between the air you breathe in and the air you breathe out (use the data in Table 1). [3 marks]
b People often say we breathe in oxygen and breathe out carbon dioxide. Use your bar chart to explain why this is wrong. [2 marks]
c Describe the adaptations of the human gas exchange system and explain how they make it as efficient as possible. [6 marks]

Key points

- The lungs are in your chest cavity, protected by your ribcage and separated from your abdomen by the diaphragm.
- The alveoli provide a very large surface area and a rich supply of blood capillaries. This means gases can diffuse into and out of the blood as efficiently as possible.

B4.6 Tissues and organs in plants

Learning objectives

After this topic, you should know:

- the roots, stem, and leaves of a plant form a plant organ system for transport of substances around the plant.

Figure 1 *The flower of the elephant yam is a plant organ made up of a number of different tissues*

Elephant yams are plants that produce a large flower that releases a disgusting stench like rotting meat that attracts carrion beetles. The beetles become trapped in the flower – its slippery, waxy walls stop them escaping. Around 24 hours after the stench is released, the flower releases pollen that coats the trapped beetles. Then the walls of the flower change texture – they become rough so the beetles can crawl out, carrying the pollen to another flower, lured again by the powerful smell of dead meat. These flowers are one type of plant organ – they are temporary and for reproduction only. But as you will see, plants have other organs, made up of combinations of many different tissues.

Plant tissues

The specialised cells in multicellular plants are organised into tissues and organs. **Epidermal** tissues cover the surfaces and protect them. These cells often secrete a waxy substance that waterproofs the surface of the leaf. **Palisade mesophyll** tissue contains lots of chloroplasts, which carry out photosynthesis. **Spongy mesophyll** tissue contains some chloroplasts for photosynthesis but also has big air spaces and a large surface area to make the diffusion of gases easier. **Xylem** and **phloem** are the transport tissues in plants. Xylem carry water and dissolved mineral ions from the roots up to the leaves and phloem carry dissolved food from the leaves around the plant. You will learn more about the role of the xylem and phloem in Topic B4.7.

The meristem tissue at the growing tips of roots and shoots is made up of rapidly dividing plant cells that grow and differentiate into all the other cell types needed.

Plant organs

Within the body of a plant, specialised tissues such as palisade, spongy mesophyll, xylem, and phloem are arranged to form organs. Each organ carries out its own particular functions. The leaves, stems, and roots are all plant organs, each of which has a very specific job to do (Figure 2).

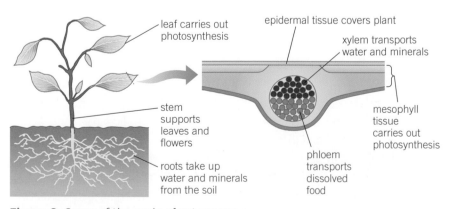

Figure 2 *Some of the main plant organs*

Within each plant organ there are collections of different tissues working together to perform specific functions for the organism.

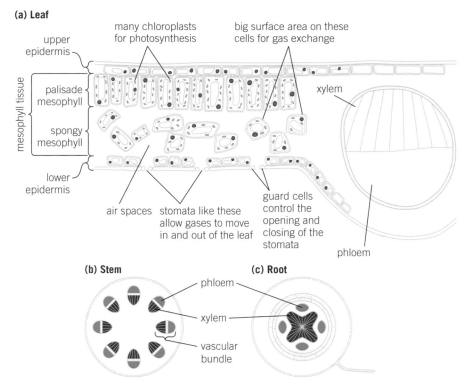

(a) Leaf

upper epidermis

mesophyll tissue

palisade mesophyll

spongy mesophyll

lower epidermis

many chloroplasts for photosynthesis

big surface area on these cells for gas exchange

xylem

air spaces

stomata like these allow gases to move in and out of the leaf

guard cells control the opening and closing of the stomata

phloem

(b) Stem

(c) Root

phloem

xylem

vascular bundle

Figure 3 *Plants have specific tissues to carry out particular functions. They are arranged in organs such as the: **a** leaf, **b** stem, and **c** roots.*

Plant organs can be very large indeed. For example, some trees, such as the giant redwood, have trunks over 40 m tall. A plant cell is about 100 µm long. The plant stem is 400 000 times bigger than an individual cell.

Plant organ systems

The whole body of the plant – the roots, stem, and leaves – form an organ system for the transport of substances around the plant. Trees form the largest and oldest land organisms, so plant organ systems are also the biggest land based organ systems in the living world.

Synoptic links

You learnt about some of the specialised plant cells that make up plant tissues and organs in Topic B1.5, and about meristems in plants in Topic B2.3.

Key points

- Plant tissues are collections of cells specialised to carry out specific functions.
- The structure of the tissues in plant organs is related to their functions.
- The roots, stem, and leaves form a plant organ system for the transport of substances around the plant.

1 Plants contain specialised tissues that are adapted to their function. Explain how:
  a epidermal tissues protect the surface of the leaf [1 mark]
  b palisade mesophyll tissue is adapted for photosynthesis [1 mark]
  c spongy mesophyll tissue is adapted for photosynthesis. [1 mark]
2 Explain how the tissues in a leaf are arranged to form an effective organ for photosynthesis. [6 marks]

B4.7 Transport systems in plants

Learning objectives

After this topic, you should know:

- the substances that are transported in plants
- how transport in the xylem tissue differs from transport in the phloem tissue.

Synoptic link

For information on phloem and xylem cells, see Topic B1.5.

Figure 1 *Aphids take the liquid full of dissolved sugars directly from the phloem*

Plants make glucose (a simple sugar) by photosynthesis in the leaves and other green parts. This glucose is needed all over the plant. Similarly, water and mineral ions move into the plant from the soil through the roots, but they are needed by every cell of the plant. Plants have two separate transport systems to move substances around the whole plant.

Phloem – moving food

The phloem tissue transports the sugars made by photosynthesis from the leaves to the rest of the plant. This includes transport to the growing areas of the stems and roots where the dissolved sugars are needed for making new plant cells. Food is also transported to the storage organs where it provides an energy store for the winter.

Phloem is a living tissue – the phloem cells are alive. The movement of dissolved sugars from the leaves to the rest of the plant is called **translocation**.

Greenfly and other **aphids** are plant pests. They push their sharp mouthparts right into the phloem and feed on the sugary fluid. If too many of them attack a plant, they can kill it by taking all of its food.

Xylem – moving water and mineral ions

The xylem tissue is the other transport tissue in plants. It carries water and mineral ions from the soil around the plant to the stem and the leaves. Mature xylem cells are dead.

Evidence for movement through xylem

You can demonstrate the movement of water up the xylem by placing leafy celery stalks in water containing a coloured dye. After a few hours, slice the stem in several places – you will see coloured circles where the water and dye have moved through the xylem. You may also see patches of dye in the leaves where the water has entered the mesophyll cells for photosynthesis.

In woody plants like trees, the xylem makes up the bulk of the wood and the phloem is found in a ring just underneath the bark. This makes young trees particularly vulnerable to damage by animals – if a complete ring of bark is eaten, transport in the phloem stops and the tree will die.

Figure 2 *Without protective collars on the trunks, deer would destroy the transport tissue of young trees like these and kill them before they could become established in the woodland*

Why is transport so important?

It is vital to move the food made by photosynthesis around the plant – all the cells need sugars for respiration as well as for providing materials for growth. The movement of water and dissolved mineral ions from the roots is equally important – the mineral ions are needed for the production of proteins and other molecules within the cells.

The plant needs water for photosynthesis, the process in which carbon dioxide and water combine to make glucose (plus oxygen). The plant also needs water to hold itself upright. When a cell has plenty of water inside it, the vacuole presses the cytoplasm against the cell walls. This pressure of the cytoplasm against the cell walls gives support for young plants and for the structure of the leaves. For young plants and soft-stem plants (although not trees) this is the main method of support.

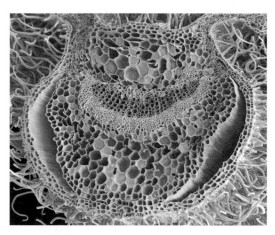

Figure 3 *The phloem and xylem are arranged in vascular bundles in the stem*

1 State why a plant needs a transport system. [1 mark]

2 Describe the main differences between xylem and phloem in a plant. [3 marks]

3 A local woodland trust has set up a scheme to put protective plastic covers around the trunks of young trees. Some local residents are objecting to this, saying it spoils the look of the woodland. Explain exactly why this protection is necessary and the impact it would have on the wood if the trees were not protected. ⬤ [6 marks]

Key points

- Plants have separate transport systems.
- Xylem tissue transports water and mineral ions from the roots to the stems and leaves.
- Phloem tissue transports dissolved sugars from the leaves to the rest of the plant, including the growing regions and storage organs.

B4.8 Evaporation and transpiration

Learning objectives

After this topic, you should know:

- what transpiration is
- the role of stomata and guard cells in controlling gas exchange and water loss.

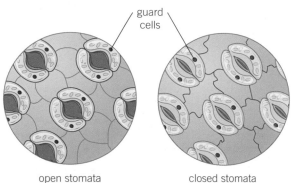

Figure 1 *The size of the opening of the stomata is controlled by the guard cells. This in turn controls the carbon dioxide going into the leaf and the water vapour and oxygen leaving it*

Figure 3 *The transpiration stream in trees can pull litres of water many metres above the ground*

The top of a tree may be many metres from the ground. Yet the leaves at the top need water just as much as those on the lower branches. So how do they get the water they need?

Water loss from the leaves

All over the leaf surface are small openings known as stomata. The stomata can be opened when the plant needs to allow air into the leaves. Carbon dioxide from the atmosphere diffuses into the air spaces and then into the cells down a concentration gradient. At the same time, oxygen produced by photosynthesis is removed from the leaf by diffusion into the surrounding air. This maintains a concentration gradient for oxygen to diffuse from the cells into the air spaces of the leaf. The size of the stomata and their opening and closing is controlled by the **guard cells** (Figure 1).

When the stomata are open, plants lose water vapour through them as well. The water vapour evaporates from the cells lining the air spaces and then passes out of the leaf through the stomata by diffusion. This loss of water vapour is known as **transpiration**.

As water evaporates from the surface of the leaves, more water is pulled up through the xylem to take its place. This constant movement of water molecules through the xylem from the roots to the leaves is known as the transpiration stream. It is driven by the evaporation of water from the leaves. So, anything that affects the rate of evaporation will also affect transpiration.

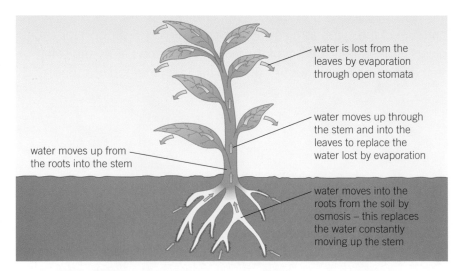

Figure 2 *The transpiration stream*

Most of the water vapour lost by plants is lost from the leaves. Most of this loss takes place by diffusion through the stomata when they are open. This is one of the main reasons why it is important that plants can close their stomata – to limit the loss of water vapour.

Finding the mean and estimating

When you carry out stomatal counts there are two bits of maths that will be useful – finding the mean and estimating.

To find the mean: The mean is the average of the numbers. To find the mean you add together all your data sets and then divide by the number of samples you have taken, for example:

A student looked at the number of stomata on the underside of a leaf. They did five counts and the results per unit area were: 10, 12, 15, 8, 10.

The mean number of stomata per unit area ($0.1\,mm^2$)

$$= \frac{10 + 12 + 15 + 8 + 10}{5}$$

$$= \frac{55}{5} = \mathbf{11}$$

Sampling: You will sample the surface area of the leaf by taking peels from randomly selected regions to make your counts. The randomness will be imposed, at least in part, by where the film actually peels successfully from the leaf.

Estimating: When you estimate, you are getting a 'ball-park' figure, not a precise measurement. For example, you might work out the mean number of stomata:

- per mm^2 of a leaf. If you know the size of the field of vision you were using, for example, $0.1\,mm^2$, you can get an estimated number of stomata per $1\,mm^2$. In this example this is $11 \times 10 = \mathbf{110}$

- per leaf. You can work out the approximate area of the entire leaf using graph paper. If, for example, the leaf has an area of $5\,cm^2$, the estimated number of stomata would be $11 \times 500 = \mathbf{5500}$ stomata per leaf.

You can find out more about finding the mean in Maths skills M2b, sampling in Maths skills M2d, and estimating numbers in Maths skills M2h.

Investigating stomata

Stomata are key to the control of transpiration. There are a number of different ways you can investigate the numbers and distribution of stomata on a leaf. You can compare the upper and lower sides of a leaf, different areas of the same leaf surface, different leaves from the same plant, and different types of leaves. The main steps are:

1   Making a stomatal peel – use nail varnish or a water-based varnish to cover an area of the leaf and then peel it off.

2   Place the peel on a microscope slide.

3   With an eyepiece graticule, use a low magnification and count the number of stomata in a random sample of squares.

4   Without an eyepiece graticule, use a higher magnification and count the number of stomata in the field of vision and repeat this with a number of sample areas of the peel to collect your data.

5   You can *calculate* the mean number of stomata on a given area of a leaf.

6   You can use this to *estimate* the number of stomata on the whole leaf.

Study tip

Remember that the transpiration stream is driven by the loss of water by evaporation out of the stomata.

Key points

- The loss of water vapour from the surface of plant leaves is known as transpiration.
- Water is lost through the stomata, which open to let in carbon dioxide for photosynthesis.
- The stomata and guard cells control gas exchange and water loss.

1   **a** What are stomata?   [1 mark]
    b Describe their role in the plant.   [2 marks]

2   Describe the process of transpiration.   [3 marks]

3   Explain how water moves up a plant in the transpiration stream.   [3 marks]

4   A student measured the numbers of stomata per mm^2 of leaf surface. Their counts were: 250, 280, 265, 245, 270, 255, 290. Calculate the mean number of stomata (to 3 significant figures).   [3 marks]

B4.9 Factors affecting transpiration

Learning objectives

After this topic, you should know:

- the factors that affect the rate of transpiration
- ways of investigating the effect of environmental factors on rates of water uptake.

Figure 1 *Dry air and high temperatures make it very hard for leafy plants to survive as they lose so much water through transpiration*

The effect of the environment on transpiration

Different conditions affect the rate of transpiration – as a result, some environments are much tougher for plants to survive in than others. Factors that affect the rate of transpiration include temperature, humidity, the amount of air movement, and light intensity.

Anything that increases the rate of photosynthesis will increase the rate of transpiration, because more stomata open up to let in carbon dioxide. When stomata are open, the rate at which water is lost by evaporation and diffusion increases. Therefore, an increase in light intensity will increase the rate of transpiration.

Conditions that increase the rate of evaporation from the leaf cells and diffusion of water from open stomata will also make transpiration happen more rapidly. Hot, dry, windy conditions increase the rate of transpiration because more water evaporates from the cells and diffusion happens quicker. Water vapour diffuses more rapidly into dry air than into humid air because the concentration gradient is steeper. Windy conditions both increase the rate of evaporation and also maintain a steep concentration gradient from the inside of the leaf to the outside by removing water vapour as it diffuses out.

Temperature affects the rate of transpiration in several ways. The molecules move faster as the temperature increases, so diffusion occurs more rapidly. The rate of photosynthesis also increases as the temperature goes up, so more stomata will be open for gas exchange to take place. Each of these conditions individually increases the rate of transpiration and, when combined, a plant will lose a lot of water in this way.

Controlling water loss

Many plants have adaptations that help them to photosynthesise as much as possible while losing as little water as possible.

Most leaves have a waxy, waterproof layer (the cuticle) to prevent uncontrolled water loss. In very hot environments, the cuticle may be very thick and shiny. Most of the stomata are found on the underside of the leaves. This protects them from the direct light and energy of the Sun.

If a plant begins to lose water faster than it is replaced by the roots, it can result in some drastic measures.

- The whole plant may wilt. Wilting is a protection mechanism against further water loss. The leaves all collapse and hang down. This greatly reduces the surface area available for water loss by evaporation.
- Stomata close, which stops photosynthesis and risks overheating. However, this prevents most water loss and any further wilting.

The plant will remain wilted until the temperature drops, the sun goes in, or it rains.

Measuring transpiration rates

There are many ways to investigate the effect of different factors on the rate of transpiration in plants. Many of them involve a piece of apparatus known as a potometer.

A potometer can be used to show how the uptake of water by a plant changes in different conditions. This gives you a good idea of the amount of water lost by the plant in transpiration. Almost all of the water taken up by a plant is lost in transpiration, but a small amount is used in the metabolism, for example, in photosynthesis.

Synoptic link

You will learn more about plant adaptations to reduce transpiration in harsh environments in Topic B15.8.

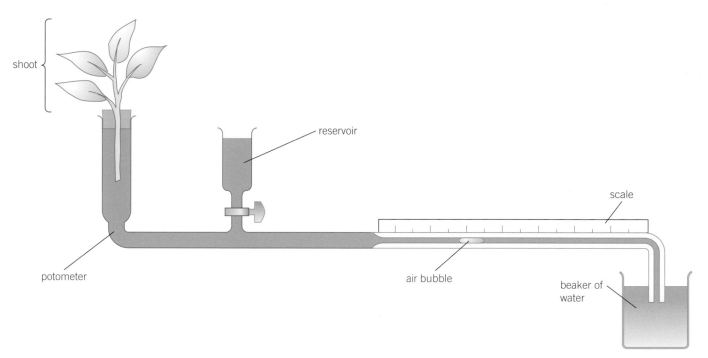

Figure 2 *A potometer is used to show the water uptake of a plant under different conditions*

1 a Name the parts of the leaf that help the plant to reduce water loss under normal conditions. [2 marks]
  b Explain the effect on transpiration of a fan blowing onto the leaves of the plant. [3 marks]
2 a Describe the effect on plant transpiration of coating the top surface of the leaves in petroleum jelly. [1 mark]
  b Describe the effect on plant transpiration of coating the bottom surface of the leaves in petroleum jelly. [1 mark]
  c Explain the difference in the responses you have described for parts **a** and **b**. [3 marks]
3 Water lilies have their stomata on the tops of their leaves.
  a Suggest why this is an important adaptation for water lilies [2 marks]
  b Controlling transpiration is not very important to water lilies. Suggest reasons for this. [2 marks]

Key points

● Factors that increase the rate of photosynthesis or increase stomatal opening will increase the rate of transpiration. These factors include temperature, humidity, air flow, and light intensity.
● Transpiration is more rapid in hot, dry, windy, or bright conditions.

B4 Organising animals and plants

Summary questions

1 Here are descriptions of three heart problems. In each case, use what you know about the heart and the circulatory system to explain the problems caused by the condition.

   a The valve that stops blood flowing back into the left ventricle of the heart after it has been pumped into the aorta becomes weak and floppy and begins to leak. [4 marks]

   b Some babies are born with a 'hole in the heart' – there is a gap in the central dividing wall of the heart. They may look blue in colour and be listless. [4 marks]

   c The coronary arteries supplying blood to the heart muscle itself may become clogged with fatty material. The person affected may get chest pain when they exercise or even have a heart attack. [4 marks]

2 In each of the following examples, explain the effect on the blood and what this means to the person involved:

   a an athlete gives blood before running a race [4 marks]

   b someone eats a diet low in iron. [4 marks]

3 If a patient has a blocked blood vessel, doctors may be able to open up the blocked vessel with a stent or replace it with bits of healthy blood vessels taken from other parts of the patient's body.

Figure 1 shows you the results of these procedures in one group of patients after one year.

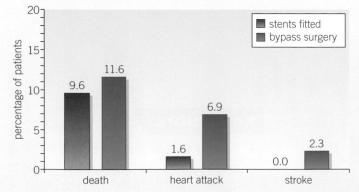

Figure 1

   a Describe a stent and explain how it works. [3 marks]

   b Determine, based on the evidence in Figure 1, which technique seems to be most successful for treating blocked coronary arteries. Explain your decision. [3 marks]

4 a Describe how the lungs are adapted to allow the exchange of oxygen and carbon dioxide between the air and the blood. [3 marks]

   b Describe how air is moved in and out of the lungs and explain how this ventilation of the lungs makes gas exchange more efficient. [4 marks]

5 Plants have specialised cells, tissues, and organs, just as animals do.

   a List **three** examples of plant tissues. [3 marks]

   b Roots, stems, and leaves are important plant organs. Describe how the structure of each is adapted to its functions. [6 marks]

   c Explain which plant tissues are common to all of the main plant organs. [2 marks]

6 The apparatus in Figure 2 (known as a potometer) is often used to give an approximate measure of the transpiration taking place in a plant.

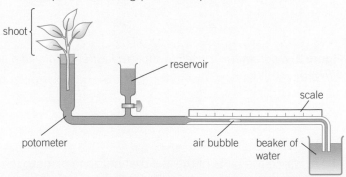

Figure 2

   a Define the term transpiration. [2 marks]

   b Explain carefully what a potometer measures and why it does *not* measure transpiration. [3 marks]

   c Readings are taken using a potometer with plants in different conditions. Explain how and why you would expect the readings to vary from the normal control shoot if:

     i a fan was set up to blow air over the plant [3 marks]

     ii the underside of all the leaves was covered with petroleum jelly. [3 marks]

Practice questions

01 **Figure 1** shows a diagram of the heart.

Figure 1

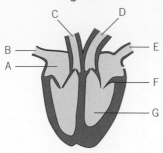

01.1 Use the correct letter from **Figure 1** to identify each of the following parts of the heart.

left ventricle
a valve
vena cava
vessel carrying blood containing the most oxygen

[4 marks]

01.2 What is the function of a valve? [1 mark]

01.3 The coronary arteries carry blood to the heart muscle cells.

In coronary heart disease layers of fatty material build up inside the coronary arteries.
Explain why this could be dangerous. [3 marks]

01.4 People who are at risk of developing coronary heart disease are often given drugs called statins.

Describe how statins reduce the risk of coronary heart disease. [2 marks]

02 **Figure 2** shows the components of the blood.

Figure 2

02.1 Name components **A**, **B,** and **C**. [3 marks]

02.2 Describe **three** ways that red blood cells are adapted to transport oxygen from the lungs to cells of the body. [3 marks]

03 A student investigated the loss of water from a plant. **Figure 3** shows the apparatus he used.

Figure 3

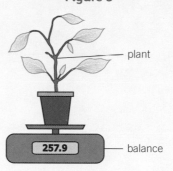

03.1 What is the loss of water through the leaves of a plant called? [1 mark]

The student measured the mass of the plant and pot at the end of each day for five days.

His results are shown in **Table 1**.

Table 1

Day	Mass of pot and plant in g
0	257.9
1	253.6
2	248.8
3	238.9
4	235.4
5	231.9

03.2 During which day did the plant lose the most water? Suggest a reason for this. [2 marks]

03.3 Calculate the mean rate of water loss in g/day.

Give your answer to 2 significant figures. [3 marks]

Marram grass grows on sand dunes where the conditions are dry and windy. The leaves are adapted to reduce the rate of water loss.

Figure 4 shows a cross section of a marram grass leaf.

Figure 4

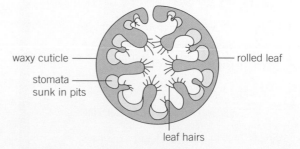

03.4 Describe how **two** features of the marram grass leaf help to reduce the rate of water loss. [2 marks]

2 Disease and bioenergetics

Communicable diseases are caused by pathogens – microorganisms that can be spread from one organism to another. In this section you will learn how we defend ourselves from the pathogens that attack us, and how our lifestyles affect our risk of developing non-communicable diseases such as heart disease and cancer.

You will also learn about photosynthesis in plants – the process where they use light to make sugar from carbon dioxide and water. You will also look at respiration – all living organisms use respiration to transfer the energy they need to carry out the reactions required for life.

Key questions

- What are communicable diseases and how can we prevent them?

- How can your lifestyle affect your risk of developing many non-communicable diseases?

- How do plants use the glucose they make during photosynthesis?

- What is the difference between aerobic and anaerobic respiration?

Making connections

- You will learn about genetic diseases, which are not infectious but can be passed from parents to their offspring, in **B12 Reproduction**

- You will discover the importance of photosynthesis in feeding relationships and ecological communities in **B15 Adaptations, interdependence, and competition** and **B16 Organising an ecosystem**.

- You will find out how pollution of a waterway by fertilisers or sewage can make it impossible for water animals to respire in **B17 Biodiversity and ecosystems**.

I already know...

The consequences of imbalances in the diet.

The importance of bacteria in the human digestive system.

The impact of exercise and smoking on the human gas exchange system.

The effects of recreational drugs on behaviour, health, and life processes.

The basic principles of photosynthesis.

The differences between aerobic and anaerobic respiration.

I will learn...

More about the impact of obesity on human health.

The role of bacteria and other pathogens in human and plant diseases, and how to calculate the effect of antibacterial chemicals by measuring the area of zones of inhibition.

How exercise and smoking can affect the health of other systems of the body.

How to interpret data to understand the effect of lifestyle factors including diet, alcohol, and smoking on the incidence of non-communicable diseases at local, national, and global levels.

How to measure and calculate the rate of photosynthesis, and how different factors affect the rate of photosynthesis.

How an oxygen debt builds up during anaerobic respiration in your muscles.

Required Practicals

Practical		Topic
5	Light intensity and the rate of photosynthesis	B8.2

B 5 Communicable diseases

5.1 Health and disease

Learning objectives

After this topic, you should know:

- what health is
- the different causes of ill health
- how different types of disease interact.

Synoptic links

Find out more about diseases in Chapter B6 and Chapter B7.

Synoptic links

You will learn more about cancer in Topic B7.2.

Synoptic link

For more help in interpreting correlations, looks at Maths skills MS2g.

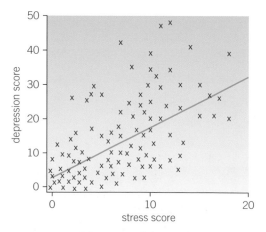

Figure 1 *Scatter graphs can show a correlation between stress and depression*

Your health is a state of physical and mental well-being, not just an absence of disease. It is at least partly based on individual perceptions. A cold or headache that might make you feel ill enough to stay in bed on a school day might be less likely to be a problem if you are on holiday.

What makes us ill?

Communicable (infectious) diseases (e.g., tuberculosis and flu) are caused by **pathogens** such as bacteria and viruses that can be passed from one person to another. **Non-communicable diseases** cannot be transmitted from one person to another (e.g., heart disease and arthritis). Both communicable and non-communicable diseases are major causes of ill health, but other factors can also affect health. Here are three examples:

- Diet – if you do not get enough to eat, or the right nutrients, you may suffer from diseases ranging from starvation to anaemia or rickets. Too much food, or the wrong type of food, can lead to problems such as obesity, some cancers, or type 2 diabetes.

- Stress – a certain level of stress is inevitable in everyone's life and is probably needed for our bodies to function properly. However, scientists are increasingly linking too much stress to an increased risk of developing a wide range of health problems. These include heart disease, certain cancers, and mental health problems.

- Life situations – these include:
    - the part of the world where you live
    - your gender
    - your financial status
    - your ethnic group
    - the levels of free health care provided where you live
    - how many children you have
    - local sewage and rubbish disposal.

People often have little or no control over their life situation, especially as children or young people. Yet such factors have a big effect on health and well-being and are responsible for many causes of ill health around the world. These include communicable diseases such as diarrhoeal diseases and malaria, through to non-communicable diseases such as heart disease and cancer.

How health problems interact

In the next three chapters you will be looking at different types of diseases in isolation. It is important to remember that in the real world, different diseases and health conditions happen at the same time. They interact and often one problem makes another worse. Here are a number of examples – you will learn more about the details of many of these conditions in later chapters.

● Viruses living in cells can trigger changes that lead to cancers – for example, the human papilloma virus can cause cervical cancer.

● The immune system of your body helps you destroy pathogens and get better. If there are defects in your immune system, it may not work effectively. This may be a result of your genetic makeup, poor nutrition, or infections such as HIV/AIDS. This means you will be more likely to suffer from other communicable diseases (Figure 2).

● Immune reactions initially caused by a pathogen, even something like the common cold, can trigger allergies to factors in the environment. These allergies may cause skin rashes, hives, or asthma.

● Physical and mental health are often closely linked. Severe physical ill health can lead to depression and other mental illness.

● Malnutrition is often linked to health problems including deficiency diseases, a weakened immune system, obesity, cardiovascular diseases, type 2 diabetes, and cancer.

The interaction between different factors, including lifestyle, environment, and pathogens, is an important principle to remember as you look at different types of disease.

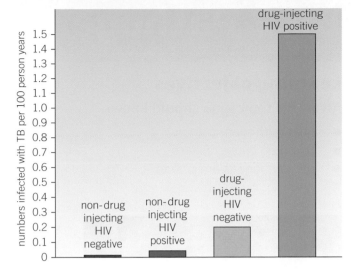

Figure 2 *Data collected in the Netherlands looks at the interaction between a number of health problems including HIV status and drug use in the incidence of tuberculosis (TB) in Amsterdam*

1 Define what is meant by good health. [1 mark]
2 a State three different factors that can cause ill health. [3 marks]
  b Give an example of ill health that each factor can produce. [3 marks]
3 a What health interactions does the data in Figure 2 cover? [2 marks]
  b i What effect does injecting drugs have on your chances of becoming infected with tuberculosis (TB)? [1 mark]
  c Which group has the greatest chance of getting TB? [1 mark]
  d How much more likely is it for an injecting drug user who is HIV positive to get tuberculosis than for an injecting drug user who is HIV negative. Give your answer to the nearest whole number. [2 marks]
4 Explain how the interactions between different types of disease can affect the prevalence of a disease around the World. 🕐 [6 marks]

Key points

● Health is a state of physical and mental well-being.
● Diseases, both communicable and non-communicable, are major causes of ill health.
● Other factors including diet, stress, and life situations may have a profound effect on both mental and physical health.
● Different types of diseases may and often do interact.

B5.2 Pathogens and disease

Learning objectives

After this topic, you should know:

- what pathogens are
- how they cause disease
- how pathogens are spread.

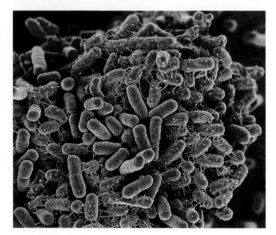

Figure 1 *Many bacteria are very useful to humans but some, such as this strain of E. coli, are pathogens and cause disease*

Synoptic link

Remind yourself about the structure of bacteria by looking back to Topic B1.3.

Communicable diseases, also known as infectious diseases, are found all over the world. **Microorganisms** that cause disease are called pathogens. Pathogens may be bacteria, viruses, protists, or fungi, and they infect animals and plants, causing a wide range of diseases.

Communicable diseases are caused either directly by a pathogen or by a toxin made by a pathogen. The pathogen can be passed from one infected individual to another individual who does not have the disease. Some communicable diseases are fairly mild, such as the common cold and tonsillitis. Others are known killers, such as tetanus, influenza, and HIV/AIDS.

Sometimes communicable diseases can be passed between different species of organisms. For example, infected animals such as dogs or bats can pass rabies on to people. Tuberculosis can be passed from badgers to cows, and from cows to people.

What are the differences between bacteria and viruses?

Bacteria and viruses cause the majority of communicable diseases in people. In plants, viruses and fungi are the most common pathogens. Bacteria are single-celled living organisms that are much smaller than animal and plant cells. Bacteria are used to make food such as yogurt and cheese, to treat sewage, and to make medicines. Bacteria are important both in the environment, as decomposers, and in your body. Scientists estimate that most people have between 1 and 2 kg of bacteria in their guts, and they are rapidly discovering that these bacteria have a major effect on our health and well-being.

Pathogenic bacteria are the minority – but they are significant because of the major effects they can have on individuals and society.

Viruses are even smaller than bacteria. They usually have regular shapes. Viruses cause diseases in every type of living organism.

How pathogens cause disease

Once bacteria and viruses are inside your body, they may reproduce rapidly.

- Bacteria divide rapidly by splitting in two (called binary fission). They may produce toxins (poisons) that affect your body and make you feel ill. Sometimes they directly damage your cells.
- Viruses take over the cells of your body. They live and reproduce inside the cells, damaging and destroying them.

Common disease symptoms are a high temperature, headaches, and rashes. These are caused by the way your body responds to the cell damage and toxins produced by the pathogens.

How pathogens are spread

The more pathogens that get into your body, the more likely it is that you will develop an infectious disease. There are a number of ways in which pathogens spread from one individual to another.

- By air (including droplet infection). Many pathogens including bacteria, viruses, and fungal spores (that cause plant diseases) are carried and spread from one organism to another in the air. In human diseases, droplet infection is common. When you are ill, you expel tiny droplets full of pathogens from your breathing system when you cough, sneeze, or talk (Figure 2). Other people breathe in the droplets, along with the pathogens they contain, so they pick up the infection. Examples include flu (influenza), tuberculosis, and the common cold.

- Direct contact. Some diseases are spread by direct contact of an infected organism with a healthy one. This is common in plant diseases, where a tiny piece of infected plant material left in a field can infect an entire new crop. In people, diseases including sexually transmitted infections, such as syphilis and chlamydia, are spread by direct contact of the skin. Pathogens such as HIV/AIDS or hepatitis enter the body through direct sexual contact, cuts, scratches, and needle punctures that give access to the blood. Animals can act as vectors of both plant and animal diseases by carrying a pathogen between infected and uninfected individuals.

- By water. Fungal spores carried in splashes of water often spread plant diseases. For humans, eating raw, undercooked, or contaminated food, or drinking water containing sewage can spread diseases such as diarrhoeal diseases, cholera, or salmonellosis. The pathogen enters your body through your digestive system.

Lifestyle factors often affect the spread of disease. For example, when people live in crowded conditions with no sewage system, infectious diseases can spread very rapidly.

Figure 2 *Droplets carrying millions of pathogens fly out of your mouth and nose at up to 100 miles an hour when you sneeze*

Synoptic link

For more information on bacteria that are resistant to antibiotics, see Topic B14.4.

Key points

- Communicable diseases are caused by microorganisms called pathogens, which include bacteria, viruses, fungi, and protists.
- Bacteria and viruses reproduce rapidly inside your body. Bacteria can produce toxins that make you feel ill.
- Viruses live and reproduce inside your cells, causing cell damage.
- Pathogens can be spread by direct contact, by air, or by water.

1 a What causes infectious diseases? [1 mark]
   b How do pathogens make you ill? [2 marks]

2 a Give two ways in which diseases are spread from one person to another. [2 marks]
   b Give two ways in which diseases are spread from one plant to another. [2 marks]
   c For each method given in part **a** and part **b**, explain how the pathogens are passed from one organism to the other. [4 marks]

3 Describe and explain the main differences between bacteria and viruses, and how they cause disease. [6 marks]

B5.3 Preventing infections

After this topic, you should know:

- how the spread of disease can be reduced or prevented.

Figure 1 *In hospitals today, simply reminding doctors, nurses, and visitors to wash their hands more often is still an important way to prevent the spread of disease*

People have recognised the symptoms of disease for many centuries. There are records of illnesses people recognise today from the ancient Egyptians and ancient Greeks. However, it is only in the past 150–200 years that people have really understood the causes of these diseases and how they are spread. The work of pioneering doctors and scientists such as Ignaz Semmelweis, Louis Pasteur, and Joseph Lister has helped develop the modern understanding of pathogens. Their work enabled people to prevent the spread of pathogens, and in some cases cure the diseases they cause.

The work of Ignaz Semmelweis

Semmelweis was a doctor in the mid-1850s. At the time, many women in hospital died from childbed fever a few days after giving birth. However, no one knew what caused it.

Semmelweis noticed that his medical students went straight from dissecting a dead body to delivering a baby without washing their hands. The women delivered by medical students and doctors rather than midwives were much more likely to die. Semmelweis wondered if they were carrying the cause of disease from the corpses to their patients.

He noticed that another doctor died from symptoms identical to childbed fever after cutting himself while working on a body. This convinced Semmelweis that the fever was caused by some kind of infectious agent. He therefore insisted that his medical students wash their hands before delivering babies. Immediately, fewer mothers died from the fever. However, other doctors were very resistant to Semmelweis's ideas.

Other discoveries

Also in the mid- to late- 19th century:

- Louis Pasteur showed that microorganisms caused disease. He developed **vaccines** against diseases such as anthrax and rabies.

- Joseph Lister started to use antiseptic chemicals to destroy pathogens before they caused infection in operating theatres.

- As microscopes improved, it became possible to see pathogens more clearly. This helped convince people that they were really there.

Understanding how communicable diseases are spread from one person to another helps us prevent it happening.

Preventing the spread of communicable diseases

There are a number of key ways to help prevent the spread of communicable diseases between people, between animals and people, and between plants.

Hygiene

Simple hygiene measures are one of the most effective ways of preventing the spread of pathogens. These include:

- Hand washing, especially after using the toilet, before cooking, or after contact with an animal or someone who has an infectious illness.

- Using disinfectants on kitchen work surfaces, toilets, etc. to reduce the number of pathogens.

- Keeping raw meat away from food that is eaten uncooked to prevent the spread of pathogens.

- Coughing or sneezing into a handkerchief, tissue, or your hands (and then washing your hands).

- Maintaining the hygiene of people and agricultural machinery to help prevent the spread of plant diseases.

Isolating infected individuals

If someone has an infectious disease, especially a serious disease such as Ebola or cholera, they need to be kept in isolation. The fewer healthy people who come into contact with the infected person, the less likely it is that the pathogens will be passed on. This is also true of plants infected with diseases but it is only possible with smaller plants that can be moved and destroyed easily.

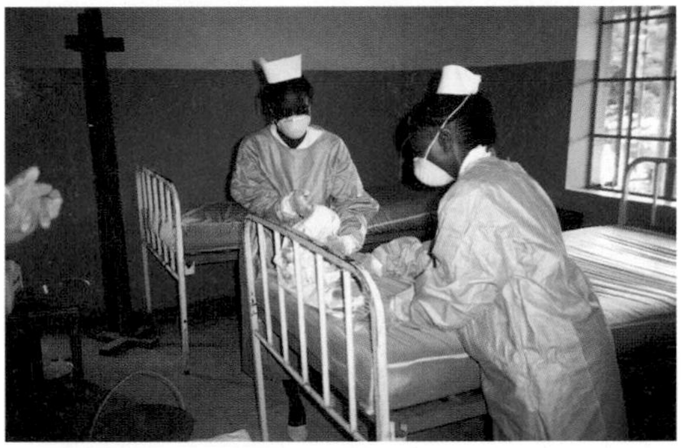

Destroying or controlling vectors

Some communicable diseases are passed on by vectors. For example, mosquitoes carry a range of diseases, such as malaria and dengue fever. Houseflies can carry over 100 human diseases, while rats also act as vectors of disease. Aphids transmit over 150 different plant diseases and different types of beetle carry disease to plants in the form of viral, bacterial, and fungal pathogens. If the vectors are destroyed, the spread of the disease can be prevented. By controlling the number of vectors, the spread of disease can be greatly reduced.

Vaccination

During vaccination, doctors introduce a small amount of a harmless form of a specific pathogen into your body. As a result, if you come into contact with the live pathogen, you will not become ill as your immune system will be prepared. Vaccination is a very successful way of protecting large numbers of humans and animals against serious diseases. However, it cannot protect plants against disease as they do not have an immune system.

Figure 2 *Isolation of infected patients played a major role in the control of the deadly disease Ebola in West Africa during the 2014 outbreak*

Synoptic link

You will learn more about preventing communicable diseases in Topic B6.1.

1 Give three examples of things people can do to reduce the spread of pathogens to lower the risk of disease. [3 marks]

2 For each example you have chosen in your answer to Question **1**, explain how it helps to prevent the spread of disease. [6 marks]

3 Suggest why other doctors were so resistant to Semmelweis's ideas. 🖊 [6 marks]

Key points

- The spread of disease can be prevented by simple hygiene measures, by destroying vectors, by isolation of infected individuals, and by vaccination.

B5.4 Viral diseases

Learning objectives

After this topic, you should know:

- some examples of plant and animal diseases caused by viruses including measles, HIV/AIDS, and tobacco mosaic virus.

Viruses can infect and damage all types of cells. The diseases they cause can be mild or potentially deadly. Scientists have not developed medicines to cure viral diseases, so it is important to stop them spreading. In people, viral diseases often start relatively suddenly. The symptoms are the result of the way the body reacts to the viruses damaging and destroying cells as they reproduce. See below for examples of viral diseases.

Measles

The main symptoms of measles are a fever and a red skin rash. The virus is spread by the inhalation of droplets from coughs and sneezes and is very infectious. Measles is a serious disease that can cause blindness and brain damage and may be fatal if complications arise. In 2013, 145 700 people globally died of measles. There is no treatment for measles, so if someone becomes infected they need to be isolated to stop the spread of the virus. Measles is now rare in the UK as a result of improved living conditions and a vaccination programme for young children. The challenge now is to vaccinate children globally and make deaths from measles a thing of the past (Figure 2).

HIV/AIDS

Around 35 million people globally are infected with HIV, a virus that can eventually lead to AIDS. In 2013, around 1.5 million people died of HIV-related illnesses. Many people do not realise they are infected with HIV, because the virus only causes a mild, flu-like illness to begin with. HIV attacks the immune cells and after the initial mild illness it remains hidden inside the immune system, sometimes for years, until the immune system is so badly damaged that it can no longer deal with infections or certain cancers. At this point the patient has developed AIDS.

The time between infection with HIV and the onset of the final stages of AIDS is affected by many factors. These include the level of nutrition and overall health of the person, as well as access to antiretroviral drugs.

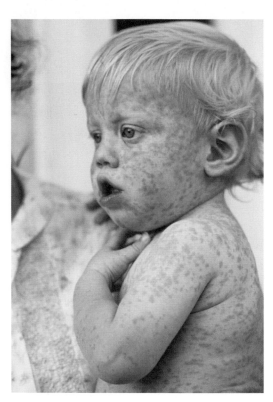

Figure 1 *A measles rash is now a rare sight in the UK*

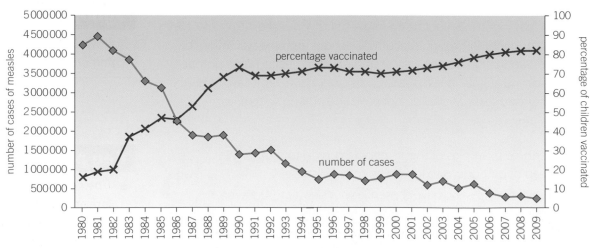

Figure 2 *World Health Organisation data on the trends in global vaccination against measles and the numbers of reported measles cases*

HIV is spread by direct sexual contact and the exchange of body fluids such as blood, which occurs when drug users share needles or when unscreened blood is used for transfusions. HIV can also be passed from mother to child in breast milk.

There is no cure for HIV/AIDS and no vaccine against it. The spread of the disease can be prevented by using condoms, not sharing needles, screening blood used for transfusions, and HIV-positive mothers bottle-feeding their children.

The regular use of antiretroviral drugs can prevent the development of AIDS for many years and give HIV positive people an almost normal life expectancy. Unfortunately, the majority of people infected with HIV live in areas such as sub-Saharan Africa where it is hard to get antiretroviral drugs. In these areas the life expectancy for people with HIV/AIDs is still very low. To have the best chance of long-term survival, antiretroviral drugs must be started as soon as possible after infection.

Tobacco mosaic virus

Tobacco mosaic virus (TMV) was the first virus ever to be isolated. It is a widespread plant pathogen that affects around 150 species of plants including tomatoes and tobacco plants. It causes a distinctive 'mosaic' pattern of discoloration on the leaves as the virus destroys the cells. This affects the growth of the plant because the affected areas of the leaf do not photosynthesise. TMV can seriously reduce the yield of a crop.

It is spread by contact between diseased plant material and healthy plants, and insects can act as vectors. The virus can remain infectious in the soil for about 50 years. There is no treatment and farmers now grow TMV-resistant strains of many crop plants. Good field hygiene and good pest control can help prevent the spread of TMV.

Figure 3 *Tobacco mosaic virus causes a typical pattern of damage in many different types of plants*

1 **a** State the main symptoms of measles. [2 marks]
  b Suggest why measles is now rare in the UK. [2 marks]
2 **a** Describe the link between HIV and AIDS. [1 mark]
  b Explain why untreated HIV is usually fatal. [4 marks]
3 Using Figure 2, calculate the following.
  a The number of cases of measles globally between:
    i 1980 and 1985 [3 marks]
    ii 2000 and 2005. [3 marks]
  b Assuming that 5% of patients (cases) will die, calculate how many people died of measles in each time period. [4 marks]
  c Discuss the apparent link between vaccination rates and cases of measles globally. ✏ [4 marks]
4 Discuss the similarities and differences between tobacco mosaic virus in plants and measles in people. ✏ [6 marks]

Key points

- Measles virus is spread by droplet infection. It causes fever and a rash and can be fatal. There is no cure. Isolation of patients and vaccination prevents spread.
- HIV initially causes flu-like illness. Unless it is successfully controlled with antiretroviral drugs the virus attacks the body's immune cells. Late stage HIV infection, or AIDS, occurs when the body's immune system becomes so badly damaged it can no longer deal with other infections or cancers. HIV is spread by sexual contact or by the exchange of body fluids, such as blood, which occurs when drug users share needles.
- Tobacco mosaic virus is spread by contact and vectors. It damages leaves and reduces photosynthesis. There is no treatment. Spread is prevented by field hygiene and pest control.

B5.5 Bacterial diseases

Learning objectives

After this topic, you should know:

- some examples of plant and animal diseases caused by bacteria, including *Salmonella* food poisoning and gonorrhoea.

Synoptic link

You can learn more about orders of magnitude in Maths skills MS2.

You will learn more about antibiotics in Topic B6.2, and more about antibiotic resistance in Topic B14.4.

Figure 1 *Handling raw poultry, undercooked food, or salads contaminated with raw meat through poor kitchen hygiene are all common sources of the Salmonella bacteria that can cause food poisoning.*

Bacterial diseases affect animals and plants. In the early 20th century, more than 30% of all deaths in the USA were due to infectious diseases. That is now an order of magnitude lower, and most of the infectious diseases that cause death are viral. Improved living standards and vaccinations have had a major effect on the incidence and death rate of communicable diseases in countries such as the USA and UK. The development of antibiotics is the other key factor in combating bacterial diseases. Antibiotics kill bacteria or stop them growing and cure bacterial diseases. Unfortunately, bacteria are becoming resistant to many antibiotics and more people are dying from bacterial diseases again.

Salmonella food poisoning

Salmonella are bacteria that live in the guts of many different animals. They can be found in raw meat, poultry, eggs, and egg products such as mayonnaise. If these bacteria get into our bodies, they disrupt the balance of the natural gut bacteria and can cause *Salmonella* food poisoning. One common cause of infection is eating undercooked food, when the bacteria have not been killed by heating. Another is, eating food prepared in unhygienic conditions where food is contaminated with *Salmonella* bacteria from raw meat.

The symptoms develop within 8–72 hours of eating infected food. Fever, abdominal cramps, vomiting, and diarrhoea are caused by the bacteria and the toxins they secrete. For many people *Salmonella* infections are unpleasant but don't last many days and no antibiotics are given. In very young children and the elderly it can be fatal, usually because of dehydration. In countries where there is malnutrition, *Salmonella* is more serious. The World Health Organisation estimates that globally around 2.2 million people, mainly children under 5 years old, are killed by sickness and diarrhoea each year, including *Salmonella* food poisoning.

In the UK, poultry are vaccinated against *Salmonella* to control the spread of the disease. *Campylobacter*, another bacterium found in chickens, still causes around 280 000 cases of food poisoning each year. To prevent food poisoning, keep raw chicken away from food that is eaten uncooked, avoid washing raw chicken (it sprays bacteria around the kitchen), wash hands and surfaces well after handling raw chicken, and cook chicken thoroughly.

Gonorrhoea

Gonorrhoea is a **sexually transmitted disease (STD)**, which are also known as sexually transmitted infections (STIs). It is spread by unprotected sexual contact with an infected person. Like many STDs, gonorrhoea has symptoms in the early stages but then becomes relatively symptomless. The early symptoms include a thick yellow or green discharge from the vagina or penis and pain on urination.

However, about 10% of infected men and 50% of infected women get no symptoms at all. Untreated gonorrhoea can cause long-term pelvic pain, infertility, and ectopic pregnancies. Babies born to infected mothers may have severe eye infections and even become blind. Gonorrhoea is bacterial, so it can be treated with antibiotics. Originally it was easily cured using penicillin but now many antibiotic-resistant strains of gonorrhoea have evolved so it is increasingly difficult to treat. All sexual partners of an infected individual must be treated with antibiotics to prevent the disease spreading in the community. The spread of gonorrhoea can also be prevented by using a barrier method of contraception such as a condom and by reducing the number of sexual partners.

Bacterial disease in plants

There are relatively few bacterial diseases of plants and these diseases are usually found in tropical and sub-tropical regions. *Agrobacterium tumefaciens* is a bacterium that causes crown galls – a mass of unspecialised cells that often grow at the join between the root and the shoot in infected plants (Figure 2). It infects many different plant types including fruit trees, vegetables, and garden flowering plants. The bacteria insert plasmids into the plant cells and cause a mass of new undifferentiated genetically modified cells to grow. For this reason, these bacteria have become a key tool for scientists when genetically modifying plant cells. Scientists make use of the way the bacteria naturally infect plant cells and give them new added genes. They manipulate the bacteria so they carry desirable genes into the cells they infect.

Figure 2 *The bacteria that cause galls like this one on a chrysanthemum plant are also widely used in genetically modifying plants*

Synoptic links

You will find out more about the use of bacteria in the production of genetically modified plants in B13.4 and B13.5.

Key points

- *Salmonella* is spread through undercooked food and poor hygiene. Symptoms include fever, abdominal cramps, diarrhoea, and vomiting caused by the toxins produced by the bacteria. In the UK, poultry are vaccinated against *Salmonella* to control the spread of disease.
- Gonorrhoea is a sexually transmitted disease. Symptoms include discharge from the penis and vagina and the pain on urination. Treatment involves using antibiotics, although many strains are now resistant. Using condoms and limiting sexual partners prevents spread.
- There are relatively few bacterial diseases of plants but *Agrobacterium tumefaciens* causes galls.

1 State one way that antibiotics work to cure bacterial infections.
[1 mark]

2 a Describe how people become infected with food poisoning caused by *Salmonella*. [2 marks]
  b Doctors in the UK rarely treat *Salmonella* food poisoning with antibiotics. Suggest reasons for this. [3 marks]

3 a Gonorrhoea is an STD. Explain what this means. [2 marks]
  b Until recently gonorrhoea was relatively easy to treat. Explain this statement. [2 marks]
  c Suggest three ways of preventing the spread of gonorrhoea. [3 marks]
  d Discuss the implications of increased antibiotic resistance in the bacteria causing gonorrhoea for the 106 million people who are infected with the disease each year. [4 marks]

4 Write a paragraph for your local newspaper on food preparation for summer barbeques to help people avoid *Salmonella* and other forms of food poisoning. 🖊 [6 marks]

B5.6 Diseases caused by fungi and protists

Learning objectives

After this topic, you should know:

- some examples of plant diseases caused by fungi, including rose black spot
- some examples of animal diseases caused by protists, including malaria
- how the spread of diseases can be reduced or prevented.

Figure 1 *Roses are beautiful flowers but fungal blackspot infections weaken the plants and reduce the flowers. Similar fungal diseases weaken and destroy crop plants around the world*

Fungi and protists are less well known than bacteria and viruses but they are also important pathogens. Some of the diseases they cause are of great significance, both in terms of global economies and of human suffering.

Fungal diseases

There are relatively few fungal diseases that affect people. Athlete's foot is a well-known, relatively minor fungal skin condition. A small number of human fungal diseases can be fatal when they attack the lungs or brains of people who are already ill. Damaged heart valves can also develop serious fungal infections. However, these conditions are rare. Antifungal drugs are usually effective against skin fungi like athlete's foot, but it can be hard to treat deep-seated tissue infections.

In plants, however, fungal diseases are common and can be devastating. Huge areas of crops, from cereals to bananas are lost every year as a result of fungal infections, including stem rusts and various rotting diseases.

Rose black spot

Rose black spot is a fungal disease of rose leaves. It causes purple or black spots to develop on the leaves and it is a nuisance in gardens and for commercial flower growers. The leaves often turn yellow and drop early. This weakens the plant because it reduces the area of leaves available for photosynthesis. As a result the plant does not flower well – and the main reason people grow roses is for the lovely flowers.

The spores of the fungus are spread in the environment, carried by the wind. They are then spread over the plant after it rains in drips of water that splash from one leaf or plant to another. The spores stay dormant over winter on dead leaves and on the stems of rose plants. Gardeners try to prevent the spread by removing and burning affected leaves and stems. Chemical fungicides can also help to treat the disease and prevent it spreading. Horticulturists have bred types of roses that are relatively resistant to black spot but the disease still cannot be prevented or cured.

Diseases caused by protists

Protists (a type of single-celled organism) cause a range of diseases in animals and plants. They are relatively rare pathogens but the diseases they cause are often serious and damaging to those infected. Diseases caused by protists usually involve a vector that transfers the protist to the host. One of the best known and globally serious protist diseases is malaria.

Malaria

Malaria is a disease caused by protist pathogens that are parasites – they live and feed on other living organisms. The life cycle of the protists includes time in the human body and time in the body of a female *Anopheles* mosquito. The protists reproduce sexually in the mosquito and asexually in the human body. The mosquitoes act as vectors of the disease. The female mosquito needs two meals of human blood before she can lay her eggs, and this is when the protists are passed into the human bloodstream. The protists travel around the human body in the circulatory system. They affect the liver and damage red blood cells. Malaria causes recurrent episodes of fever and shaking when the protists burst out of the blood cells, and it can be fatal. It weakens the affected person over time even if it does not kill them. Globally several hundred million cases of malaria occur each year, and around 660 000 people die from the disease.

Synoptic links

You will learn more about asexual and sexual reproduction in Topics B12.1 and B12.2.

Figure 2 *The malaria protist causes disease and mosquitoes act as very effective vectors*

Figure 3 *Simple control measures such as insecticide-treated mosquito nets have reduced the incidence of malaria by as much as 75% in some countries*

If malaria is diagnosed quickly, it can be treated using a combination of drugs, but this is not always available in the countries most affected by malaria. The protists have also become resistant to some of the most commonly used medicines. The spread of malaria can be controlled in a number of ways, most of which target the mosquito vector. These include:

- Using insecticide-impregnated insect nets to prevent mosquitoes biting humans and passing on the protists (Figure 3).
- Using insecticides to kill mosquitoes in homes and offices.
- Preventing the vectors from breeding by removing standing water and spraying water with insecticides to kill the larvae.
- Travellers can take antimalarial drugs that kill the parasites in the blood if they are bitten by an infected mosquito.

1 a Describe three ways in which fungal diseases such as black spot or stem rust can be spread from plant to plant. [3 marks]
  b Explain why roses affected by black spot produce fewer, smaller flowers than healthy plants. [3 marks]

2 a Describe how malaria is passed from one person to another. [2 marks]
  b Insecticide-treated mosquito nets help prevent the spread of malaria in two ways. Explain how. [2 marks]

3 For travellers from the UK going to an area with malaria, doctors suggest the ABCD approach. This stands for Awareness, Bite prevention, Chemoprophylaxis (antimalarial medicines), and Diagnosis. Discuss how each of these points would reduce the chance of becoming seriously ill with malaria. [6 marks]

Key points

- Rose black spot is a fungal disease spread in the environment by wind and water. It damages leaves so they drop off, affecting growth as photosynthesis is reduced. Spread is controlled by removing affected leaves and chemical sprays, but is not very effective.
- Malaria is caused by parasitic protists and is spread by the bite of female mosquitos. It damages blood and liver cells, causes fevers and shaking, and can be fatal. Some drugs are effective if given early but protists are becoming resistant. Spread is reduced by preventing the vectors from breeding and by using mosquito nets to prevent people from being bitten.

B5.7 Human defence responses

Learning objectives

After this topic, you should know:

- how your body stops pathogens getting in
- how your white blood cells protect you from disease.

The mucus produced from your nose turns green when you have a cold. Why does this happen? It is all part of the way your body defends itself against disease.

Preventing microorganisms getting into your body

Each day, you meet millions of disease-causing microorganisms. Every body opening as well as any breaks in the skin give pathogens a way in. The more pathogens that get into your body, the more likely it is that you will get an infectious disease. Fortunately, your body has many defence mechanisms that work together to keep the pathogens out.

Skin defences

- Your skin covers your body and acts as a barrier. It prevents bacteria and viruses reaching the tissues beneath. If you damage or cut your skin, the barrier is broken but your body restores it. You bleed, and the platelets in your blood set up a chain of events to form a clot that dries into a scab (Figure 1). This forms a seal over the cut, stopping pathogens getting in. It also stops you bleeding to death.

- Your skin produces antimicrobial secretions to destroy pathogenic bacteria.

- Healthy skin is covered with microorganisms that help keep you healthy and act as an extra barrier to the entry of pathogens.

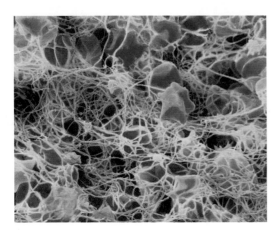

Figure 1 *The scabs that restore the protective barrier of the skin and prevent pathogens getting in are made of red blood cells tangled in protein strands formed by platelets*

Defences of the respiratory and digestive systems

Your respiratory system is a weak link in your body defences. Every time you breathe in, you draw air full of pathogens into the airways of the lungs. In the same way, you take food and drink, as well as air, into your digestive system through your mouth. Both systems have good defences to help prevent pathogens constantly causing infections.

- Your nose is full of hairs and produces a sticky liquid, called mucus. The hairs and mucus trap particles in the air that may contain pathogens or irritate your lungs. If you spend time in an environment with lots of air pollution, the mucus you produce when you blow your nose is blackened, showing that the system works.

- The trachea and bronchi also secrete mucus that traps pathogens from the air. The lining of the tubes is covered in cilia – tiny hair-like projections from the cells. The cilia beat to waft the mucus up to the back of the throat where it is swallowed.

- The stomach produces acid and this destroys the microorganisms in the mucus you swallow, as well as the majority of the pathogens you take in through your mouth in your food and drink.

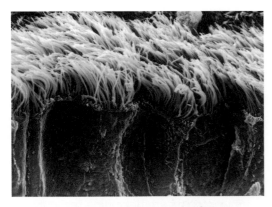

Figure 2 *The cilia of the airways beat together to move mucus containing trapped pathogens away from the lungs*

The immune system – internal defences

In spite of your body's defence mechanisms, some pathogens still get inside your body. Once there, they will meet your second line of defence – the white blood cells of your immune system. The immune system will try to destroy any pathogens that enter the body in several ways.

Table 1 *Ways in which your white blood cells destroy pathogens and protect you against disease*

Role of white blood cell	How it protects you against disease
Ingesting microorganisms bacterium — white blood cell	Some white blood cells ingest (take in) pathogens, digesting and destroying them so they cannot make you ill.
Producing antibodies antibody — antigen — bacterium — white blood cell — antibody attached to antigen	Some white blood cells produce special chemicals called antibodies. These target particular bacteria or viruses and destroy them. You need a unique antibody for each type of pathogen. When your white blood cells have produced antibodies once against a particular pathogen, they can be made very quickly if that pathogen gets into the body again. This stops you getting the disease twice.
Producing antitoxins white blood cell — antitoxin molecule — toxin and antitoxin joined together — toxin molecule — bacterium	Some white blood cells produce antitoxins. These counteract (cancel out) the toxins released by pathogens.

The different body systems work together to help protect you from disease. For example, some white blood cells contain green-coloured enzymes. These white blood cells destroy the cold viruses and any bacteria trapped in the mucus of your nose when you have a cold. The dead white blood cells, along with the dead bacteria and viruses, are removed in the mucus, making it look green.

1 State how each action can prevent the spread of disease.
  a Washing your hands before preparing a salad. [1 mark]
  b Throwing away tissues after you have blown your nose. [1 mark]
  c Making sure that sewage does not get into drinking water. [1 mark]

2 Explain why the following symptoms of certain diseases increase your risk of getting infections.
  a Your blood won't clot properly. [2 marks]
  b The number of white cells in your blood falls. [3 marks]

3 Discuss how your white blood cells help to prevent you from suffering from communicable diseases. ✏ [6 marks]

B5 Communicable diseases

Summary questions

1 a Describe the difference between a communicable disease and a non-communicable disease and give one example of each. [4 marks]

b Give three factors that can affect the health of a person. [3 marks]

c For each factor you chose in part **b**, explain how this factor can affect the health of an individual. [6 marks]

d Often a disease is the result of several different factors interacting.
Explain how these interactions can work, giving three examples. [6 marks]

e As well as human diseases, we also study plant diseases caused by pathogens. Suggest reasons why plant diseases are so important to human well-being. [5 marks]

2 a Explain how communicable diseases spread from one person to another. [4 marks]

b What steps can individuals take to reduce the spread of communicable diseases from one person to another? [4 marks]

c Suggest how companies and organisations might reduce the spread of communicable diseases between employees. [4 marks]

3 Use the data in Figure 1 and your knowledge of HIV/AIDS to answer the following questions.

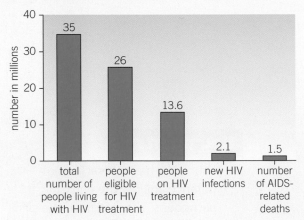

Figure 1 *Data on the global HIV/AIDs epidemic in one year*

a What is the difference between HIV and AIDS? [2 marks]

b Approximately 70% of the people living with HIV/AIDS and 70% of the deaths from AIDS are in sub-Saharan Africa. Using data from Figure 1:
i Calculate the approximate numbers of people living with HIV and dying of AIDS in sub-Saharan Africa in 2013. [4 marks]
ii Explain why your answer is only approximate. [3 marks]

c i What percentage of the people suitable for treatment with antiretroviral therapy (ART) actually get treatment? [2 marks]
ii Suggest reasons for this. [3 marks]

d i Give three ways of preventing the spread of HIV. [3 marks]
ii Explain how each method works. [6 marks]

4 Table 1 shows the percentage of crops lost to disease from a study published in 2006, both with and without crop protection measures:

a Display these results graphically or in a chart. [5 marks]
b Which crop benefitted most from the protective measures? Explain your answer. [3 marks]
c Which crop was least affected by crop protection measures? Explain your answer [3 marks]

Table 1

	% crop loss without protection	% crop loss with protection
wheat	49.8	28.2
rice	75.0	35.4
maize	68.8	31.2
potatoes	84.8	40.3
soybeans	60.1	26.3
cotton	82.0	28.8

Source: Oerke *et al.*, *Journal of Agricultural Science* 2006

d i State three different types of plant diseases [3 marks]
ii Explain how the diseases might affect the yield of a crop. [6 marks]
e Suggest three ways in which farmers can help to protect their crops from plant diseases. [3 marks]
f Explain, using your answer to part **a**, why research into the prevention of plant diseases is so important to human health and well being. [6 marks]

Practice questions

01 Pathogens cause infectious diseases.

01.1 Match each disease to the type of pathogen that causes the disease.

Disease		Type of pathogen
AIDS		bacterium
malaria		virus
salmonella food poisoning		protist
		fungus

[3 marks]

Salmonella food poisoning is caused by ingesting food contaminated with the pathogen.
Figure 1 shows the number of laboratory confirmed cases of *Salmonella* food poisoning in Wales from 1981 to 2013.

Figure 1

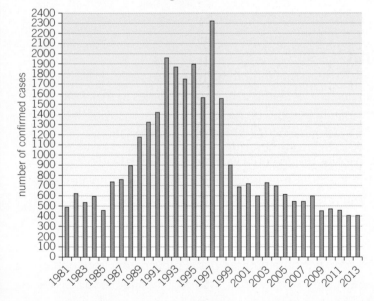

01.2 In which year was the highest number of confirmed cases of *Salmonella*? [1 mark]

01.3 How many confirmed cases were there in that year? [1 mark]

01.4 Suggest **one** reason why this value may not be the total number of *Salmonella* cases in Wales that year. [1 mark]

01.5 The high number of cases that year was due to people eating eggs from infected chickens. The number of cases of *Salmonella* has rapidly decreased since then.
Suggest **two** reasons for the decrease. [2 marks]

02 Rose black spot is a fungal infection of roses.
Figure 2 shows a rose leaf infected with the rose black spot fungus.

Figure 2

Black spots appear on the leaves. The leaves may turn yellow and drop from the plant.

02.1 Explain how rose black spot disease might affect the growth of a rose bush. [3 marks]

02.2 Suggest **two** things a gardener could do to kill the fungus and reduce the chance of another rose black spot infection. [2 marks]

03 Pathogens can enter the body through cuts in the skin, through the breathing system, and through the digestive system.

03.1 Describe how the body is adapted to reduce the entry of pathogens through these three routes. [6 marks]

If pathogens do enter the body, white blood cells work in different ways to defend against the pathogens.

03.2 Describe **two** ways that white blood cells can kill pathogens. [2 marks]

03.3 If pathogens reproduce to form a large population inside the body they cause an infection. The person may feel ill.
Describe how white blood cells can reduce the symptoms of an infection. [1 mark]

6.1 Vaccination

Figure 1 *No one likes having a vaccination very much – but they save millions of lives around the world every year!*

Every cell has unique proteins on its surface called antigens. The antigens on the microorganisms that get into your body are different to the ones on your own cells. Your immune system recognises that they are different.

Your white blood cells then make specific antibodies, which join up with the antigens and inactivate or destroy that particular pathogen.

Some of your white blood cells (the memory cells) 'remember' the right antibody needed to destroy a particular pathogen. If you meet that pathogen again, these memory cells can make the same antibody very quickly to kill the pathogen, so you become immune to the disease.

The first time you meet a new pathogen you get ill because there is a delay while your body sorts out the right antibody needed. The next time, your immune system destroys the invaders before they can make you feel unwell.

Vaccination

Some pathogens, such as meningitis, can make you seriously ill very quickly. In fact, you can die before your body manages to make the right antibodies. Fortunately, you can be protected against many of these serious diseases by vaccination (also known as immunisation).

Immunisation involves giving you a **vaccine** made of a dead or inactivated form of a disease-causing microorganism. It stimulates your body's natural immune response to invading pathogens.

A small amount of dead or inactive forms of a pathogen is introduced into your body. This stimulates the white blood cells to produce the antibodies needed to fight the pathogen and prevent you from getting ill. Then, if you meet the same, live pathogen, your white blood cells can respond rapidly. They can make the right antibodies just as if you had already had the disease, so that you are protected against it.

Doctors use vaccines to protect us both against bacterial diseases, such as tetanus and diphtheria, and viral diseases such as polio, measles and mumps. For example, the MMR vaccine protects against measles, mumps, and rubella. Vaccines have saved millions of lives around the world. One disease – smallpox – has been completely wiped out by vaccinations. Doctors hope polio will also disappear in the next few years.

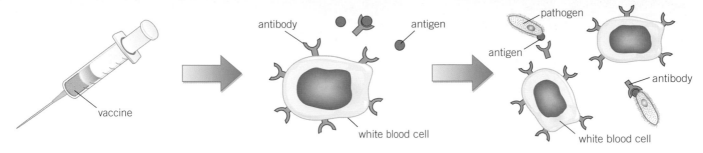

Small amounts of dead or inactive pathogen are put into your body, often by injection.

The antigens in the vaccine stimulate your white blood cells into making antibodies. The antibodies destroy the antigens without any risk of you getting the disease.

You are immune to future infections by the pathogen. That's because your body can respond rapidly and make the correct antibody as if you had already had the disease.

Figure 2 *This is how vaccination protects you against dangerous infectious diseases*

Herd immunity

If a large proportion of the population is immune to a disease, the spread of the pathogen in the population is very much reduced and the disease may even disappear. This is known as herd immunity. If, for any reason, the number of people taking up a vaccine falls, the herd immunity is lost and the disease can reappear. This is what happened in the UK in the 1970s when there was a scare about the safety of the whooping cough vaccine. Vaccination rates fell from over 80% to around 30% (Figure 3). In the following years, thousands of children got whooping cough again and a substantial number died. Yet the vaccine was as safe as any medicine. Eventually people realised this and enough children were vaccinated for herd immunity to be effective again. There are global vaccination programmes to control a number of diseases, including tetanus in mothers and new-born babies, polio, and measles. The World Health Organisation want 95% of children to have two doses of measles vaccine to give global herd immunity. Current global figures show that 85% of children get the first dose and 56% get the second. It will take money and determination to get global herd immunity against a range of different diseases, but the advantages both to individuals and to global economies are huge.

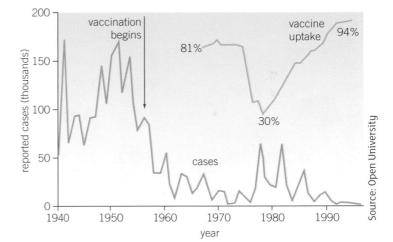

Figure 3 *Graph showing the effect of the whooping cough scare on both uptake of the vaccine and the number of cases of the disease*

1 a Describe what an antigen is. [1 mark]
  b Describe what an antibody is. [1 mark]
  c Give an example of one bacterial and one viral disease that you can be immunised against. [2 marks]

2 Explain, using diagrams if they help you:
  a how the immune system of your body works [5 marks]
  b how vaccines use your natural immune system to protect you against serious diseases. [5 marks]

3 Explain why vaccines can be used against both bacterial and viral diseases. [5 marks]

Key points

- If a pathogen enters the body the immune system tries to destroy the pathogen.
- Vaccination involves introducing small amounts of dead or inactive forms of a pathogen into your body to stimulate the white blood cells to produce antibodies. If the same live pathogen re-enters the body, the white blood cells respond quickly to produce the correct antibodies, preventing infection.
- If a large proportion of the population is immune to a pathogen, the spread of the pathogen is much reduced.

B6.2 Antibiotics and painkillers

Learning objectives

After this topic, you should know:

- what medicines are and how some of them work
- that painkillers and other medicines treat disease symptoms but do not kill pathogens
- the ways in which antibiotics can and cannot be used.

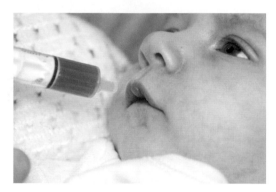

Figure 1 *Giving this baby a painkiller will make him feel better, but he will not actually get better any faster as a result*

Study tip

Don't confuse antiseptics, antibiotics and antibodies.

- Antiseptics kill microorganisms in the environment.
- Antibiotics kill bacteria (*not* viruses) in the body.
- Antibodies are made by white blood cells to destroy pathogens (both bacteria *and* viruses).

When you have an infectious disease, you generally take medicines that contain useful drugs. Often the medicine doesn't affect the pathogen that is causing the problems – it just eases the symptoms and makes you feel better.

Treating the symptoms

Drugs such as aspirin and paracetamol are very useful painkillers. When you have a cold, they will help relieve your headache and sore throat. On the other hand, they will have no effect on the viruses that have entered your tissues and made you feel ill.

Many of the medicines you can buy at a chemists or supermarket relieve your symptoms but do not kill the pathogens, so they do not cure you any faster. You have to wait for your immune system to overcome the pathogens before you actually get well again.

Antibiotics – drugs to cure bacterial diseases

Drugs that make you feel better are useful, but in some cases what you really need are drugs that can cure you. You can use antiseptics and disinfectants to kill bacteria outside the body, but they are far too poisonous to use inside your body. They would kill you and your pathogens at the same time.

The drugs that have really changed the treatment of communicable diseases are antibiotics. These are medicines that can work inside your body to kill bacterial pathogens. The impact of antibiotics on deaths from communicable diseases has been enormous. Antibiotics first became widely available in the 1940s. They were regarded as wonder drugs. For example, the number of women who died of infections in the first days after giving birth dropped dramatically (Figure 2).

How antibiotics work

Antibiotics, such as penicillin, work by killing the bacteria that cause disease whilst they are inside your body. They damage the bacterial cells without harming your own cells. Bacterial diseases that killed millions of people in the past can now be cured using antibiotics. They have had an enormous effect on our society.

If you need antibiotics, you usually take a pill or syrup, but if you are very ill antibiotics may be put straight into your bloodstream. This makes sure that they reach the pathogens in your cells as quickly as possible. Some antibiotics kill a wide range of bacteria. Others are very specific and only work against particular bacteria. It is important that the right antibiotic is chosen and used. Specific bacteria should be treated with the specific antibiotic that is effective against them.

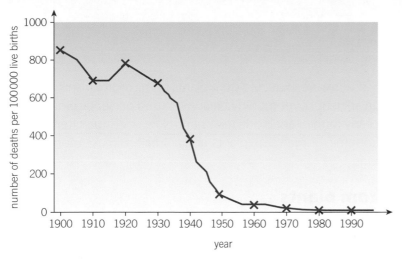

Figure 2 *The introduction of antibiotics in the 1940s had an enormous impact on deaths from maternal septicaemia*

Figure 3 *Penicillin was the first antibiotic. Now there are many different ones that kill different types of bacterium. Here, several different antibiotics are being tested on an agar plate.*

Unfortunately, antibiotics are not the complete answer to the problem of infectious diseases:

● Antibiotics cannot kill viral pathogens so they have no effect on diseases caused by viruses. Viruses reproduce inside the cells of your body. It is extremely difficult to develop drugs that will kill the viruses without damaging the cells and tissues of your body at the same time.

● Strains of bacteria that are resistant to antibiotics are evolving. This means that antibiotics which used to kill a particular type of bacteria no longer have an effect, so they cannot cure the disease. There are some types of bacteria that are resistant to all known antibiotics. The emergence of antibiotic-resistant strains of bacteria is a matter of great concern. Unless scientists can discover new antibiotics soon, we may no longer be able to cure bacterial diseases. This means that many millions of people in the future will die of bacterial diseases that we can currently cure. You will learn more about the discovery of antibiotics in Topic B6.3.

1 Describe the main difference between drugs such as paracetamol and drugs such as penicillin. [2 marks]

2 Explain why it is more difficult to develop medicines against viruses than it has been to develop antibiotics. [4 marks]

3 Use Figure 2 to answer the following.
a State how many women died in childbirth or shortly afterwards in 1930, 1940, and 1950 from maternal septicaemia. [3 marks]
b Calculate the percentage fall or rise in the death rates of mothers around the time of birth in:
  i 1930 and 1940 [3 marks]
  ii 1940 and 1950. [3 marks]
c Suggest reasons for these observations. [3 marks]
d Based on this evidence, explain why the emergence of antibiotic-resistant bacteria is such a cause of concern. [3 marks]

Synoptic links

You will learn more about the development of antibiotic resistance in bacteria in Topic B14.4.

Key points

● Painkillers and other medicines treat the symptoms of disease but do not kill the pathogens that cause it.
● Antibiotics cure bacterial diseases by killing the bacterial pathogens inside your body.
● The use of antibiotics has greatly reduced deaths from infectious diseases.
● The emergence of strains of bacteria resistant to antibiotics is a matter of great concern.
● Antibiotics do not destroy viruses because viruses reproduce inside the cells. It is difficult to develop drugs that can destroy viruses without damaging your body cells.

B6.3 Discovering drugs

Learning objectives

After this topic, you should know:

- some of the drugs traditionally extracted from plants
- how penicillin was discovered
- how scientists look for new drugs.

Figure 1 *Chewing on the glands of beaver tail brought unexplained relief to people in the days before clean effective painkillers such as aspirin or paracetamol were available. The tail glands contained concentrated pain-relieving chemicals from willow bark chewed by the beaver*

Traditionally drugs were extracted from plants or microorganisms such as moulds. In ancient Egypt mouldy bread was used on septic wounds, perhaps an early form of antibiotic treatment. Now scientists often adapt chemicals from microorganisms, plants, and animals to make more effective drugs.

Drugs from plants

There are a number of drugs used today that are based on traditional medicines extracted from plants.

Digitalis is one of several drugs extracted from foxgloves, and the drug digoxin is another. They have been used since the 18th century to help strengthen the heartbeat. There are many more modern drugs but doctors still use digoxin, especially for older patients with heart problems. Large amounts of these chemicals can act as poisons.

The painkiller aspirin originates from a compound found in the bark of willow trees. The anti-inflammatory and pain-relieving properties were first recorded in 400BC. In 1897, Felix Hoffman synthesised acetyl salicylic acid (aspirin), which not only relieves pain and inflammation better than willow bark but has fewer side effects. Aspirin is still commonly used to treat a wide range of health problems.

Drugs from microorganisms – discovering penicillin

In the early 20th century, scientists were looking for chemicals that might kill bacteria and cure infectious diseases. In 1928, Alexander Fleming was growing bacteria for study purposes. He was rather careless, often leaving the lids off his culture plates – health and safety procedures were not as good in those days.

After one holiday, Fleming saw that lots of his culture plates had mould growing on them. He noticed a clear ring in the jelly around some of the spots of mould and realised something had killed the bacteria covering the gel. Fleming recognised the importance of his observations. He called the substance that killed bacteria 'penicillin' after the *Penicillium* mould that produced it. He tried unsuccessfully for several years to extract an active juice from the mould before giving up and moving on to other work.

About 10 years after Fleming's discovery, Ernst Chain and Howard Florey set about trying to extract penicillin, and they succeeded. They gave some penicillin to a man dying of a blood infection and he recovered almost miraculously – until the penicillin ran out. Even though their patient died, Florey and Chain demonstrated that penicillin could cure bacterial infections in people. Eventually, working with the company Pfizer in the USA, Florey and Chain made penicillin on an industrial scale, producing enough to supply the demands of World War II. It is still used today.

Medicines for the future

There is a continuing drive to find new medicines but it is difficult. For example, it is not easy to find chemicals that kill bacteria without damaging human cells. Most drugs are now synthesised by research chemists working in the pharmaceutical industry using chemical banks and computer models. However, the starting point may still be a chemical extracted from a plant or microorganism. Compounds showing promise as antibiotics can be modified to produce more powerful molecules that can be synthesised easily and cheaply.

For example, the noni fruit is widely used in traditional medicine in Costa Rica and many other countries to treat both infections and non-communicable diseases. People have also used it for food and drink for centuries with no apparent problems. Recent research shows that it has antibiotic properties. More research is taking place to see if this traditional healing plant might be the source of new antibiotics or other medicines.

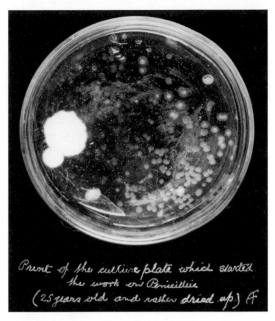

Figure 2 *Alexander Fleming was on the lookout for something that would kill bacteria. Because he noticed the effect of this mould on his cultures, millions of lives have been saved around the world*

Figure 3 *The noni fruit looks strange and smells stranger, but will it provide us with medicines for the future?*

Scientists are also collecting soil samples globally and searching for microorganisms to produce a new antibiotic against antibiotic-resistant bacteria. Only about 1% of soil microorganisms can be cultured in the lab. Scientists have developed a special unit that enables them to grow microorganisms in the soil in a controlled way. Using this technology, in 2015 they announced a completely new type of antibiotic from some soil bacteria. In tests so far, this antibiotic has destroyed all bacteria including MRSA and other antibiotic resistant pathogens. It worked in mice – will it work in humans?

1 Describe how Alexander Fleming discovered penicillin. [3 marks]

2 Suggest reasons for using synthetic forms of drugs rather than using plant extracts directly. [4 marks]

3 Discuss the advantages and disadvantages of looking for new antibiotic compounds in living organisms based on the example of the noni fruit. [6 marks]

Synoptic links

You will learn more about the development of antibiotic resistant bacteria in Topic B14.4.

Key points

- Traditionally drugs were extracted from plants, for example, digitalis, or from microorganisms, for example, penicillin.
- Penicillin was discovered by Alexander Fleming from the *Penicillium* mould.
- Most new drugs are synthesised by chemists in the pharmaceutical industry. However, the starting point may still be a chemical extracted from a plant.

B6.4 Developing drugs

Learning objectives

After this topic, you should know:

- the stages involved in testing and trialling new drugs
- why testing new drugs is so important.

Figure 1 *The development of a new medicine costs millions of pounds and involves many people and lots of equipment*

New medicines are being developed all the time, as scientists and doctors try to find ways of curing more diseases. Scientists test new medicines in the laboratory. Every new medical treatment has to be extensively tested and trialled in a series of stages before it is used. This process makes sure that it works well and is as safe as possible.

A good medicine is:

- Effective – it must prevent or cure a disease or at least make you feel better.
- Safe – the drug must not be too toxic (poisonous) or have unacceptable side effects for the patient
- Stable – you must be able to use the medicine under normal conditions and store it for some time.
- Successfully taken into and removed from your body – it must reach its target and be cleared from your system once it has done its work.

Developing and testing a new drug

When scientists research a new medicine they have to make sure all these conditions are met. It can take up to 12 years to bring a new medicine into your doctor's surgery and costs around £1700 million, including failures and capital costs.

Researchers target a disease and make lots of possible new drugs. These are tested in the laboratory to find out if they are toxic (toxicity) and if they seem to do their job (efficacy). In the laboratory they are tested on cells, tissues, and even whole organs. Many chemicals fail at this stage.

The small numbers of chemicals, which pass the earlier tests, are then laboratory tested on animals, to find out how they work in a whole living organism. It also gives information about possible doses and side effects. The tissues and animals are used as models to predict how the drugs may behave in humans.

Up to this point the chemicals are undergoing **preclinical testing**. This always takes place in the laboratory using cells, tissues, and live animals.

Drugs that pass animal testing move on to **clinical trials**. Clinical trials use healthy volunteers and patients. First, very low doses are given to healthy people to check for side effects. If the drug is found to be safe, it is tried on a small number of patients to see if it treats the disease. If it seems to be safe and effective, bigger clinical trials take place to find the optimum dose for the drug.

If the medicine passes all the legal tests, it is licenced so your doctor can prescribe it. Its safety will be monitored for as long as it is used.

Double blind trials

In human trials, scientists use a double blind trial to see just how effective the new medicine is. A group of patients with the target disease agree to take part in the trials. Some are given a **placebo** that does not contain the drug and some are given the new medicine. Patients are randomly allocated to the different groups. Then neither the doctor nor the patients know who has received the real drug or the placebo until the trial is complete. The patients' health is monitored carefully.

Often the placebo will contain a different drug that is already used to treat the disease. This means the patient is not deprived of treatment whilst taking part in the trial.

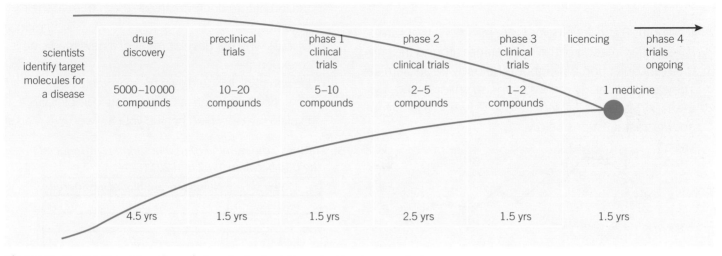

Figure 2 *An enormous number of chemicals start the selection process but few actually become a new, useful medicine*

Publishing results

The results of drug tests and trials, like all scientific research, are published in journals after they have been scrutinised in a process of peer review. This means other scientists working in the same area can check the results over, helping to prevent false claims. National bodies such the National Institute for Clinical Excellence (NICE) look at the published results of drugs trials and decide which drugs give good value for money and should be prescribed by the NHS.

1 All new drugs are extensively tested for efficacy, toxicity, and dosage. Define these three terms. [3 marks]

2 a Testing a new medicine costs a lot of money and can take up to 12 years. Draw a flow chart to show the main stages in testing new drugs. [6 marks]

 b Explain why an active drug is often used as the placebo in a clinical trial instead of a sugar pill that has no effect. [3 marks]

3 Florey and Chain tried penicillin on a human volunteer without preclinical trials. Thalidomide was used without any proper testing and caused limb deformities in developing foetuses. The outcomes were very different. Discuss the need for full trialling of all drugs and consider how and why the process has changed over time. ✏ [6 marks]

B6 Preventing and treating disease

Summary questions

1 Vaccination uses your body's natural defence system to protect you against disease.
   a Describe how vaccination works. [4 marks]
   b Produce a flow chart to summarise the process of developing active immunity after:
      i a natural infection [4 marks]
      ii a vaccination. [4 marks]
   c There are vaccines for diseases such as diphtheria, polio, tetanus, and meningitis, but not for the common cold or tonsillitis. Suggest reasons for this. [2 marks]

2 Meningitis B and meningitis C are infections that can cause inflammation of the membranes around the brain and infection throughout the body (septicaemia). They are particularly serious in young children and teenagers, and can kill rapidly. Use Figure 1 to help you answer the questions below.

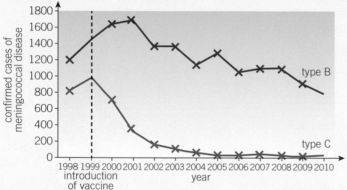

Figure 1 *Annual confirmed cases of meningococcal disease 1998–2010*

   a How many cases of meningitis B and meningitis C were recorded in 1999? [2 marks]
   b How many cases of meningitis B and meningitis C were recorded in 2005 and 2009? [2 marks]
   c Suggest whether the introduction of the meningitis C vaccine in 1999 alone is responsible for the reduction in meningitis cases in the UK between 1998 and 2010. Explain your answer. [4 marks]
   d Suggest one argument for and one argument against the introduction of a new vaccine against meningitis B. [2 marks]

3 a There are no medicines to cure measles, mumps or and rubella. What does this tell you about the pathogens that cause these diseases? [1 mark]
   b Suggest a medicine that might be used to make people feel more comfortable. [1 mark]
   c Doctors hope to get levels of MMR vaccination against measles, mumps, and rubella up to 95% of the population. Why is it important to get vaccination levels so high? [3 marks]

4 a Explain why new medicines need to be tested and trialled before doctors can use them to treat their patients. [5 marks]
   b Discuss why the development of a new medicine is so expensive. [4 marks]
   c Do you think it would ever be acceptable to use a new medicine before all the trials are completed? Explain the reasoning behind your answer. [5 marks]

5 **Table 1** *The number of new antibiotics introduced in the USA between 1983 and 2011*

Year	Number of antibiotics introduced
1983–1987	16
1988–1992	14
1993–1997	10
1998–2002	7
2003–2007	5
2008–2011	2

   a Use Table 1 to help you explain why there is such pressure to discover new medicines. [2 marks]
   b Some people argue for the conservation of biodiversity because the living world is a potential source of new medicines. Explain why this argument is not completely accurate. [4 marks]
   c Describe the main steps in the development of a new medicine to the point where it can be used by your local GP or hospital. [6 marks]

6 Discuss the importance of herd immunity in the elimination of an infectious disease in a population of people. [5 marks]

Practice questions

01 A scientist investigated how effective five different antibiotics were at killing two types of bacteria, *E.coli* and *S.aureus*.

- The scientist grew the bacteria on agar in two different Petri dishes.
- He placed paper discs soaked in the 5 different antibiotic solutions, **A**, **B**, **C**, **D**, and **E** onto the agar.
- He used the same concentration of each antibiotic and the same sized paper discs.
- The Petri dishes were incubated at 25 °C for three days.

A clear area around the paper disc means that the antibiotic has killed the bacteria there.

The results are shown in **Figure 1**.

Figure 1

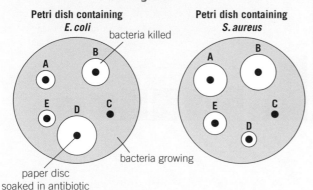

01.1 Give one variable the scientist controlled. [1 mark]

Use the results shown in the diagram to help you to answer the following questions.

01.2 Which antibiotic, **A**, **B**, **C**, **D**, or **E**, was the most effective at killing *E.coli*? [1 mark]

01.3 Which antibiotic, **A**, **B**, **C**, **D**, or **E**, did not kill either *E.coli* or *S.aureus*? [1 mark]

01.4 Which antibiotic, **A**, **B**, **C**, **D**, or **E**, would be the best to use to kill both *E.coli* **and** *S.aureus*?
Give a reason for your answer. [2 marks]

01.5 MRSA is a strain of *S.aureus*. MRSA cannot be killed by most antibiotics.
Use the correct word from the box to complete the sentence.

| immune | powerful | resistant |

Bacteria that cannot be killed by antibiotics are . . . . [1 mark]

AQA, 2013

02 New drugs have to be tested before they can be sold. **Figure 2** shows how much time the different stages of testing took for a new drug.

Figure 2

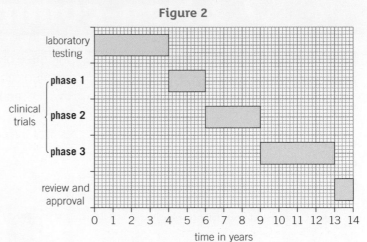

02.1 How much more time did the clinical trials take than the laboratory testing? [1 mark]

02.2 Apart from the time taken, what other difference is there between laboratory testing and clinical trials? [1 mark]

02.3 During **Phase 1** clinical trials, the drug is tested on healthy volunteers using low doses.
Suggest why **only** healthy volunteers and **only** low doses are used at this stage of drug testing. [2 marks]

02.4 In **Phase 2** and **Phase 3** clinical trials, a double blind trial is usually done. Explain what a double blind trial is and why a double blind trial is good practice. [3 marks]

AQA, 2013

03 Whooping cough is a highly infectious disease that can have severe complications.
In 2012 there was a nationwide outbreak of whooping cough in the UK. Fourteen babies under three months of age died. In October 2012 a programme was introduced, offering pregnant women a whooping cough vaccine.

03.1 Babies under three months old are at the greatest risk of developing complications from whooping cough. Suggest why. [1 mark]

03.2 Suggest **one** reason why there was a sudden increase in the number of cases of whooping cough in 2012. [1 mark]

03.3 Suggest **one** reason why vaccinating pregnant women may reduce the incidence of whooping cough in babies. [2 marks]

Learning objectives

After this topic, you should know:

- what is meant by a non-communicable disease
- what a lifestyle factor is
- how scientists consider risk
- the human and financial costs involved
- what a causal mechanism is.

Only three of the top 10 killer diseases in the world in 2012 were communicable – lower respiratory tract infections such as pneumonia, HIV/AIDS, and diarrhoeal diseases. The other seven were non-communicable diseases. These are diseases that are not infectious and affect people as a result of their genetic makeup, their lifestyle, and factors in their environment.

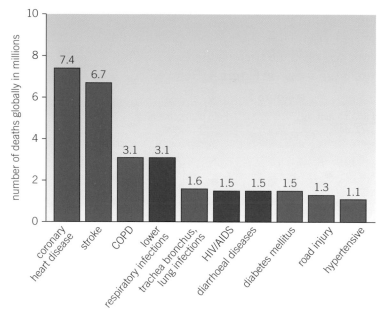

Figure 1 *The leading causes of death globally in 2012 (WHO). Non-communicable diseases (blue bars) contribute to more deaths than communicable diseases (pink bars)*

Risk factors for disease

There are many risk factors for disease, including the genes you inherit from your parents and your age, which you cannot change. Risk factors for disease also include:

- aspects of your lifestyle such as smoking, lack of exercise, or overeating
- substances that are present in the environment or in your body such as ionising radiation, UV light from the sun, or second-hand tobacco smoke.

Certain lifestyle factors, or environmental substances, have been shown to increase your risk of developing particular diseases. Risk factors for non-communicable diseases vary from one disease to another and some may affect more than one disease. Examples of risk factors for a number of non-communicable diseases include diet, obesity, fitness levels, smoking, drinking alcohol, and exposure to **carcinogens** in the environment such as **ionising radiation**. We have the power to influence, change, or remove many of these risk factors.

Causal mechanisms

Scientists often see similarities in the patterns between non-communicable diseases such as cardiovascular disease or lung cancer with lifestyle factors such as lack of exercise or smoking. These similarities may suggest a link or relationship between the two, known as a **correlation**. However, a correlation does not prove that one thing is the cause of another (Figure 2).

It is useful to find correlations between lifestyle factors and particular diseases, but this is only the first step. Doctors and scientists then need to do lots of research to discover if there is a **causal mechanism**. A causal mechanism explains how one factor influences another through a biological process. If a causal mechanism can be demonstrated, there is a link between the two. For example, there is a clear causal link between smoking tobacco and lung cancer. Anyone can get lung cancer, but smoking increases that risk because you take carcinogens into your lungs.

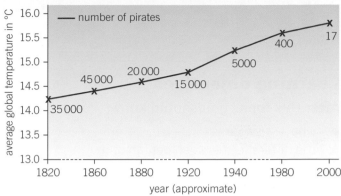

Figure 2 *Falling levels of piracy and rising global temperatures between the years 1820 and 2000 look to be closely linked on this very unscientific graph but the apparent correlation does not mean one causes the other*

The impact of non-communicable disease

Every serious disease has a human impact on the individual affected and their family. It will often have a financial cost as well if a wage-earner becomes ill and cannot work. Local communities often bear the cost of supporting people who are ill, whether formally through taxes or informally by taking care of affected families.

Diseases cost nations huge sums of money both in the expense of treating ill people and in the loss of money earned when large numbers of the population are ill. The global economy suffers too, especially when diseases affect younger, working-age populations. Non-communicable diseases affect far more people than communicable diseases, so they have the greatest effect at both human and economic levels.

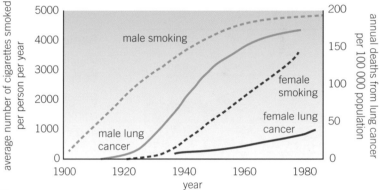

Figure 3 *This graph shows the number of deaths from lung cancer and the average number of cigarettes smoked. In the UK, 86% of lung cancer cases are linked to smoking*

1 **a** What is a non-communicable disease? [1 mark]
  b Describe the main differences between a communicable disease and a non-communicable disease. [2 marks]

2 **a** From Figure 1, identify which of the 10 leading causes of death in the world is *not* a disease? [1 mark]
  b Approximately how many people died as a result of a non-communicable disease? [2 marks]
  c What percentage of people died as a result of a non-communicable disease? [2 marks]

3 Explain the difference between risk factors, correlations, and causal mechanisms. 🖉 [6 marks]

Key points

- A non-communicable disease cannot be passed from one individual to another.
- Risk factors are aspects of a person's lifestyle, or substances present in a person's body or environment, that have been shown to be linked to an increased rate of a disease.
- For some non-communicable diseases, a causal mechanism for some risk factors has been proven, but not in others.

B7.2 Cancer

Learning objectives

After this topic, you should know:

- what a tumour is
- the difference between benign and malignant tumours
- how cancer spreads.

Synoptic link

You learnt about mitosis and the cell cycle in Topic B2.1.

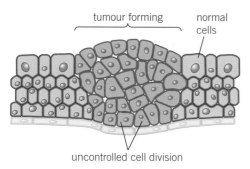

Figure 1 *A tumour forms when there is uncontrolled cell division*

Cancer is a disease that affects people in many families. The cells in your body divide on a regular basis in a set sequence known as the cell cycle that involves several stages. A **tumour** forms when control of this sequence is lost and the cells grow in an abnormal, uncontrolled way.

Tumour formation

Tumour cells do not respond to the normal mechanisms that control the cell cycle. They divide rapidly with very little non-dividing time for growth in between each division. This results in a mass of abnormally growing cells called a tumour (Figure 1). Some tumours are caused by communicable diseases. For example, the bacteria *agrobacterium tumefaciens* can cause crown galls in plants, and the human papilloma virus (HPV) can cause cervical cancer in humans.

Benign tumours are growths of abnormal cells contained in one place, usually within a membrane. They do not invade other parts of the body but a benign tumour can grow very large, very quickly. If it causes pressure or damage to an organ, this can be life-threatening. For example, benign tumours on the brain can be very dangerous because there is no extra space for them to grow into.

Malignant tumour cells can spread around the body, invading neighbouring healthy tissues. A malignant tumour is often referred to as **cancer**. The initial tumour may split up, releasing small clumps of cells into the bloodstream or lymphatic system. They circulate and are carried to different parts of the body where they may lodge in another organ. Then they continue their uncontrolled division and form secondary tumours. Cancer cells not only divide more rapidly than normal cells, they also live longer. The growing tumour often completely disrupts normal tissues and, if left untreated, will often kill the person. Because of the way malignant tumours spread around the body, it can be very difficult to treat them.

The causes of cancer

Scientists still do not understand what triggers the formation of many cancers, but some of the causes are well known.

- There are clear genetic risk factors for some cancers including early breast cancer and ovarian cancer.

- Most cancers are the result of mutations – changes in the genetic material. Chemicals such as asbestos and the tar found in tobacco smoke can cause mutations that trigger the formation of tumours. These cancer-causing agents are called carcinogens.

- Ionising radiation, such as UV light and X-rays, can also interrupt the normal cell cycle and cause tumours to form. For example, melanomas (Figure 2) appear when there is uncontrolled growth of pigment-forming cells in the skin as a result of exposure to UV light from the Sun.

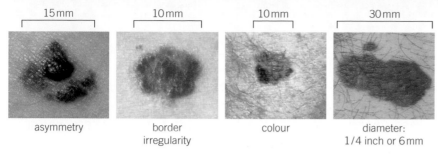

Figure 2 *Melanomas are malignant tumours often triggered by exposure to UV radiation. Over 2000 people a year die from melanomas in the UK alone, so it is important to know the signs to look out for*

● About 15% of human cancers are caused by virus infections. For example, cervical cancer is almost always the result of infection by HPV. Teenagers in the UK are now routinely vaccinated against the virus.

Treating cancer

Because of the way cancer can spread through the body it can be difficult to treat. In recent years, treatments have become increasingly successful. Scientists are working hard to develop new treatments, and are also finding that combining some older treatments makes them more successful too. The two main ways we can treat cancer at the moment are:

● Radiotherapy, when the cancer cells are destroyed by targeted doses of radiation. This stops mitosis in the cancer cells but can also damage healthy cells. Methods of delivering different types of radiation in very targeted ways are improving cure rates.

● Chemotherapy, where chemicals are used to either stop the cancer cells dividing or to make them 'self-destruct'. There are many different types of chemotherapy and scientists are working to make them as specific to cancer cells as possible.

1 **a** What is a tumour? [3 marks]
  b Describe the difference between a benign tumour and a malignant tumour. [3 marks]
  c Suggests ways in which both types of tumour can cause serious health problems. [4 marks]

2 One of the most common methods of treating cancers is chemotherapy. Chemotherapy drugs often affect other parts of the body, particularly hair follicles, skin cells, cells lining the stomach, and blood cells as well as the cancer cells.
  a Explain how the drugs used in chemotherapy might work. [2 marks]
  b Suggest reasons why healthy hair, skin, blood, and stomach lining cells are particularly badly affected by the drugs used to treat cancer. [4 marks]

3 Describe and explain the different treatments that are used to treat cancer. ✍ [4 marks]

Key points

● Benign and malignant tumours result from abnormal, uncontrolled cell division.
● Benign tumours form in one place and do not spread to other tissues.
● Malignant tumour cells are cancers. They invade neighbouring tissues and may spread to different parts of the body in the blood where they form secondary tumours.
● Lifestyle risk factors for various types of cancer include smoking, obesity, common viruses, and UV exposure. There are also genetic risk factors for some cancers.

B7.3 Smoking and the risk of disease

Learning objectives

After this topic, you should know:

- how smoking affects the risk of developing cardiovascular disease
- how smoking affects the risk of developing lung disease and lung cancer
- the effect of smoking on unborn babies.

Synoptic links

You can remind yourself of the structure of the breathing system in Topic B4.5 and of the way oxygen is carried around the body in Topic B4.1.

There are around 1.1 billion smokers world-wide, smoking around 6000 billion cigarettes each year, so smoking is big business. Every cigarette smoked contains tobacco leaves which, as they burn, produce around 4000 different chemicals that are inhaled into the throat, trachea, and lungs. At least 150 of these are linked to disease. Some of these chemicals are absorbed into the bloodstream to be carried around the body to the brain.

Nicotine and carbon monoxide

Nicotine is the addictive but relatively harmless drug found in tobacco smoke. It produces a sensation of calm, well-being, and 'being able to cope' and this is why people like smoking. Unfortunately some of the other chemicals in tobacco smoke can cause lasting and often fatal damage to the body cells. Carbon monoxide is a poisonous gas found in tobacco smoke and it takes up some of the oxygen-carrying capacity of your blood. After smoking a cigarette, up to 10% of the blood will be carrying carbon monoxide rather than oxygen. This can lead to a shortage of oxygen, one reason why many smokers get more breathless when they exercise than non-smokers.

Smoking during pregnancy

Oxygen shortage is a particular problem in pregnant women who smoke. During pregnancy a woman is carrying oxygen for her developing fetus as well as herself. If the mother's blood is carrying carbon monoxide, the fetus may not get enough oxygen to grow properly. This can lead to premature births, low birthweight babies and even stillbirths, where the baby is born dead. There are around 3500 stillbirths in the UK each year. Scientists estimate that around 20% result from the mother smoking during her pregnancy. In other words, 700 babies a year are born dead due to smoking.

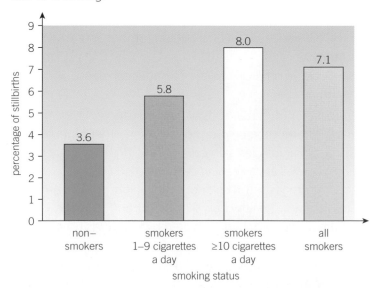

Figure 1 *Smoking during pregnancy has a dramatic effect on the risk of stillbirths*

Carcinogens

The cilia in the trachea and bronchi that move mucus, bacteria, and dirt away from the lungs are anaesthetised by some of the chemicals in tobacco smoke. They stop working for a time, allowing dirt and pathogens down into the lungs and increasing the risk of infections. Mucus also builds up over time and causes coughing.

Other toxic compounds in tobacco smoke include tar. This is a sticky, black chemical that accumulates in the lungs, turning them from pink to grey. Along with other chemicals in the smoke, tar makes smokers much more likely to develop bronchitis (inflammation and infection of the bronchi). The build-up of tar in the delicate lung tissue can lead to a breakdown in the structure of the alveoli, causing chronic obstructive pulmonary disease (COPD). This reduces the surface area to volume ratio of the lungs, leading to severe breathlessness and eventually death.

Tar is also a carcinogen. It acts on the delicate cells of the lungs and greatly increases the risk of lung cancer developing. Tar also causes other cancers of the breathing system, for example, the throat, larynx, and trachea.

Smoking and the heart

The chemicals in tobacco smoke also affect the heart and blood vessels. Scientists have data showing that smokers are more likely to suffer from cardiovascular problems than non-smokers (Table 1). They have also worked out the mechanisms that show it is a causal link, not just a correlation.

Smoking narrows the blood vessels in your skin, ageing it. Nicotine makes the heart rate increase whilst other chemicals damage the lining of the arteries. This makes coronary heart disease more likely, and it increases the risk of clot formation. The mixture of chemicals in cigarette smoke also lead to an increase in blood pressure. This combination of effects increases the risk of suffering cardiovascular disease including heart attacks and strokes.

Synoptic links

You can find out more about coronary heart disease and how it can be treated in Topic B4.3, and about the clotting of the blood in Topic B5.7.

Table 1 *The number of deaths by cardiovascular disease (CVD) by average number of cigarettes smoked a day*

Cigarettes smoked per day	CVD deaths per 100 000 men per year
0	572
10 (range 1–14)	802
20 (range 15–24)	892
30 (range >24)	1025

1 a State three components of tobacco smoke. [4 marks]
   b State what effect the three components you gave in part **a** have on the human body. [4 marks]
2 a Display the data from Table 1 as a bar chart. [4 marks]
   b Summarise what this data shows. [1 mark]
   c Describe possible causal mechanisms to explain the trend shown in the data. [3 marks]
3 Summarise the information given that cigarette smoking increases the risk of developing lung cancer and explain how scientists think this effect is caused. [4 marks]
4 Many women continue to smoke when they are pregnant. Explain why smoking during pregnancy is so harmful. 🕭 [5 marks]
5 Discuss the human and financial cost of smoking to individuals and to nations around the world. 🕭 [6 marks]

Key points

- Smoking can cause cardiovascular disease including coronary heart disease, lung cancer, and lung diseases such as bronchitis and COPD.
- A fetus exposed to smoke has restricted oxygen, which can lead to premature birth, low birthweight, and even stillbirth.

B7.4 Diet, exercise, and disease

Learning objectives

After this topic, you should know:

- the effect of diet and exercise on the development of obesity
- how diet and exercise affect the risk of developing cardiovascular disease
- that obesity is a risk factor for type 2 diabetes.

The evidence is building that your weight and the amount of exercise you take affects your risk of developing various diseases. These diseases can be life-changing and even life-threatening.

Diet, exercise, and obesity

If you eat more food than you need, the excess is stored as fat. You need some body fat to cushion your internal organs and act as an energy store. However, over time regularly eating too much food will make you overweight and then obese.

Carrying too much weight is often inconvenient and uncomfortable. Far worse, obesity can lead to serious health problems, such as type 2 diabetes (high blood sugar levels, which are hard to control), high blood pressure, and heart disease.

Exercise and health

The food you eat transfers energy to your muscles as they work from respiration, so the amount of exercise you do affects the amount of respiration in your muscles, and the amount of food you need. People who exercise regularly are usually much fitter than people who take little exercise. People who take regular exercise make bigger muscles, up to 40% of their body mass, and muscle tissue needs much more energy to be transferred from food than body fat. People who exercise regularly have fitter hearts and bigger lungs than people who don't exercise. But exercise doesn't always mean time spent training or 'working out' in the gym. Walking to school, running around the house looking after small children, or doing a physically active job all count as exercise too. Between 60 and 75% of your daily food intake is needed for the basic reactions that keep you alive. About 10% is needed to digest your food so only the final 15–30% is affected by your physical activity!

Scientists and doctors have collected lots of evidence that people who exercise regularly are less likely to develop cardiovascular disease than people who do not exercise. They are less likely to suffer from many other health problems too, including type 2 diabetes (see later).

Synoptic link

You can find out about the effect of fitness on the way the body reacts to exercise in Topic B9.2.

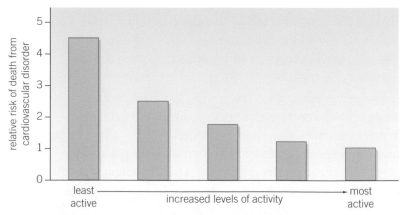

Figure 1 *The effect of exercise on the risk of death associated with cardiovascular disease in men and women*

Here are some of the causal mechanisms that explain why exercise helps to keep you healthy. You will have more muscle tissue, increasing your metabolic rate, so you are less likely to be overweight. This reduces the risk of developing arthritis, diabetes, and high blood pressure, for example. Your heart will be fitter and develop a better blood supply. Regular exercise lowers your blood cholesterol levels and helps the balance of the different types of cholesterol. This reduces your risk of fatty deposits building up on your coronary arteries, so lowering your risk of heart disease and other health problems.

Obesity and type 2 diabetes

In type 2 diabetes, either your body doesn't make enough insulin to control your blood sugar levels or your cells stop responding to insulin. This can lead to problems with circulation, kidney function, and eyesight, which may eventually lead to death. Type 2 diabetes gets more common with age and some people have a genetic tendency to develop it. The evidence is now overwhelming that being overweight or obese and not doing much exercise are risk factors for type 2 diabetes at any age. Type 2 diabetes is becoming increasingly common in young people. By 2025, an estimated 4 million people in the UK will have diabetes and 90% of those cases will be type 2. Fortunately most people can restore their normal blood glucose balance simply by eating a balanced diet with controlled amounts of carbohydrate, losing weight, and doing regular exercise.

Synoptic links

You will learn more about both type 1 and type 2 diabetes in Topic B11.2 and Topic B11.3.

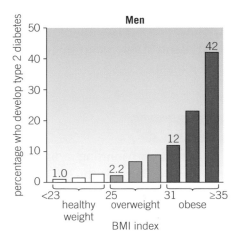

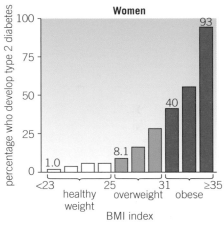

Figure 2 *The effect of obesity on the risk of developing type 2 diabetes in men and women*

1 Explain why people who exercise regularly are usually healthier than people who take little exercise. [5 marks]

2 Exercise levels and obesity levels are often linked. Suggest reasons for this. [4 marks]

3 Based on the data in Figure 1, what is the relative risk of suffering cardiovascular disease in men who exercise least compared to men who exercise most? [2 marks]

4 Type 2 diabetes has been described as an epidemic. It was observed that if it is an epidemic, it is an epidemic that particularly affects women. Look at the data in Figure 2 and discuss these statements, taking into account the scientific evidence. Suggest both the reasons for the observations and how the 'epidemic' might be controlled. ✪ [6 marks]

Key points

- Diet affects your risk of developing cardiovascular and other diseases directly through cholesterol levels and indirectly through obesity.
- Exercise levels affect the likelihood of developing cardiovascular disease.
- Obesity is a strong risk factor for type 2 diabetes.

B7.5 Alcohol and other carcinogens

There are many different agents that increase your risk of developing non-communicable diseases. Some of these you may take into your body willingly, and some you may be unaware of.

Alcohol and health

Alcohol (ethanol) is a commonly used social drug in many parts of the world. It is poisonous but the liver can usually remove it before permanent damage or death results. Alcohol is also very addictive.

After an alcoholic drink, the ethanol is absorbed into the blood from the gut and passes easily into the body tissues, including the brain. It affects the nervous system, making thought processes, reflexes, and many reactions slower than normal. In small amounts alcohol makes people feel relaxed, cheerful, and reduces inhibitions. Larger amounts lead to lack of self-control and lack of judgement. If the dose of alcohol is too high, it can sometimes lead to unconsciousness, coma, and death.

Brain and liver damage

People can easily become addicted to alcohol, needing the drug to function, and they may drink heavily for many years. Their liver and brain may suffer long-term damage and eventually the alcohol can kill them:

- They may develop cirrhosis of the liver, a disease that destroys the liver tissue. The active liver cells are replaced with scar tissue that cannot carry out vital functions.

- Alcohol is a carcinogen so heavy drinkers are at increased risk of developing liver cancer. This usually spreads rapidly and is difficult to treat.

- Long-term heavy alcohol use also causes damage to the brain. In some alcoholics the brain becomes so soft and pulpy that the normal brain structures are lost and it can no longer function properly. This too can cause death.

The damage to the liver and brain associated with heavy drinking usually develops over years, but short bouts of very heavy drinking risk the same symptoms appearing relatively fast, even in young people.

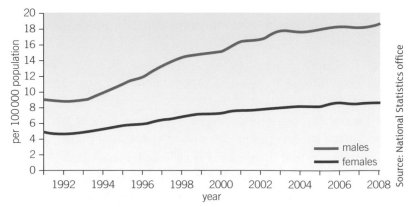

Figure 1 *This graph shows the increase in alcohol-related deaths in the UK between 1990 and 2008*

Alcohol and pregnancy

If a pregnant woman drinks alcohol, it passes across the placenta into the developing baby. Miscarriage, stillbirths, premature births, and low birthweight are all risks linked to drinking alcohol during pregnancy. The developing liver cannot cope with alcohol, so the development of the brain and body of an unborn baby can be badly affected, especially in the early stages of pregnancy.

The baby may have facial deformities, problems with its teeth, jaw or hearing, kidney, liver, and heart problems, and may have learning and other developmental problems. This is known as fetal alcohol syndrome (FAS). Doctors are not sure how much alcohol is safe during pregnancy. The best advice is not to drink at all to avoid fetal alcohol syndrome. The more you drink, the higher the risk to the unborn baby.

Ionising radiation

Ionising radiation in the form of different types of electromagnetic waves is a well-known carcinogen (risk factor for cancer). Radioactive materials are a source of ionising radiation. The radiation penetrates the cells and damages the chromosomes, causing mutations in the DNA. The more you are exposed to ionising radiation, the more likely it is that mutations will occur and that cancer will develop.

Ionising radiation is particularly dangerous when taken directly into your body. For example, breathing radioactive materials into the lungs enables the ionising radiation to penetrate directly into the cells. Well-known sources of ionising radiation include:

- Ultraviolet light from the sun – this increases the risk of skin cancers such as melanoma (protection includes sunscreen and sensible clothing).

- Radioactive materials found in the soil, water, and air (including radon gas in granite-rich areas such as Cornwall and the Pennines).

- Medical and dental X-rays.

- Accidents in nuclear power generation, especially accidents such as the one in Chernobyl, Ukraine in 1986, can spread ionising radiation over wide areas.

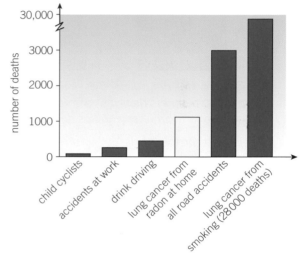

Figure 2 *Living in a home contaminated by radon gas can increase your risk of dying from lung cancer, especially if you are also a smoker. However, the number of people directly affected is small compared to the number who die each year as a result of smoking tobacco, or from road accidents*

---

1 **a** Define a carcinogen. [1 mark]
   b Name three different carcinogens. [3 marks]

2 Use data from Figure 1 to help you answer the following.
   a How many men and women died of alcohol-related diseases per 100 000 of the population in 1992? [2 marks]
   b How many men and women died of alcohol-related diseases per 100 000 of the population in 2008? [2 marks]
   c Suggest reasons for this increase in alcohol-related deaths. [3 marks]
   d Explain why you think alcohol remains a legal drug when it causes so many deaths. [3 marks]

Key points

- Alcohol can damage the liver and cause cirrhosis and liver cancer.
- Alcohol can cause brain damage and death.
- Alcohol taken in by a pregnant woman can affect the development of her unborn baby.

Summary questions

1 a What is a non-communicable disease? [1 mark]
   b Define a lifestyle factor and give three examples. [3 marks]
   c Explain what is meant by:
      i a correlation between a lifestyle factor and a non-communicable disease [2 marks]
      ii a causal link between a lifestyle factor and a particular disease. [3 marks]

2 **(a)** **(b)**

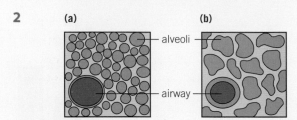

Figure 1 *This diagram shows areas of lung tissue from two people. One is a non-smoker* **a**, *the other is a long-term smoker who has developed COPD,* **b**

   a Which sample of tissue is from the non-smoker? Explain your choice. [3 marks]
   b What are the main symptoms of COPD? [2 marks]
   c Explain how smoking causes COPD and the reason for the symptoms. [5 marks]
   d The risk of having a number of diseases (like COPD) increases with the number of cigarettes smoked a day and the length of time someone has been a smoker.
      i Name two more of these smoking-related diseases. [2 marks]
      ii Explain how both the number of cigarettes smoked and the time someone has been a smoker affects their risk of developing a smoking-related disease. [4 marks]

3 The following chart shows the effect of smoking on the annual death rate in men.

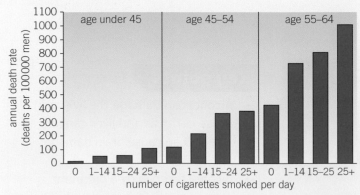

Figure 2 *Effect of smoking on the annual death rate for men*

a i What is the annual death rate per 100 000 men aged 55–64 who smoke 1–14 cigarettes a day? [1 mark]
   ii The death rate of men aged 45–54 who smoke more than 25 cigarettes a day is higher than the death rate for non-smokers. How much higher is it per 100 000 men? [3 marks]
   b Explain why the death rate for smokers is higher than the death rate for non-smokers in each age group. [3 marks]

4 a Which two body organs are most affected by a large amount of alcohol over a long period of time? [2 marks]
   b Explain why it is unsafe for someone to drive when they have been drinking alcohol. [4 marks]
   c Alcohol is closely linked to violence and crime. Suggest why alcohol has this effect. [3 marks]
   d Pregnant women are advised not to drink alcohol, even though scientists do not have clear evidence of exactly what level of alcohol consumption is safe for the unborn baby. Discuss why this is sensible advice. [6 marks]

5 a Look at Figure 3. Describe the trend in obesity in England between 1996 and 2005. [1 mark]
   b Suggest reasons for the observed trend in obesity. [4 marks]
   c A similar graph shows a rise in type 2 diabetes over the same time scale. Suggest how the two sets of data may be linked. [4 marks]

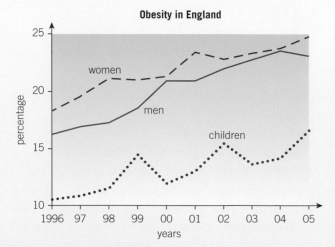

Figure 3 *Trends in obesity in England between 1996 and 2005*

Practice questions

01 Non-communicable diseases can be linked to a range of risk factors. Match each disease to its main risk factor.

Disease	Main risk factor
liver disease	alcohol
skin cancer	ionising radiation
type 2 diabetes	obesity
	smoking

[3 marks]

02 Body mass index (BMI) is a measure of whether a person is a healthy mass for their height.
BMI is calculated using the equation:

$$BMI = \frac{body\ mass\ in\ kg}{(height\ in\ m)^2}$$

02.1 A woman has a mass of 92 kg and a height of 1.7 m. Use the equation to calculate her BMI. [2 marks]
Figure 1 shows the relative risk of developing Type 2 diabetes in women for different BMI values.

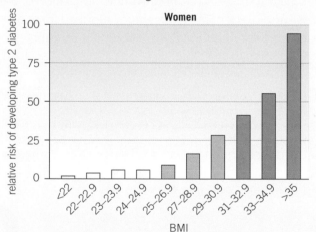

Figure 1

02.2 Use your answer from **02.1** to determine the relative risk for this woman of developing type 2 diabetes. [1 mark]
02.3 Describe the trend shown in **Figure 1**. [2 marks]
02.4 Suggest **one** other factor that can affect a person's risk of developing Type 2 diabetes. [1 mark]
02.5 A man went to a doctor with symptoms that could indicate diabetes.
What blood test would confirm that the man had diabetes? [1 mark]

02.6 If it was confirmed the man had diabetes the doctor would recommend some changes the man could make to his lifestyle.
Suggest **two** changes the man could make to help control his diabetes. [2 marks]

03 Cancers are the result of uncontrolled cell growth and division.
03.1 What type of cell division occurs in cancer cells? [1 mark]
03.2 Describe how a cancer can spread to different parts of the body. [2 marks]
Figure 2 shows the number of deaths from lung cancer and the average number of cigarettes smoked for men and women.
03.3 Give **three** conclusions that can be made from **Figure 2**. [3 marks]

Figure 2

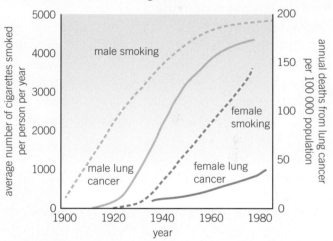

03.4 Give **one** other non-communicable disease associated with smoking cigarettes. [1 mark]

04 Health is a state of physical and mental well-being.
Describe some things a person could do **and** things they could avoid, in order to stay healthy.
Explain how these actions would help the person to stay healthy. [6 marks]

Learning objectives

After this topic, you should know:

- the raw materials and energy source for photosynthesis
- that photosynthesis is an endothermic reaction
- the equations that summarise photosynthesis.

All organisms, including plants and algae, need food for respiration, growth, and reproduction. However, plants don't need to eat – they can make their own food by photosynthesis. This takes place in the green parts of plants (especially the leaves) when it is light. Algae can also carry out photosynthesis.

The process of photosynthesis

The cells in algae and plant leaves are full of small green parts called chloroplasts, which contain a green substance called chlorophyll. During photosynthesis, energy is transferred from the environment to the chloroplasts by light. This energy is then transferred to convert carbon dioxide (CO_2) from the air, plus water (H_2O) from the soil into a simple sugar called **glucose** ($C_6H_{12}O_6$). The chemical reaction also produces oxygen gas (O_2) as a by-product. The oxygen is released into the air, which you can then use when you breathe it in. Every year plants produce about 368 000 000 000 tonnes of oxygen, so there is plenty to go around.

Photosynthesis is an **endothermic reaction** – it needs an input of energy from the environment. The energy transferred from the environment when the bonds holding carbon dioxide and water are broken is more than that transferred back to the environment with the formation of the new bonds in glucose and oxygen. The extra energy required for the reaction to take place is transferred from the environment by light.

Photosynthesis can be summarised as follows:

$$\text{carbon dioxide} + \text{water} \xrightarrow{\text{light}} \text{glucose} + \text{oxygen}$$

$$6CO_2 + 6H_2O \xrightarrow{\text{light}} C_6H_{12}O_6 + 6O_2$$

Some of the glucose produced during photosynthesis is used immediately by the cells of the plant for respiration. However, a lot of the glucose is converted into insoluble starch and stored.

Producing oxygen

You can show that a plant is photosynthesising by the oxygen gas it gives off as a by-product. Oxygen is a colourless gas, so in land plants it isn't easy to show that it is being produced. However, if you use water plants such as *Cabomba* or *Elodea*, you can see and collect the bubbles of gas they give off when they are photosynthesising. The gas will relight a glowing splint, showing that it is rich in oxygen.

Figure 1 *The oxygen produced during photosynthesis is vital for life on Earth. You can demonstrate that it is produced using water plants such as this Cabomba*

Leaf adaptations

For photosynthesis to be successful, a plant needs plenty of carbon dioxide, light, and water. The leaves of plants are perfectly adapted as organs of photosynthesis because:

- most leaves are broad, giving them a big surface area for light to fall on
- most leaves are thin so diffusion distances for the gases are short
- they contain chlorophyll in the chloroplasts to absorb light
- they have veins, which bring plenty of water in the xylem to the cells of the leaves and remove the products of photosynthesis in the phloem
- they have air spaces that allow carbon dioxide to get to the cells, and oxygen to leave by diffusion
- they have guard cells that open and close the stomata to regulate gas exchange.

These adaptations mean that the plant can photosynthesise as much as possible whenever there is light available.

 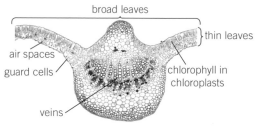

Figure 2 *Leaves are well-adapted for photosynthesis*

Algae are aquatic so they are adapted to photosynthesising in water. They have a large surface area and absorb carbon dioxide dissolved in the water around them. The oxygen they produce also dissolves in the water around them as it is released.

1 a State where a plant gets the carbon dioxide and water that it needs for photosynthesis. [2 marks]
   b State where algae get the carbon dioxide, water, and light they need for photosynthesis. [3 marks]

2 Describe the path taken by a carbon atom as it moves from being part of the carbon dioxide in the air to being part of a starch molecule in a plant. [5 marks]

3 Explain why a leaf kept in the light for 24 hours will turn an iodine solution blue-black, whereas a leaf kept in the light for 24 hours and then in the dark for 24 hours will have no effect on an iodine solution. [4 marks]

4 a Give the word equation for photosynthesis. [2 marks]
   b Explain why photosynthesis is an endothermic reaction. [3 marks]

B8.2 The rate of photosynthesis

Learning objectives

After this topic, you should know:

- which factors limit the rate of photosynthesis in plants.

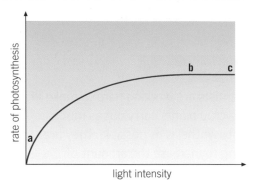

Figure 1 *Investigating the effect of light intensity on the rate of photosynthesis*

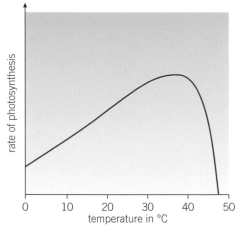

Figure 2 *The effect of increasing temperature on the rate of photosynthesis*

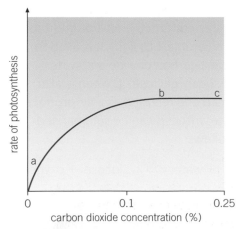

Figure 3 *The effect of increasing carbon dioxide concentration on the rate of photosynthesis*

You may have noticed that plants grow quickly in the summer, yet they hardly grow at all in the winter. Plants need light, warmth, and carbon dioxide if they are going to photosynthesise and grow as fast as they can. Sometimes any one or more of these things can be in short supply and limit the amount of photosynthesis a plant can manage. This is why they are known as **limiting factors**.

Light

The most obvious factor affecting the rate of photosynthesis is light intensity. If there is plenty of light, lots of photosynthesis can take place. If there is very little or no light, photosynthesis will stop, whatever the other conditions are around the plant. For most plants, the brighter the light, the faster the rate of photosynthesis (Figure 1).

Temperature

Temperature affects all chemical reactions, including photosynthesis. As the temperature rises, the rate of photosynthesis increases as the reaction speeds up. However, photosynthesis is controlled by enzymes. Most enzymes are denatured once the temperature rises to around 40–50 °C. If the temperature gets too high, the enzymes controlling photosynthesis are denatured and the rate of photosynthesis will fall (Figure 2).

Carbon dioxide concentration

Plants need carbon dioxide to make glucose. The atmosphere is only about 0.04% carbon dioxide. This means carbon dioxide often limits the rate of photosynthesis. Increasing the carbon dioxide concentration will increase the rate of photosynthesis (Figure 4).

On a sunny day, carbon dioxide concentration is the most common limiting factor for plants. The carbon dioxide concentrations around a plant tend to rise at night, because in the dark a plant respires but doesn't photosynthesise. As light intensity and temperature increase in the morning, most of the carbon dioxide around the plant gets used up.

In a science lab or a greenhouse the levels of carbon dioxide can be increased artificially. This means that carbon dioxide is no longer the limiting factor. Then the rate of photosynthesis increases with the rise in carbon dioxide concentration.

In a garden, woodland or field (rather than a lab or greenhouse, where conditions can be controlled), light intensity, temperature, and carbon dioxide concentrations interact, and any one of them might be the factor that limits photosynthesis.

Light intensity and rate of photosynthesis

You can investigate the effect of light intensity on the rate of photosynthesis (Figure 4).

At the beginning of the investigation, the rate of photosynthesis in the water plant increases as the light is moved closer to the plant, which increases the light intensity. This tells us that light is acting as a limiting factor. When the light is moved further away from the water plant, the rate of photosynthesis falls, shown by a slowing down in the stream of bubbles produced. If the light is moved closer again the stream of bubbles becomes faster, showing an increased rate of photosynthesis. We often add a heat shield to the apparatus in Figure 4. This helps keep the temperature of the water around the plant constant regardless of the position of the light source.

Eventually, no matter how close the light, the rate of photosynthesis stays the same. At this point, light is no longer limiting the rate of photosynthesis. Something else has become the limiting factor.

The results can be plotted on a graph showing the effect of light intensity on the rate of photosynthesis (Figure 1).

Safety: keep electrical equipment dry and do not handle if hands are wet.

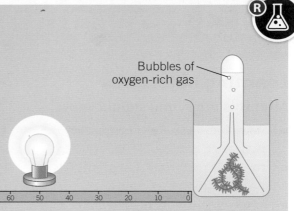
Bubbles of oxygen-rich gas

Figure 4 *Simple apparatus for investigating the effect of light intensity on the rate of photosynthesis*

Chlorophyll levels in the leaf

If the amount of chlorophyll in a leaf is limited in any way, less photosynthesis will take place. The leaves of some ornamental plants have white, chlorophyll-free areas. These plants grow less vigorously than plants with all green leaves. If they are permanently in dim light, variegated leaves often turn completely green. If a plant does not have enough minerals, especially magnesium, it cannot make chlorophyll. The rate of photosynthesis drops and eventually the plant may die.

1 State the three main limiting factors that affect the rate of photosynthesis in a plant. [3 marks]

2 Look at the graph in Figure 1.
   a Explain what is happening between points **a** and **b** on the graph. [2 marks]
   b Explain what is happening between points **b** and **c** on the graph. [2 marks]
   c Now look at Figure 2. Explain why it is a different shape to the graphs in Figures 1 and 3. [4 marks]

3 Explain in terms of limiting factors why the plants growing in a tropical rainforest are so much bigger than the plants that grow in a UK woodland, and why both are bigger than the plants on the Arctic tundra. [6 marks]

Higher
Light intensity and the inverse square law

The relationship between light intensity and the rate of photosynthesis is not a simple one. This is because light intensity involves the inverse square law.

As the distance of the light from the plant increases, the light intensity decreases. That is an inverse relationship – as one goes up the other goes down. However the relationship between distance and light intensity is not linear. The light intensity decreases or increases in inverse proportion to the square of the distance.

Light intensity $\propto \dfrac{1}{distance^2}$

For example, if you double the distance between the light and your plant, light intensity falls by a quarter.

Light intensity $\propto \dfrac{1}{2^2} = \dfrac{1}{4}$

Key points

- The rate of photosynthesis may be affected by light intensity, temperature, level of carbon dioxide, and the amount of chlorophyll.

B8.3 How plants use glucose

Learning objectives

After this topic, you should know:

- how plants use the glucose they make
- the extra materials that plant cells need to produce proteins
- some practical tests for starch, sugars, and proteins.

Plants make food by photosynthesis in their leaves and other green parts. Algae also photosynthesise. So how do they use the glucose they make?

Using glucose

Plant cells and algal cells, like any other living cells, respire all the time. They use some of the glucose produced during photosynthesis as they respire. The glucose is broken down using oxygen to provide energy for the cells. Carbon dioxide and water are the waste products of the reaction. Chemically, respiration is the reverse of photosynthesis.

Cellulose for strength, starch for storage

Energy transferred in respiration may be used to build smaller molecules into bigger molecules. For example, plants build up glucose into complex carbohydrates such as cellulose. They use this to strengthen their cell walls.

Plants convert some of the glucose produced in photosynthesis into starch to be stored. Glucose is soluble in water. If it were stored in plant cells, it could affect the way water moves into and out of the cells by osmosis. Lots of glucose stored in plant cells could affect the water balance of the whole plant.

Starch is insoluble in water. It has no effect on the water balance of the plant so plants can store large amounts of starch in their cells. Starch is the main energy store in plants and it is found in cells all over a plant:

- Starch is stored in the cells of the leaves. It provides an energy store for when it is dark or when light levels are low.

- Starch is also kept in special storage areas of a plant. For example, many plants produce tubers and bulbs that are full of stored starch, to help them survive through the winter. Humans often take advantage of these starch stores, found as vegetables such as potatoes and onions.

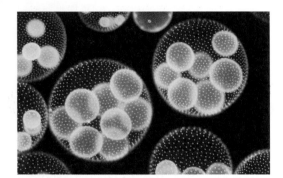

Figure 1 *Worldwide, photosynthesis in algae like this* Volvox *produces more oxygen and biomass than photosynthesis in plants, but they are often forgotten about*

Synoptic links

For more information on the cellulose wall in plant cells, see Topic B1.2, and Topic B1.7.

To find out more about respiration in cells, see Topic B9.1.

Synoptic links

For more information on transport in plants, see Topic B4.7.

For more on osmosis in plants, see Topic B1.7.

Testing for starch

The presence of starch in a leaf is evidence that photosynthesis has taken place. You can test for starch using the iodine solution test and also use this test to show that light is vital for photosynthesis to take place.

Take a leaf from a plant kept in the light and a plant kept in the dark for at least 24 hours. Just adding iodine solution to a leaf does not work, because the waterproof cuticle keeps the iodine out. Leaves have to be specially prepared so the iodine solution can reach the cells and react with any starch stored there. Also, the green chlorophyll would mask any colour changes if the iodine did react with the starch. You therefore need to treat the leaves by boiling them in ethanol, to destroy the waxy cuticle and then to remove the colour. The leaves are then rinsed in hot water to soften them. After treating the leaves, add iodine solution to them both. Iodine solution turns blue-black in the presence of starch.

The leaf that has been in the light will turn blue-black. The iodine solution on the leaf kept in the dark remains orange-red (Figure 2).

Safety: Take care when using ethanol. It is volatile, highly flammable, and harmful. Always wear eye protection. No naked flames – use a hot water bath to heat ethanol.

Figure 2 *The results of the iodine test for starch on a leaf kept in the light (on the left) and a leaf kept in the dark (on the right).*

Nitrates, proteins and carnivorous plants

Plants use some of the glucose from photosynthesis to make amino acids. They do this by combining sugars with nitrate ions and other mineral ions from the soil. These amino acids are then built up into proteins to be used in the plant cells in many ways, including as enzymes. This uses energy from respiration.

Algae also make amino acids. They do this by taking the nitrate ions and other materials they need from the water they live in.

Very few plants can survive well if the soil they are growing in is low in minerals. For example, bogs are wet and their peaty soil has very few nutrients in it. This makes it a difficult place for plants to live.

Some carnivorous plants, such as pitcher plants, Venus flytraps, and sundews are especially adapted to live in nitrate-poor soil. They can survive because they obtain most of their nutrients from the animals, such as insects, that they catch. The plants produce enzymes to digest the insects they trap. They then use nitrates and other minerals from the digested bodies of their victims in place of the nutrients they cannot get from the bog soil in which they grow. After an insect has been digested, the trap reopens, ready to try again.

Making lipids

Plants and algae use some of the glucose from photosynthesis and energy transferred from respiration to build up fats and oils. These may be used in the cells as an energy store. They are sometimes used in the cell walls to make them stronger. In addition, plants often use fats or oils as an energy store in their seeds. Seeds provide food for the new plant to respire as it germinates. Some algal cells are also very rich in oils. They are even being considered as a possible source of biofuels for the future.

1   State three ways that a plant uses the glucose produced by photosynthesis. [3 marks]

2   a   Describe where you might find starch in a plant. [4 marks]
    b   Explain why some of the glucose made by photosynthesis is converted to starch. [3 marks]
    c   Suggest how you could demonstrate that a potato is a store of starch. [3 marks]

3   Look at Figure 2. Explain fully why the two leaves look so different. [4 marks]

4   Explain why pitcher plants, sundews, and Venus fly traps are often found growing in bogs globally, an environment where not many other plants can survive. 🖊 [6 marks]

Study tip

Two important points to remember:

- Plants **respire** 24 hours a day to transfer useable energy for the cells.
- Glucose is soluble in water, but starch is insoluble.

Synoptic link

You can remind yourself of the Biuret test for proteins in Topic B3.3.

Figure 3 *Sundews trap insects on their sticky hairs and digest them where they are stuck to make use of the valuable nitrates and other minerals*

Key points

- Plant and algal cells use the glucose produced during photosynthesis for respiration, to convert into insoluble starch for storage, to produce fats or oils for storage, to produce cellulose to strengthen cell walls, and to produce amino acids for protein synthesis.
- Plants and algal cells also need nitrate ions absorbed from the soil or water to make the amino acids used to make proteins.

B8.4 Making the most of photosynthesis

Learning objectives

After this topic, you should know:

- how the different factors affecting the rate of photosynthesis interact
- how humans can manipulate the environment in which plants grow.

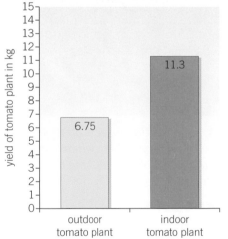

Figure 1 *One piece of American research showed the crop yield for tomatoes was almost doubled in a greenhouse*

The more a plant photosynthesises, the more biomass it makes and the faster it grows. It's not surprising that farmers want their plants to grow as fast and as big as possible. It makes more food for people to eat and helps them to make a profit.

In theory, if you give plants a warm environment with plenty of light and carbon dioxide, and water, they should grow as fast as possible. Out in the fields it is almost impossible to manipulate any of these factors. However, people have found a number of ways that they can artificially manipulate the environment of their plants – and get a number of benefits from doing so.

The garden greenhouse

Lots of people have a greenhouse in their garden. The first recorded greenhouse was built in about AD 30 for Tiberius Caesar, a Roman emperor who wanted to eat cucumbers out of season. Glass hadn't been invented so they used sheets of the mineral mica to build it. Now farmers use huge plastic 'polytunnels' as giant greenhouses for growing crops from tomatoes to strawberries and potatoes.

How does a greenhouse affect the rate of photosynthesis? The glass or plastic building means that the environment is much more controllable than in an ordinary garden or field. Most importantly, the atmosphere is warmer inside than out. This affects the rate of photosynthesis, speeding it up so plants grow faster, flower and fruit earlier, and crop better. In the UK, greenhouses can be used to grow fruit like peaches, lemons, and oranges which don't grow well outside.

Controlling everything

More and more farmers are taking the idea of the greenhouse a bit further. In the laboratory you can isolate different factors and see how they limit the rate of photosynthesis. However, for most plants it is usually a mixture of these factors that affects them. Early in the morning, light levels and temperature probably limit the rate of photosynthesis. Then as the levels of light and temperature rise, carbon dioxide levels become limiting. On a bright, cold winter day, temperature probably limits the rate of the process. There is a constant interaction between the different factors. Remind yourself of the effects of limiting factors on the rates of photosynthesis in Topic B8.2.

Big commercial greenhouses now take advantage of what is known about limiting factors. They control not only the temperature but also the levels of light and carbon dioxide to get the fastest possible rates of photosynthesis. As a result the plants grow as quickly as possible. The plants are even grown in water with a perfect balance of nutrients instead of soil to make sure nothing slows down their growth. This type of system is known as hydroponics.

These greenhouses are enormous and conditions are controlled using computer software. It costs a lot of money but manipulating the environment has many benefits. You can change the carbon dioxide levels in the greenhouses during the day as well as the temperature and the light levels. Furthermore, you can change the mineral content of the water as the plants get bigger.

Turnover is fast which means profits can be high. The crops are clean and soil-free. There is no need to plough or prepare the land in these systems, and crops can be grown where the land is poor.

Greenhouse economics

It takes a lot of planning to keep conditions in the greenhouses just right. Electricity and gas are used to maintain the lighting and temperatures and to control the carbon dioxide levels. Expensive monitoring equipment and computers are needed to maintain conditions inside the greenhouse within narrow boundaries, and alarms are vital if things go wrong. On the other hand, less staff are needed, the time from seed to harvest is much shorter, and the final crop is larger and cleaner. All of these factors, along with the size of the business, have to be considered when deciding whether an enclosed system will increase or reduce profits. The increased income from a larger crop and the ability to grow more crops each year has to be balanced against the cost of setting up and maintaining the system.

There are many decisions for growers to make, but for plants grown hydroponically, limiting factors are a thing of the past.

1   **a** State the main differences between a garden greenhouse and a hydroponics growing system. [3 marks]
    b State the main benefits of artificially manipulating the environment in which food plants are grown. [3 marks]

2   **a** In each of these situations, identify the one factor that is most likely to be limiting photosynthesis. In *each* case explain why the rate of photosynthesis is limited.
      i A wheat field first thing in the morning [3 marks]
      ii The same field later on in the day [3 marks]
      iii Plants growing on a woodland floor in winter [3 marks]
      iv Plants growing on a woodland floor in summer. [3 marks]
    b Explain why is it impossible to be certain which factor is involved in each of these cases. [3 marks]

3   Use Figure 3 to answer this question.
    The cost of running a large greenhouse is particularly affected by the levels of light and temperature. If plants are provided with plenty of carbon dioxide the grower can get two crops a year at 20 °C and three crops a year at 30 °C. Explain why many growers use 20 °C rather than 30 °C in their greenhouses. In your explanation suggest why many growers are also investing in insulation systems that can be added and removed electronically. ✔ [6 marks]

Figure 2 *By controlling the temperature, light, and carbon dioxide levels in a greenhouse like this you can produce the biggest possible crops.*

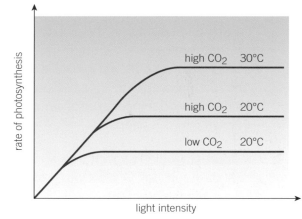

Figure 3 *Growers need to look at this type of data to help them decide the best economic condition for growing their plants. The cost of providing the conditions that give the very highest yields may be too expensive and may wipe out the profits from the bigger, cleaner crop*

Key points

- The factors that limit the rate of photosynthesis interact and any one of them may limit photosynthesis.
- Limiting factors are important in the economics of enhancing the conditions in greenhouses to gain the maximum rate of photosynthesis, while still maintaining profit.

B8 Photosynthesis

Summary questions

1 a Complete the word equation for photosynthesis.

_____ + water $\xrightarrow{\text{light}}$ glucose + _____

[2 marks]

  b Geraniums are green plants that grow in gardens.

  i   Where does the light for photosynthesis in the geranium come from? [1 mark]

  ii How does the geranium absorb this light? [2 marks]

  c On a cold morning, the rate of photosynthesis in the geranium plant is very slow. Suggest which factors may be limiting and why. [2 marks]

  d Some of the glucose produced by the geranium plant is used for respiration. Give three other ways in which the plant uses the glucose produced in photosynthesis. [3 marks]

  e Plants grown in pure water will die, even if they are supplied with light, carbon dioxide, and a growing temperature of around 20 °C. Explain why this happens. [4 marks]

2 The figures in Table 1 show the mean growth of two sets of oak seedlings. One set was grown in 85% full sunlight and the other set in only 35% full sunlight.

Table 1

Year	Mean height of seedlings grown in 85% full sunlight in cm	Mean height of seedlings grown in 35% full sunlight in cm
2000	12	10
2001	16	12.5
2002	18	14
2003	21	17
2004	28	20
2005	35	21
2006	36	23

  a Plot a graph to show the growth of both sets of oak seedlings. [4 marks]

  b Using what you know about photosynthesis and limiting factors, explain the difference in the growth of the two sets of seedlings. [4 marks]

3 Plants make food in one organ and take up water from the soil in another organ. But both the food and the water are needed all over the plant.

  a Where do plants make their food? [2 marks]

  b Where do plants take in water? [1 mark]

  c Describe how you would demonstrate that photosynthesis had taken place in the leaves of a plant. ✔ [6 marks]

4 Palm oil is made from the fruit of oil palms. Large areas of tropical rainforests have been destroyed to make space to plant these oil palms, which grow rapidly.

  a Explain why you think that oil palms can grow rapidly in the conditions that support a tropical rainforest. [3 marks]

  b Where does the oil in the oil palm fruit come from? [1 mark]

  c What is it used for in the plant? [2 marks]

  d How else is glucose used in the plant? [3 marks]

5 **H** Table 2 shows the yields of some different plants grown in Bengal. The yields per acre when grown normally in the field and when grown hydroponically are compared.

Table 2

Name of crop	Hydroponic crop per acre in kg	Ordinary soil crop per acre in kg
wheat	3629	2540
rice	5443	408
potatoes	70760	8164
cabbage	8164	5896
peas	63503	11340
tomatoes	181 437	9072
lettuce	9525	4080
cucumber	12700	3175

  a Explain why yields are always higher when the crops are grown hydroponically. [2 marks]

  b Which crops would be most economically sensible to grow hydroponically? Explain your choice. [4 marks]

  c Which crops would it be least sensible to grow hydroponically? Explain your choice. [3 marks]

  d State the benefits and problems of growing crops in:

  i   the natural environment [3 marks]

  ii an artificially manipulated environment. [3 marks]

Practice questions

01.1 What is the correct word equation for photosynthesis?

A carbon dioxide + glucose → oxygen + water

B light + carbon dioxide → glucose + oxygen

C water + carbon dioxide → glucose + oxygen

D water + oxygen → carbon dioxide + glucose

[1 mark]

01.2 Write down the chemical symbol for glucose.

[1 mark]

01.3 Photosynthesis is an endothermic reaction. What does this statement mean? [2 marks]

01.4 Give **two** reasons why photosynthesis in plants is essential for the survival of animals. [2 marks]

02 **Figure 1** shows the apparatus used to measure the rate of photosynthesis.

Figure 1

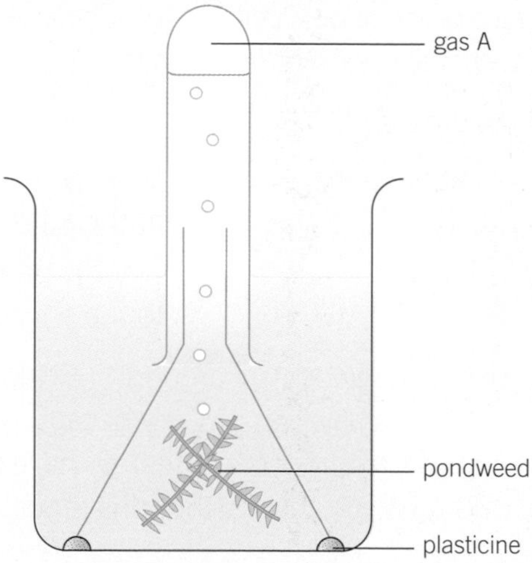

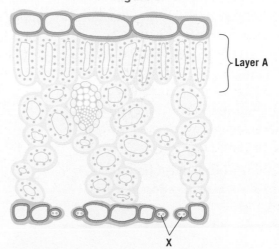

gas A

pondweed

plasticine

02.1 Name gas A? [1 mark]

02.2 Suggest why the funnel is supported on pieces of plasticine. [2 marks]

02.3 Describe how the apparatus is used to measure the rate of photosynthesis. [2 marks]

02.4 Give **three** factors that could affect the rate of photosynthesis in the pondweed. [3 marks]

03 Read the following method used to test a leaf for the presence of starch.

Step 1 Put the leaf in boiling water for 1 minute.

Step 2 Transfer the leaf into boiling ethanol for 5 minutes.

Step 3 Wash the leaf in hot water.

Step 4 Spread the leaf on a white tile and cover it with iodine solution.

03.1 Explain the purpose of each step in the method. [4 marks]

03.2 Describe **two** safety precautions you should take in **Step 2**. [2 marks]

Figure 2 shows a leaf that is part green and part white. It has been removed from a plant that has been in bright light.

Figure 2

03.3 The leaf is tested for the presence of starch. The green part of the leaf is stained black. The white part of the leaf is stained orange. What conclusion could you make from this result? [2 marks]

04 Plants need to make starch. Starch is used as a food storage product. Describe how plants make starch from simple raw materials. [6 marks]

05 **Figure 3** shows a section through a leaf.

Figure 3

Layer A

X

Layer A is the main photosynthetic layer in the leaf.

05.1 What is the name of **Layer A**? [1 mark]

05.2 Describe how **Layer A** is adapted for photosynthesis. [2 marks]

05.3 What are the cells labelled **X** called? [1 mark]

05.4 Describe the function of these cells. [2 marks]

Learning objectives

After this topic, you should know:

- the chemistry of aerobic respiration
- why cellular respiration is so important.

Investigating respiration

Animals, plants, and microorganisms all respire. It is possible to show that cellular respiration is taking place. You can either deprive a living organism of the things it needs to respire, or show that waste products are produced from the reaction.

Depriving a living thing of food and/or oxygen would kill it. So you should concentrate on the products of respiration. Carbon dioxide is the easiest product to identify. You can also measure the energy transferred to the surroundings.

Limewater goes cloudy when carbon dioxide bubbles through it. The higher the concentration of carbon dioxide, the quicker the limewater goes cloudy. This gives us an easy way of showing that carbon dioxide has been produced. You can also look for a rise in temperature to show that energy is being transferred to the environment during respiration.

- Plan an ethical investigation into aerobic respiration in living organisms.

Study tip

Make sure you know the word equation for aerobic respiration.

Remember that aerobic respiration takes place in the mitochondria.

One of the most important enzyme-controlled processes in living things is aerobic respiration. It takes place all the time in plant and animal cells.

Your digestive system, lungs, and circulation all work to provide your cells with the glucose and oxygen they need for respiration.

During **aerobic respiration**, glucose (a sugar) reacts with oxygen. This reaction transfers energy that your cells can use. This energy is vital for everything that goes on in your body.

Carbon dioxide and water are produced as waste products of the reaction. The process is called aerobic respiration because it uses oxygen from the air.

Aerobic respiration is an **exothermic reaction**. Exothermic reactions transfer energy to the environment – more energy is transferred to the environment when new bonds are formed in the products than is taken in to break the bonds in the reactants. Some of the energy transferred in respiration is used for all of the reactions that take place inside a cell. The rest of the energy is transferred to the environment, making it slightly warmer.

Aerobic respiration can be summarised:

glucose + oxygen $\rightarrow$ carbon dioxide + water (energy transferred to the environment)

$$C_6H_{12}O_6 + 6O_2 \rightarrow 6CO_2 + 6H_2O$$ (energy transferred to the environment)

The average energy needs of a teenage boy are around 11 510 kJ daily, but teenage girls only need 8830 kJ a day. This is partly because, on average, girls are smaller than boys, and also because boys have more muscle cells. More muscle cells mean more mitochondria requiring fuel for aerobic respiration.

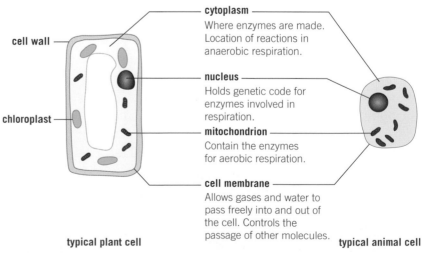

cytoplasm
Where enzymes are made. Location of reactions in anaerobic respiration.

cell wall

nucleus
Holds genetic code for enzymes involved in respiration.

chloroplast

mitochondrion
Contain the enzymes for aerobic respiration.

cell membrane
Allows gases and water to pass freely into and out of the cell. Controls the passage of other molecules.

typical plant cell

typical animal cell

Figure 1 *Aerobic respiration takes place in the mitochondria, but other parts of the cell play vital roles*

Mitochondria – the site of respiration

Aerobic respiration involves lots of chemical reactions. Each reaction is controlled by a different enzyme. Most of these reactions take place in the mitochondria of your cells.

Mitochondria are tiny rod-shaped parts (organelles) that are found in almost all plant and animal cells as well as in fungal and algal cells. They have a folded inner membrane that provides a large surface area for the enzymes involved in aerobic respiration. The number of mitochondria in a cell shows you how active the cell is.

The need for respiration

The energy transferred during respiration supplies all the energy needs for living processes in the cells:

- Living cells need energy to carry out the basic functions of life. They build up large molecules from smaller ones to make new cell material. Much of the energy transferred in respiration is used for these 'building' activities (synthesis reactions). Energy is also transferred to break down larger molecules to smaller ones, both during digestion and within the cells themselves.

- In animals, energy from respiration is transferred to make muscles contract. Muscles are working all the time in your body. Even when you sleep, your heart beats, you breathe, and your stomach churns. All muscular activities require energy.

- Mammals and birds maintain a constant internal body temperature almost regardless of the temperature of their surroundings. On cold days energy transferred from respiration helps you to stay warm, while on hot days you sweat and transfer energy to your surroundings to keep your body cool.

- In plants, energy from respiration is transferred to move mineral ions such as nitrates from the soil into root hair cells. It is also transferred to convert sugars, nitrates, and other nutrients into amino acids, which are then built up into proteins.

Synoptic links

You learnt about mitochondria in Topic B1.2, and about adaptations of active cells in Topic B1.4. You can find out more about active transport and the movement of mineral ions into root hair cells in Topic B1.9.

Figure 2 *When the weather is cold birds like this American robin use up a lot of food for respiration just to keep warm. Giving them extra food supplies during the winter can therefore mean the difference between life and death*

1 **a** Give the word equation for aerobic respiration. [2 marks]
   b Give the symbol equation for aerobic respiration. [2 marks]
   c Explain why muscle cells have many mitochondria while fat cells have very few. [4 marks]

2 You need a regular supply of food to provide energy for your cells. If you don't get enough to eat, you become thin and stop growing. As a result, you don't want to move around and you start to feel cold.
   a State the three main uses of the energy transferred in your body during aerobic respiration. [3 marks]
   b Suggest how this explains the symptoms of starvation described above. [4 marks]

3 Plan an experiment to show that oxygen is taken up and carbon dioxide is released during aerobic respiration. ✔ [6 marks]

Key points

Cellular respiration is an exothermic reaction that occurs continuously in living cells.

- Aerobic respiration is summarised as: glucose + oxygen → carbon dioxide + water (energy transferred to the environment).
- The energy transferred supplies all the energy needed for living processes.

B9.2 The response to exercise

Learning objectives

After this topic, you should know:

- how your body responds to the increased demands for energy during exercise.

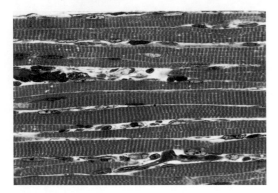

Figure 1 *All the work done by your muscles is based on these special protein fibres, which need energy from respiration to contract*

Go further

Slow twitch muscle fibres rely on aerobic respiration and give endurance. Fast twitch muscle fibres rely on anaerobic respiration and are good for sprinting.

Study tip

You need to be clear about:

- the difference between the rate and the depth of breathing
- the difference between the breathing rate and the rate of respiration.

Your muscles require a lot of energy to carry out their functions. They move you around and help support your body against gravity. Your heart is made of muscle and pumps blood around your body. The movement of food through your digestive system depends on muscles too.

Muscle tissue is made up of protein fibres that contract when energy is transferred from respiration. Muscle fibres need a lot of energy to contract. They contain many mitochondria to carry out aerobic respiration and transfer the energy needed. Muscle fibres usually occur in big blocks or groups, which contract to cause movement. They then relax, which allows other muscles to work.

Your muscles also store glucose as the carbohydrate **glycogen**. Glycogen can be converted rapidly back to glucose to use during exercise. The glucose is used in aerobic respiration to transfer the energy needed to make your muscles contract:

glucose + oxygen → carbon dioxide + water (energy transferred to the environment)

$$C_6H_{12}O_6 + 6O_2 \rightarrow 6CO_2 + 6H_2O \quad \text{(energy transferred to the environment)}$$

The response to exercise

Even when you are not moving about, your muscles use up a certain amount of oxygen and glucose. However, when you begin to exercise, many muscles start contracting harder and faster. As a result, they need more glucose and oxygen for respiration. During exercise the muscles also produce increased amounts of carbon dioxide. This needs to be removed for muscles to keep working effectively.

During exercise, when muscular activity increases, several changes take place in your body:

- Your heart rate increases and the arteries supplying blood to your muscles dilate (widen). These changes increase the flow of oxygenated blood to your exercising muscles. This in turn increases the rate of supply of oxygen and glucose for the increased cellular respiration rate needed. It also increases the rate that carbon dioxide is removed from the muscles.

- Your breathing rate increases and you breathe more deeply. This means you breathe more often and also bring more air into your lungs each time you breathe in. The rate at which oxygen is brought into your body and picked up by your red blood cells is increased, and this oxygen is carried to your exercising muscles. It also means that carbon dioxide can be removed more quickly from the blood in the lungs and breathed out.

- Glycogen stored in the muscles is converted back to glucose, to supply the cells with the fuel they need for increased cellular respiration.

In this way, the heart rate and breathing rate increase during exercise to supply the muscles with what they need and remove the extra waste produced. Cellular respiration increases to supply the muscle cells with the increased levels of energy needed for contraction during exercise. The increase in your breathing and heart rate is to keep up with the demands of the cells.

Synoptic links

You learned about the heart in Topic B4.3, and about the lungs and breathing in Topic B4.5.

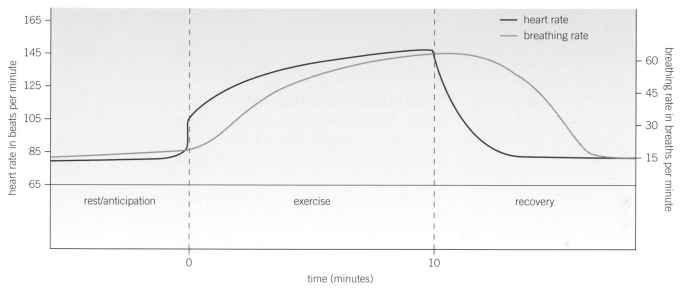

Figure 2 *The changes measured in the heart and breathing rate before, during, and after a period of exercise*

Table 1 *The heart and lung functions change during exercise whether you are fit or not*

	Unfit person	Fit person
amount of blood pumped out of the heart during each beat at rest in cm³	64	80
volume of the heart at rest in cm³	120	140
resting breathing rate in breaths per min	14	12
resting pulse rate in beats per min	72	63

1 Using Figure 2 and Table 1:
  a Describe the effect of exercise on the:
    i heart rate of a fit person [3 marks]
    ii breathing rate of a fit person. [3 marks]
  b Use Table 1 to explain the difference in heart rate between a fit person and an unfit person. [4 marks]
2 a State the function of glycogen. [2 marks]
  b Explain why muscles contain a store of glycogen but most other tissues of the body do not. [5 marks]
3 a Describe an experiment to test the fitness levels of your classmates. Include a method that your classmates could follow. 🖊 [6 marks]
  b Explain what you would expect the results to be and why. [4 marks]

Key points

- The energy that is transferred during respiration is used to enable muscles to contract.
- During exercise the human body responds to the increased demand for energy.
- Body responses to exercise include:
  - an increase in the heart rate, in the breathing rate and in the breath volume
  - glycogen stores in the muscles are converted to glucose for cellular respiration
  - the flow of oxygenated blood to the muscles increases.
- These responses act to increase the rate of supply of glucose and oxygen to the muscles and the rate of removal of carbon dioxide from the muscles.

B9.3 Anaerobic respiration

Learning objectives

After this topic, you should know:

- why less energy is transferred by anaerobic respiration than by aerobic respiration
- **(H)** what is meant by an oxygen debt
- that anaerobic respiration takes place in lots of different organisms, including plants, bacteria, and fungi.

Your everyday muscle movements use energy transferred by aerobic respiration. However, when you exercise hard, your muscle cells may become short of oxygen. Although you increase your heart and breathing rates, sometimes the blood cannot supply oxygen to the muscles fast enough. When this happens, energy from the breakdown of glucose can still be transferred to the muscle cells. They use **anaerobic respiration**, which takes place without oxygen.

Anaerobic respiration

Anaerobic respiration is not as efficient as aerobic respiration because the glucose molecules are not broken down completely. In animal cells the end product of anaerobic respiration is **lactic acid** instead of carbon dioxide and water. Because the breakdown of glucose is incomplete, far less energy is transferred than during aerobic respiration.

Anaerobic respiration:

glucose $\rightarrow$ lactic acid (energy transferred to the environment)

Muscle fatigue

Using your muscle fibres vigorously for a long time can make them become fatigued and they stop contracting efficiently. One cause of this muscle fatigue is the build-up of lactic acid, produced by anaerobic respiration in the muscle cells. The build up of lactic acid in the muscles as a result of anaerobic respiration creates an **oxygen debt**.

For example, repeated movements can soon lead to anaerobic respiration in your muscles – particularly if you're not used to the exercise. If you are fit, your heart and lungs will be able to keep a good supply of oxygen going to your muscles while you exercise for a relatively long time. If you are unfit, your muscles will run short of oxygen much sooner.

Figure 1 *Training hard is the simplest way to avoid anaerobic respiration. When you are fit, you can get oxygen to your muscles and remove carbon dioxide more efficiently*

Making lactic acid

Repeat a single action many times. For example, you could step up and down, lift a weight or clench and unclench your fist. You will soon feel the effect of a build-up of lactic acid in your muscles as they begin to ache.

Oxygen debt

If you have been exercising hard, you often carry on puffing and panting for some time after you stop. The length of time you remain out of breath depends on how fit you are. Why do you carry on breathing fast and deeply when you have stopped using your muscles?

The waste lactic acid you produce during anaerobic respiration is a problem. You cannot simply get rid of lactic acid by breathing it out as you can with carbon dioxide. As a result, when the exercise is over, lactic acid has to be broken down to produce carbon dioxide and water. This needs oxygen.

The amount of oxygen needed to break down the lactic acid to carbon dioxide and water is known as the oxygen debt. After a race, your heart rate and breathing rate stay high to supply the extra oxygen needed to pay off the oxygen debt. The bigger the debt (the larger the amount of lactic acid), the longer you will puff and pant.

Higher

Oxygen debt repayment:

$$lactic\ acid\ +\ oxygen\ \rightarrow\ carbon\ dioxide\ +\ water$$

In a 100 m sprint, some athletes do not breathe at all. This means that the muscles use the oxygen taken in before the start of the race and then don't get any more oxygen until the race is over. Although the race only takes a few seconds, a tremendous amount of energy is used up, so a big oxygen debt can develop, even if the athletes are very fit.

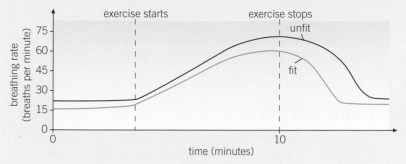

Figure 2 *Everyone gets an oxygen debt if they exercise hard, but if you are fit you can pay it off faster*

Anaerobic respiration in other organisms

Humans and other animals are not the only living organisms that can respire anaerobically. Plants and microorganisms can also respire without oxygen. However, when plant cells respire anaerobically they do not form lactic acid – they form ethanol and carbon dioxide. Some microorganisms form lactic acid during anaerobic respiration – the bacteria used to form yoghurts, for example. Other microorganisms, including yeast, form ethanol and carbon dioxide. Anaerobic respiration in yeast cells is known as fermentation. People have made use of this for thousands of years. It is a very economically important reaction because it is used globally in the manufacture of bread and alcoholic drinks.

$$glucose\ \rightarrow\ ethanol\ +\ carbon\ dioxide \quad \text{(energy transferred to the environment)}$$

1 If you exercise very hard or for a long time, your muscles begin to ache and do not work effectively. Explain why. [4 marks]

2 **a** Define anaerobic respiration. [1 mark]
   b Explain how anaerobic respiration differs between animals, plants, and yeast. In each case, explain the benefits to the organism of being able to respire in this way. [6 marks]
   c Write the word equation for anaerobic respiration in animals, plants, and yeast. [3 marks]

3 **H** If you exercise vigorously, you often puff and pant for some time after you stop. Explain what is happening. [4 marks]

Testing fitness

A good way of telling how fit you are is to measure your resting heart rate and breathing rate. The fitter you are, the lower they will be. This is because the heart and lungs of fit people are bigger and have a better blood supply than those of less-fit people. They are more efficient. Then see what happens when you exercise. The increase in your heart rate and breathing rate, along with how quickly they return to normal, are also ways of finding out how fit you are – or aren't!

Key points

- If muscles work hard for a long time, they become fatigued and don't contract efficiently. If they don't get enough oxygen, they will respire anaerobically.
- Anaerobic respiration is respiration without oxygen. When this takes place in animal cells, glucose is incompletely broken down to form lactic acid.
- The anaerobic breakdown of glucose transfers less energy than aerobic respiration.
- **H** After exercise, oxygen is still needed to break down the lactic acid that has built up. The amount of oxygen needed is known as the oxygen debt.
- Anaerobic respiration in plant cells and some microorganisms, such as yeast, results in the production of ethanol and carbon dioxide.

B9.4 Metabolism and the liver

Learning objectives

After this topic, you should know:

- that metabolism is the sum of all the reactions in a cell or the body of an organism
- **H** how the liver is involved in repaying the oxygen debt

Synoptic links

You learnt about the reactions that build up carbohydrates, proteins, and lipids in Topic B3.3.

You learnt about the reactions of photosynthesis in Topic B8.1 and Topic B8.3.

The metabolism of an organism is the sum of all the reactions that take place in a cell or in the body. Some of the energy transferred by respiration reactions in cells simply heats the environment. However, some of it is used by the organism for the continual enzyme-controlled processes of metabolism that make molecules or break them down. It is used to bring about change and movement where needed. As you saw in Topic B9.1, the energy transferred by respiration in animals is also used to enable the muscles to contract. In mammals and birds it is also used to maintain a constant body temperature.

Metabolic reactions

There are hundreds of thousands of metabolic reactions but you don't have to learn them all. Some of the most common metabolic reactions include:

- the conversion of glucose to starch, glycogen, and cellulose
- the formation of lipid molecules from a molecule of glycerol and three fatty acid molecules
- the use of glucose and nitrate ions to form amino acids that are then used to make proteins
- the reactions of respiration
- the reactions of photosynthesis
- the breakdown of excess proteins in the liver to form urea for excretion in the urine by the kidneys.

You have already met a number of these reactions in some detail.

Figure 1 *Many of the metabolic reactions of plants and animals, such as aerobic respiration, are exactly the same, but some, for example photosynthesis, are very different*

Higher

The role of the liver

Your liver is a large reddish-brown organ that carries out many different functions in your body. Liver cells grow and regenerate themselves very rapidly. The liver is a very active organ with many different metabolic functions. These include:

● detoxifying poisonous substances such as the ethanol from alcoholic drinks

● passing the breakdown products into the blood so they can be excreted in the urine via the kidneys

● breaking down old, worn out blood cells and storing the iron until it is needed to synthesise more blood cells.

Removing lactic acid

One important role of the liver is in dealing with the lactic acid produced by the muscles during anaerobic respiration. Blood flowing through the muscles transports the lactic acid to the liver where it is converted back into glucose. The oxygen debt is repaid once the lactic acid has been converted back to glucose and the glucose has been completely broken down in aerobic respiration to form carbon dioxide and water. If it isn't needed, the glucose made from the lactic acid may be converted to glycogen and stored in the liver until it is needed.

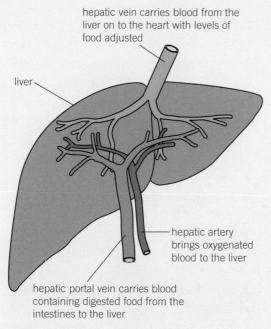

hepatic vein carries blood from the liver on to the heart with levels of food adjusted

liver

hepatic artery brings oxygenated blood to the liver

hepatic portal vein carries blood containing digested food from the intestines to the liver

Figure 2 *Your liver weighs about 1.5 kg and plays a vital role in removing poisons from your body*

Synoptic links

You can remind yourself about the role of the liver in digestion in Topic B3.7.

Key points

● Metabolism is the sum of all the reactions in the body.

● The energy transferred by respiration in cells is used by the organism for the continual enzyme-controlled processes of metabolism that synthesise new molecules

● Metabolism includes the conversion of glucose to starch, glycogen and cellulose. Metabolism also includes the formation of lipid molecules, and the use of glucose and nitrate ions to form amino acids, which are used to synthesise proteins, and breakdown excess proteins to form urea.

● **H** Blood flowing through the muscles transports lactic acid to the liver where it is converted back to glucose.

---

1 What is metabolism?   [1 mark]

2 Give four examples of metabolic reactions in the body.   [4 marks]

3 **H** The liver is an organ of respiration. Discuss this statement.
                                                                [5 marks]

B9 Respiration

Summary questions

1 Some students investigated the process of cellular respiration. They set up three vacuum flasks. One contained live, soaked peas. One contained dry peas. One contained peas that had been soaked and then boiled. They took daily observations of the temperature in each flask for a week. The results are shown in Table 1.

Table 1

Day	Room temperature in °C	Temperature in flask A containing live, soaked peas in °C	Temperature in flask B containing dry peas in °C	Temperature in flask C containing soaked, boiled peas in °C
1	20.0	20.0	20.0	20.0
2	20.0	20.5	20.0	20.0
3	20.0	21.0	20.0	20.0
4	20.0	21.5	20.0	20.0
5	20.0	22.0	20.0	20.0
6	20.0	22.2	20.0	20.5
7	20.0	22.5	20.0	21.0

a Plot a graph to show these results. [6 marks]

b Explain the results in flask A containing the live, soaked peas. [3 marks]

c Why were the results in flask B the same as the room temperature readings? [3 marks]

d Why was the room temperature in the lab recorded every day? [2 marks]

e Look at the results for flask C.

   i Why is the temperature at 20 °C for the first five days? [3 marks]

   ii After five days the temperature increases. Suggest **two** possible explanations for why the temperature increases. [4 marks]

2 It is often said that taking regular exercise and getting fit is good for your heart and your lungs.

Table 2

	Before getting fit	After getting fit
amount of blood pumped out of the heart during each beat in cm³	64	80
heart volume in cm³	120	140
breathing rate in breaths per min	14	12
pulse rate in beats per min	72	63

a Table 2 shows the effect of getting fit on the heart and lungs of one person. Display this data in four bar charts. [5 marks]

b Use the information on your bar charts to help you explain exactly what effect increased fitness has on:

   i your heart [3 marks]

   ii your lungs. [2 marks]

3 a What is aerobic respiration? [3 marks]

b What is anaerobic respiration? [3 marks]

c How does anaerobic respiration differ between a human and a yeast cell? [4 marks]

d **H** Explain what is meant by the term 'oxygen debt'. [4 marks]

e **H** Explain the difference in the responses of a fit and an unfit individual to exercise in terms of their muscles, heart and lung function, and oxygen debt. **✔** [6 marks]

4 Athletes want to be able to use their muscles aerobically for as long as possible when they compete. They train to develop their heart and lungs. Many athletes also train at altitude. There is less oxygen in the air so your body makes more red blood cells, which helps to avoid oxygen debt. Some athletes remove some of their own blood, store it and then just before a competition transfuse it back into their system. This is called blood doping and it is illegal. Other athletes use hormones to stimulate the growth of extra red blood cells. This is also illegal.

a Why do athletes want to be able to use their muscles aerobically for as long as possible? [3 marks]

b How does developing more red blood cells by training at altitude help athletic performance? [3 marks]

c How does blood doping help performance? [2 marks]

d Explain in detail what happens to the muscles if the body cannot supply glucose and oxygen quickly enough when they are working hard. **✔** [4 marks]

Practice questions

01.1 Complete the word equation for aerobic respiration.
glucose + ... → carbon dioxide + ...
[1 mark]

01.2 What type of reaction is aerobic respiration?
A endothermic
B exothermic
C fermentation
D reversible [1 mark]

01.3 In which part of a cell does aerobic respiration take place?
A chloroplast
B nucleus
C mitochondrion
D ribosome [1 mark]

02 Glycogen is stored in the liver.
Describe how stored glycogen can provide glucose to respiring cells in a different part of the body.
[4 marks]

03 Muscle cells can respire aerobically and anaerobically.
Compare these two types of respiration in human muscle cells.
You should include both similarities and differences.
[6 marks]

04 Yeast cells can respire anaerobically.
04.1 What is anaerobic respiration in yeast cells called?
[1 mark]

04.2 A student measured the rate of anaerobic respiration in yeast using the apparatus shown in **Figure 1**.

Figure 1

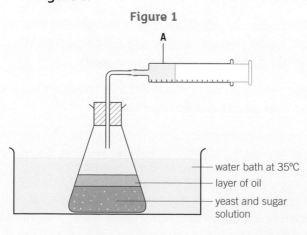

water bath at 35ºC
layer of oil
yeast and sugar solution

The student's method was as follows.
● Mix yeast and sugar solution together in a flask.
● Cover the surface of the liquid with oil.
● Put the flask into the water bath and leave it for 15 minutes.
● After the 15 minutes, connect the apparatus labelled **A** to the flask.
● Measure the volume of gas collected in **A** every two minutes for 20 minutes.
Suggest why a layer of oil was put on top of the yeast and sugar solution. [1 mark]

04.3 What is the piece of apparatus labelled **A** called?
[1 mark]

04.4 Give **one** reason why the flask was left in the water bath for 15 minutes before apparatus **A** was connected. [1 mark]

The student's results are shown in **Table 1**.

Table 1

Time in minutes	Volume of gas produced in cm³
0	0
2	3
4	7
6	14
8	24
10	34
12	45
14	56
16	65
18	74
20	82

04.5 Plot the results on graph paper.
Choose suitable scales, label both axes, and draw a line of best fit. [4 marks]

04.6 Calculate the rate of anaerobic respiration over the 20 minutes in cm³/min. [1 mark]

04.7 Predict how the results would be different if the investigation was done at 65 °C.
Explain your prediction. [3 marks]

04.8 Give **two** ways anaerobic respiration in yeast is of economic importance to humans. [2 marks]

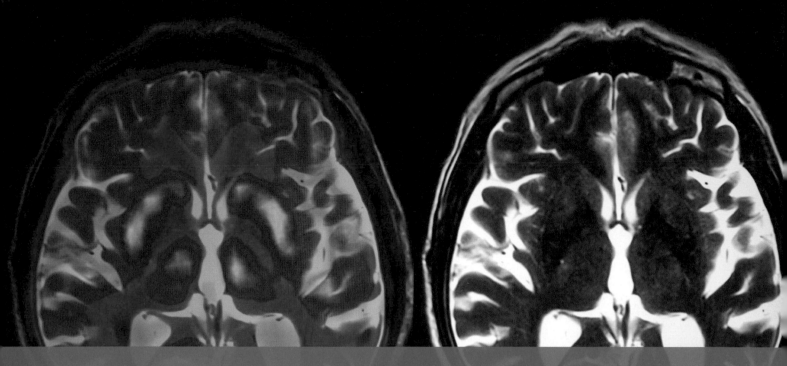

3 Biological responses

The ability to respond to the world around you is key to survival. Your body needs awareness of the conditions outside and inside your body to coordinate your responses. Everything works towards keeping your internal conditions stable and within very narrow ranges.

In humans, the nervous system controls most of our rapid responses, and our hormones control everything from our reproduction to the rate of our metabolism. We have found ways to produce hormones, such as insulin, which we can use to treat diseases such as type 1 and type 2 diabetes. We have also found ways to use artificial hormones and other methods to control our own fertility.

Key questions

- What is homeostasis and why is it so important?

- Why are reflex actions so important for survival?

- How do hormones control responses, such as the release of a mature egg in the human menstrual cycle?

- How are blood sugar levels controlled?

Making connections

- You learnt about the specialised cells and tissues that are important in the nervous and hormonal responses of animals in **B1 Cell structure and transport**.

- You learnt how glucose is absorbed into the blood in **B3 Organisation and the digestive system**.

- You discovered how important the control of heart and breathing rate are in exercise in **B4 Organising animals and plants**.

- You will learn more about the adaptations of organisms to maintain homeostasis in challenging environments in **B16 Adaptations, interdependence and competition**.

I already know...

The basic structure of neurones.

That tissues can be organised into organs with particular functions in the body.

That enzymes act as biological catalysts.

The basic processes of human reproduction.

The male and female reproductive organs.

I will learn...

The differences between sensory and motor neurones and their roles in coordination and control.

About the arrangement of tissues in the endocrine organs and how they are adapted to their functions.

How the structures of enzymes are related to their functions and how different factors affect the rate of enzyme-controlled reactions.

How reproduction is controlled by hormones and how hormones can be used in the artificial control of fertility.

How hormones work together to control the menstrual cycle, and how they can be used in the artificial control of fertility.

Required Practicals

Practical		Topic
6	Measuring reaction times	B10.2

10.1 Principles of homeostasis

Learning objectives

After this topic, you should know:

- why it is important to control your internal environment
- the key elements of control systems.

Synoptic links

You learnt about the effect of temperature and pH changes on enzyme activity in Topic B3.5.

Synoptic links

You will learn more about the role of the hormonal system in Chapter B11.

Human beings live everywhere from the equator to the Antarctic. People survive wearing no clothes or many clothes, running a marathon, or never moving from the computer screen. Conditions can change dramatically around us and even inside us and yet we survive. How do we do it?

Figure 1 *Human beings are not the only organisms who can survive in extremes of temperatures. Animals and plants all have complex coordination and control mechanisms that enable them to cope with dramatic changes in the external environment*

Go further

Scientists are investigating homeostasis in organisms from whales to plants. More and more astonishing mechanisms are being discovered. They range from colour changes to regulate body temperature in invertebrates, to protecting the immune system from responding at the wrong time in plants.

Homeostasis in action

The conditions inside your body are known as its internal environment. Your organs cannot work properly if this keeps changing. Many of the processes that go on inside your body aim to keep everything as constant as possible. As well as the body as a whole, this includes the regulation of the internal conditions of cells to maintain optimum conditions for functioning, in response to internal and external changes. This balancing act is called **homeostasis**.

You know that enzymes only work at their best in specific conditions of temperature and pH. Enzymes control all the functions of a cell. The functioning of individual cells is vital for the way tissues, organs, and whole organisms work. It is important to respond to changes in the internal or external environment to maintain optimum conditions for the cellular enzymes.

Internal conditions that are controlled include:

● body temperature

● the water content of the body

● blood glucose concentration.

Working together

Homeostasis involves coordination and control. Organisms need to be aware of changes in the world around them, such as changes in temperature or levels of sunlight. They also need to respond to changes in the internal environment. When you exercise your muscles get hotter, when you have eaten a meal your blood sugar levels go up, and in hot weather you lose water and salt through sweating.

Detecting changes and responding to them involves automatic control systems. These automatic systems include nervous responses in your nervous system and chemical responses in your hormone system. They also involve many of your body organs.

Figure 2 *During the month of Ramadan, Muslims fast from dawn to sunset. Homeostatic mechanisms maintain the blood sugar levels and the ion and water balance of the body during the hours of fasting*

The demands of a control system

All control systems in the body need certain key features to function:

● **Receptors**: cells that detect changes in the internal or external environment. These changes are known as **stimuli**. Receptors may be part of the nervous or the hormonal control systems of the body.

● **Coordination centres**: areas that receive and process the information from the receptors. They send out signals and coordinate the response of the body. They include the brain, which acts as a coordination centre for both the nervous system and parts of the hormonal system, the spinal cord, and some organs such as the pancreas.

● **Effectors**: muscles or glands that bring about responses to the stimulus that has been received. These responses restore conditions in the body to the optimum levels.

1 Describe homeostasis. [3 marks]

2 Compare receptors, coordination centres, and effectors. [6 marks]

3 **a** Describe three ways in which your external environment might vary. [3 marks]

b Explain how each of your answers to part a affects your body. ⊘ [6 marks]

B10.2 The structure and function of the human nervous system

Learning objectives

After this topic, you should know:

- why you need a nervous system
- how the structure of the nervous system is adapted to its function
- how receptors enable you to respond to changes in your surroundings.

Figure 1 *Your body is made up of millions of cells that have to work together. Whatever you do with your body, whether it's walking to school, learning to drive, or playing on the computer, your movements need to be coordinated*

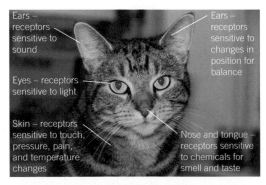

Ears – receptors sensitive to sound

Ears – receptors sensitive to changes in position for balance

Eyes – receptors sensitive to light

Skin – receptors sensitive to touch, pressure, pain, and temperature changes

Nose and tongue – receptors sensitive to chemicals for smell and taste

Figure 2 *This cat relies on its sensory receptors to detect changes in the external environment*

Synoptic link

You learnt about specialised nerve cells in Topic B1.4.

You need to know what is going on in the world around you. Your nervous system makes this possible. It enables you to react to your surroundings and coordinate your behaviour.

Your nervous system carries electrical signals (impulses) that travel fast – between 1 and 120 metres per second. This means you can react to changes in your surroundings very quickly.

The nervous system

As with most animals, you need to avoid danger, find food, and eventually find a mate. This is where your nervous system helps. Your body is particularly sensitive to changes in the world around you. Any changes (known as stimuli) are picked up by cells called receptors.

Receptor cells, such as the light receptor cells in your eyes, are similar to most animal cells. They have a nucleus, cytoplasm, and a cell membrane. These receptors are usually found clustered together in special sense organs, such as your eyes and your skin. You have many different types of sensory receptor. Some male moths have receptors so sensitive they can detect the scent of a female several kilometres away and follow the scent trail to find her.

How your nervous system works

Once a sensory receptor detects a stimulus, the information is sent as an electrical impulse that passes along special cells called **neurones**. These are usually found in bundles of hundreds or even thousands of neurones known as **nerves**.

The impulse travels along the neurone until it reaches the **central nervous system** (CNS). The CNS is made up of the brain and the spinal cord. The cells that carry impulses from your sense organs to your CNS are called **sensory neurones**.

Your brain gets huge amounts of information from all the sensory receptors in your body. It coordinates the response to the information, and sends impulses out along special cells. These cells, called **motor neurones**, carry information from the CNS to the rest of your body. They carry impulses to make the right bits of your body – the **effectors** – respond.

Effectors may be muscles or glands. Your muscles respond to the arrival of impulses by contracting. Your glands respond by releasing (secreting) chemical substances. For example, your salivary glands produce and release extra saliva when you smell food cooking, and your pancreas releases the hormone insulin when your blood sugar levels go up after a meal.

The way your nervous system works can be summed up as:

$$\text{stimulus} \rightarrow \text{receptor} \xrightarrow{\;\text{coordinator (CNS)}\;} \text{effector}$$

The receptor sends an impulse along a sensory neurone, carrying information about a change in the environment to the coordinator (CNS). Once all the incoming information has been processed, the coordinator sends impulses down motor neurones. These motor impulses stimulate the effectors to bring about the responses needed in any particular situation.

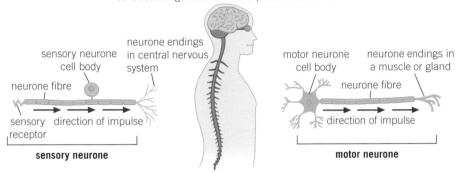

Sensory nerves carry impulses to the CNS. the information is processed and impulses are sent out along motor nerves to produce an action.

sensory neurone cell body
neurone endings in central nervous system
neurone fibre
sensory receptor
direction of impulse
sensory neurone

motor neurone cell body
neurone endings in a muscle or gland
neurone fibre
direction of impulse
motor neurone

Figure 3 *The rapid responses of our nervous system allow us to respond to our surroundings quickly – and in the right way*

Measuring reaction times

There are many ways to investigate how quickly nerve impulses travel in your body. Two simple methods are:

- use the ruler drop test or digital sensors to measure how quickly you react to a visual stimulus

- stand in a circle holding hands with your eyes closed and measure how long it takes a hand-squeeze to pass around the circle.

People claim that activities such as drinking cola, talking on the phone, and listening to music affect our reaction times. Choose a factor that interests you and use these simple techniques to investigate the effect it has – or does not have – on human reaction times.

Safety: Do not drink or eat in the laboratory.

1 a State the main function of the nervous system. [3 marks]
   b Describe the difference between a neurone and a nerve. [2 marks]
   c Describe the difference between a sensory neurone and a motor neurone. [3 marks]
2 a State the different types of sense receptor. [5 marks]
   b For each of the above receptors, give one example of what it responds to. [5 marks]
3 Explain what happens in your nervous system when you see a piece of fruit, pick it up, and eat it. [6 marks]

Key points

- The nervous system uses electrical impulses to enable you to react quickly to your surroundings and coordinate your behaviour.
- Cells called receptors detect stimuli (changes in the environment).
- Impulses from receptors pass along sensory neurones to the brain or spinal cord (CNS). The brain coordinates the response, and impulses are sent along motor neurones from the brain (CNS) to the effector organs.

B10.3 Reflex actions

Learning objectives

After this topic, you should know:

- what reflexes are
- how reflexes work
- why reflexes are important in your body.

Your nervous system lets you take in information from your surroundings and respond in the right way. However, some of your responses are so fast that they happen without giving you time to think.

When you touch something hot, or sharp, you pull your hand back before you feel the pain. If something comes near your face, you blink. Automatic responses like these are known as **reflexes**.

What are reflexes for?

Reflexes are very important both for human and other animals. They help you to avoid danger or harm because they happen so fast. There are also lots of reflexes that take care of your basic body functions. These functions include breathing and moving food through your digestive system.

Reflexes are automatic and rapid – they do not involve the conscious part of your brain. It would make life very difficult if you had to think consciously about those things all the time – and it would be fatal if you forgot to breathe!

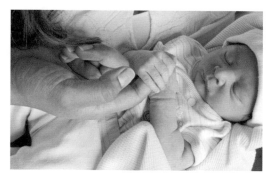

Figure 1 *Newborn babies have a number of special reflexes that disappear as they grow. This grasp reflex is one of them*

How do reflexes work?

Simple reflex actions such as the pain withdrawal reflex we are all familiar with often involve just three types of neurone. These are:

- sensory neurones
- motor neurones
- relay neurones – these connect a sensory neurone and a motor neurone, and are found in the CNS.

An electrical impulse passes from the receptor along the sensory neurone to the CNS. It then passes along a relay neurone (usually in the spinal cord) and straight back along the motor neurone. From there, the impulse arrives at the effector organ. The effector organ will be a muscle or a gland. We call this pathway a **reflex arc**.

The key point in a reflex arc is that the impulse bypasses the conscious areas of your brain. The result is that the time between the stimulus and the reflex action is as short as possible.

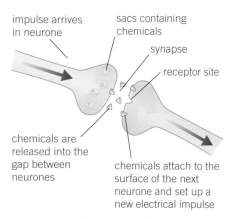

impulse arrives in neurone

sacs containing chemicals

synapse

receptor site

chemicals are released into the gap between neurones

chemicals attach to the surface of the next neurone and set up a new electrical impulse

Figure 2 *When an impulse arrives at the junction between two neurones, chemicals are released that cross the synapse and arrive at receptor sites on the next neurone. This starts up a new electrical impulse in the next neurone.*

How do synapses work?

Your neurones are not joined up directly to each other. There are junctions between them called synapses which form physical gaps between the neurones. The electrical impulses travelling along your neurones have to cross these synapses. They cannot leap the gap. The diffusion of the chemical across the synapse is slower than the electrical impulse in the neurones, but it makes it possible for the impulse to cross the gap between them. Figure 2 shows you how this happens.

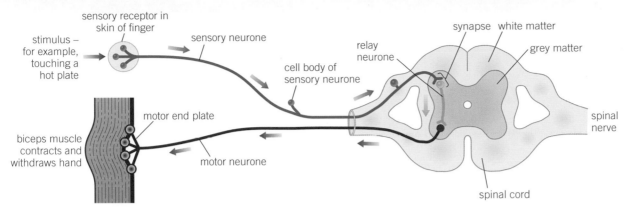

Figure 3 *The reflex action that moves your hand away from something hot can save you from being burned. Reflex actions are quick and automatic – you do not think about them*

The reflex arc in detail

Look at Figure 3. It shows what would happen if you touched a hot object.

- When you touch the object, a receptor in your skin is stimulated. An electrical impulse from a receptor passes along a sensory neurone to the CNS – in this case, the spinal cord.

- When an impulse from the sensory neurone arrives at the synapse with a relay neurone, a chemical is released. The chemical diffuses across the synapse to the relay neurone where it sets off a new electrical impulse that travels along the relay neurone.

- When the impulse reaches the synapse between the relay neurone and a motor neurone returning to the arm, another chemical is released. Again, the chemical diffuses across the synapse and starts a new electrical impulse travelling down the motor neurone to the effector.

- When the impulse reaches the effector organ, it is stimulated to respond. In this example, the impulses arrive in the muscles of the arm, causing them to contract. This action moves the hand rapidly away from the source of pain. If the effector organ is a gland, it will respond by releasing (secreting) chemical substances.

The reflex pathway is not very different from a normal conscious action. However, in a reflex action the coordinator is a relay neurone either in the spinal cord or in the unconscious areas of the brain. The whole reflex is very fast indeed.

An impulse also travels up the spinal cord to the conscious areas of your brain. You know about the reflex action, but only after it has happened.

1 a Why are reflexes important? [2 marks]
  b Why is it important that reflexes don't go to the conscious areas of your brain? [1 mark]

2 Explain why some actions such as breathing and swallowing are reflex actions, while others such as speaking and eating are under your conscious control. [4 marks]

3 Describe what happens when you step on a pin. Make sure you indicate when an electrical impulse or a chemical impulse is involved. [6 marks]

> **Study tip**
>
> Learn the reflex pathway off by heart.
>
> **stimulus → receptor →
> sensory neurone → relay neurone →
> motor neurone → effector →
> response.**

> **Key points**
>
> - Reflex actions are automatic and rapid and do not involve the conscious parts of the brain.
> - Reflexes involve sensory, relay and motor neurones.
> - Reflex actions control everyday bodily functions, such as breathing and digestion, and help you to avoid danger.
> - The main stages of a reflex arc are:
>   **stimulus → receptor →
>   sensory neurone → relay neurone →
>   motor neurone → effector →
>   response**

B10 The human nervous system

Summary questions

1 **a** What is homeostasis? [3 marks]

   b Explain why the control of conditions inside
   your body is so important. 🖊 [4 marks]

2 Match up **A–F** with **U–Z** to make complete sentences
   about animal responses.

   A Many processes in
   the body …

   U … effector organs.

   B The nervous system
   allows you …

   V … secreted by glands.

   C The cells that are
   sensitive to light …

   W … to react to your
   surroundings and
   coordinate your
   behaviour.

   D Hormones are chemical
   substances …

   X … are found in the
   eyes.

   E Muscles and glands
   are known as …

   Y … are known as
   nerves.

   F Bundles of neurones …

   Z … are controlled by
   hormones.

   [6 marks]

3 **a** State the job of your nervous system. [2 marks]

   b State where in your body you would find nervous
   receptors that respond to:

   i light [1 mark]

   ii sound [1 mark]

   iii temperature [1 mark]

   iv touch. [1 mark]

   c Draw and label a simple diagram of a reflex arc.
   [4 marks]

   Explain carefully how a reflex arc works and why
   it allows you to respond quickly to danger. [4 marks]

4
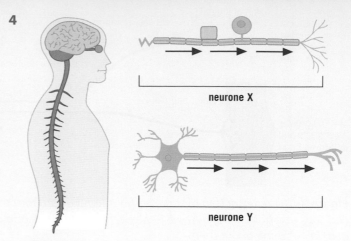

Figure 1

a State the difference between neurone X and
neurone Y. [2 marks]

b State the difference between a neurone
and a nerve. [3 marks]

c Draw and label a synapse. [3 marks]

d Describe how a nerve impulse passes across
a synapse. Explain the importance of synapses in
the nervous system. [4 marks]

5 Lots of people drink coffee or caffeinated drinks
because they think it keeps them awake or speeds up
their reaction times.
In a class investigation to see if caffeine has an effect
on simple reaction times, two groups of students
had their mass reaction times measured by standing
in a ring and squeezing hands. Their responses were
measured before and 10 minutes after having a drink.
One group was given a caffeinated drink, the other
was not.

Table 1

	Mass reaction times (mins)			Mass reaction times (min) 10 minutes after a drink		
Group A	4.0	3.8	3.6	3.8	3.8	3.6
Group B	4.2	4.0	3.8	3.6	3.4	3.2

a Calculate the mean mass reaction times for both
groups before having a drink. [2 marks]

b Calculate the mean mass reaction times for both
groups 10 minutes after having a drink. [2 marks]

c Which group do you think was given a drink
containing caffeine? Explain your answer. [4 marks]

d Suggest a reason why the mass reaction times of
both groups fell during the test. [2 marks]

Practice questions

01 A student did an investigation to see if reaction time was affected by the sense organ stimulated. A computer measured how quickly she clicked the mouse when she:

- saw a shape appear on the screen

or

- heard a man shout 'Stop!'

or

- felt a bar vibrate in her hand.

Each sense organ was tested five times. **Figure 1** shows her results.

Figure 1

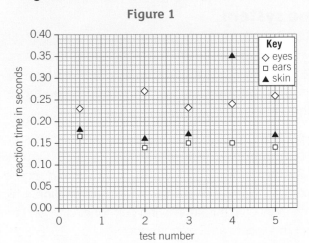

01.1 The data is shown as a scatter graph rather than a line graph.
Suggest why. [1 mark]

01.2 The results shown in the scatter graph might be easier to understand if they were drawn as a bar chart. Describe what would have to be done with these results before they could be shown in a bar chart. [2 marks]

01.3 Give **one** conclusion that can be made from these results. [1 mark]

AQA, 2013

02 Some students investigated the effect of caffeine on a person's reaction time. **Figure 2** shows how they measured reaction time using a computer program.

Figure 2

When the woman saw Stop! appear on screen she clicked the mouse.
She did this five times.
She then drank a cup of strong coffee. Coffee contains caffeine.
After drinking the coffee the woman sat for 30 minutes.
The students then tested her reaction time another five times using the same computer program.

02.1 What was the independent variable in this investigation? [1 mark]

02.2 Why did the woman sit for 30 minutes after drinking the coffee, before the students measured her reaction time? [1 mark]

Table 1 shows the results.

Table 1

Test number	Reaction time in seconds	
	Before drinking coffee	After drinking coffee
1	0.343	0.301
2	0.348	0.303
3	0.354	0.303
4	0.352	0.307
5	0.348	0.301
Mean	0.349	

02.3 What would be the best way to display these results?
A bar chart
B histogram
C line graph
D scatter diagram [1 mark]

02.4 Calculate the mean reaction time after drinking coffee. [2 marks]

02.5 What conclusion can be made from these results? [1 mark]

02.6 **H** Suggest **two** improvements the students could make to the investigation to produce more valid results. [2 marks]

03 If you touch a hot object you pull your hand away. This is an example of a reflex action.

03.1 Describe fully how the different structures of the nervous system bring about this reflex action. [6 marks]

03.2 Explain why reflex actions are important to the body. [2 marks]

11.1 Principles of hormonal control

Learning objectives

After this topic, you should know:

- what a hormone is
- the main organs of the endocrine system
- the role of the pituitary gland.

Figure 1 *Many aspects of the growth of children from birth to adulthood is controlled by hormones*

Go further

Hormones have their effects on the body by their interaction with DNA and the process of protein synthesis.

In Chapter B 10 you discovered how the nervous system acts to coordinate and control your body, reacting in seconds to changes in your internal and external environments. However, it is very important that your body acts as a coordinated whole, not just from minute to minute but from day to day and year to year throughout your life. You have a second coordination and control system to help with this – the **endocrine system**.

The endocrine system

The endocrine system is made up of glands that secrete chemicals called **hormones** directly into the bloodstream. The blood carries the hormone to its target organ (or organs) where it produces an effect. The target organ has receptors on the cell membranes that pick up the hormone molecules, triggering a response in the cell.

Many processes in your body are coordinated by these hormones. Hormones can act very rapidly but, compared to the nervous system, many hormonal effects are slower but longer lasting. Hormones that give a rapid response include **insulin**, which controls your blood glucose, and **adrenaline**, which prepares your body for fight or flight. Slow-acting hormones with long-term effects include growth hormones and sex hormones.

The endocrine glands

Hormones provide chemical coordination and control for the body and are produced by the endocrine glands. Many endocrine glands around the body are themselves coordinated and controlled by one very small but powerful endocrine gland found in the brain – the **pituitary gland**. The pituitary gland acts as a master gland. It secretes a variety of different hormones into the blood in response to changes in body conditions. Some hormones produced by the pituitary in response to changes in the internal environment have a direct effect on the body. Examples include **ADH**, which affects the amount of urine produced by the kidney, and growth hormone, which controls the rate of growth in children.

Other hormones released by the pituitary affect specific endocrine glands, stimulating them to release hormones that bring about the required effect on the body. These include:

- **follicle stimulating hormone (FSH)**, which stimulates the **ovaries** to make the female sex hormone **oestrogen**, and
- TSH, which stimulates the thyroid gland to make thyroxine, a hormone that helps control the rate of your metabolism.

Each of the endocrine glands produces hormones that have a major effect on the way your body works. The levels of the hormones vary depending on changes in the internal environment of your body. You will learn more about many of these hormones in the next few pages.

Table 1 *The main roles of hormones produced by the different endocrine glands*

Endocrine gland	Role of the hormones
Pituitary	Controls growth in children
	Stimulates the thyroid gland to make thyroxine to control the rate of metabolism
	In women – stimulates the ovaries to produce and release eggs and make the female sex hormone oestrogen
	In men – stimulates the testes to make sperm and the male sex hormone testosterone
Thyroid	Controls the metabolic rate of the body
Pancreas	Controls the levels of glucose in the blood
Adrenal	Prepares the body for stressful situations – 'fight or flight' response
Ovaries	Controls the development of the female secondary sexual characteristics and is involved in the menstrual cycle
Testes	Controls the development of the male secondary sexual characteristics and is involved in the production of sperm

Figure 2 *It isn't just humans who need hormones – without the hormones from their thyroid glands, these tadpoles will never become frogs*

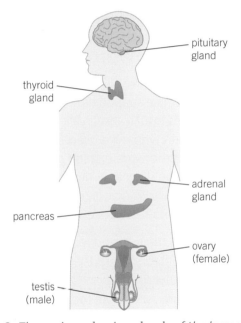

Figure 3 *The main endocrine glands of the human body*

1 **a** What is a hormone? [1 mark]
  b What is an endocrine gland? [1 mark]

2 Explain how coordination and control by hormones differs from coordination and control by the nervous system. [6 marks]

3 Suggest why the pituitary gland is sometimes called the master gland of the endocrine system. [3 marks]

4 Suggest what would happen if the pituitary gland:
  a did not produce enough growth hormone in a child [2 marks]
  b continued to produce lots of growth hormone in an adult. [2 marks]

Study tip

Find a way to help you remember the main endocrine organs of the body – for example, learn them in the order they appear in your body from the head downwards.

Key points

- The endocrine system is composed of glands that secrete chemicals called hormones directly into the blood stream. The blood carries the hormone to a target organ where it produces an effect.
- Compared to the nervous system, the effects of hormones are often slower but longer lasting.
- The pituitary gland is the master gland and secretes several hormones into the blood in response to body conditions. Some of these hormones act on other glands and stimulate them to release hormones to bring about specific effects.
- Key endocrine glands are the pituitary, thyroid, pancreas, adrenal glands, ovaries, and testes.

B11.2 The control of blood glucose levels

Learning objectives

After this topic, you should know:

- the role of the pancreas in monitoring and controlling blood glucose concentration
- how insulin controls blood glucose levels in the body
- **H** how glucagon and insulin interact to control blood glucose levels
- what causes diabetes.

Study tip

Know the difference between:

- glucose – the sugar used in respiration
- glycogen – a storage carbohydrate found in the liver and muscles
- **H** glucagon – a hormone that stimulates the liver to break down glycogen to glucose.

It is essential that your cells have a constant supply of the glucose they need for respiration. To achieve this, one of your body systems responds to changes in your blood glucose levels and controls it to within very narrow limits. This is an example of homeostasis in action.

Insulin and the control of blood glucose levels

When you digest a meal, large amounts of glucose pass into your blood. Without a control mechanism, your blood glucose levels would vary significantly. They would range from very high after a meal to very low several hours later – so low that cells would not have enough glucose to respire.

This situation is prevented by your pancreas. The pancreas is a small pink organ found under your stomach. It constantly monitors and controls your blood glucose concentration using two hormones. The best known of these is **insulin**.

When your blood glucose concentration rises after you have eaten a meal, the pancreas produces insulin. Insulin allows glucose to move from the blood into your cells where it is used. Soluble glucose is also converted to an insoluble carbohydrate called glycogen. Insulin controls the storage of glycogen in your liver and muscles. Stored glycogen can be converted back into glucose when it is needed. As a result, your blood glucose stays stable within a narrow concentration range.

When the glycogen stores in the liver and muscles are full, any excess glucose is converted into lipids and stored. If you regularly take in food that results in having more glucose than the liver and muscles can store as glycogen, you will gradually store more and more of it as lipids. Eventually you may become obese.

Glucagon and control of blood glucose levels

The control of your blood glucose doesn't just involve insulin. When your blood glucose concentration falls below the ideal range, the pancreas secretes another hormone called **glucagon**. Glucagon makes your liver break down glycogen, converting it back into glucose. In this way, the stored glucose is released back into the blood.

By using two hormones and the glycogen store in your liver, your pancreas keeps your blood glucose concentration fairly constant. It does this using negative feedback control, which involves switching between the two hormones (Figure 1).

Figure 2 shows a model of your blood glucose control system where the blood glucose is a tank. It has both controlled and uncontrolled inlets and outlets. In every case, any control is given by the hormones insulin and glucagon.

Higher

Figure 1 *Negative feedback control of blood glucose levels using insulin and glucagon*

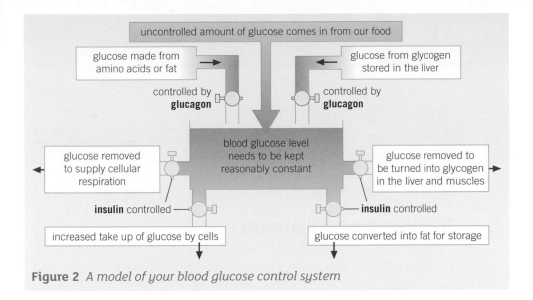

Figure 2 *A model of your blood glucose control system*

What causes diabetes?

If your pancreas does not make enough (or any) insulin, your blood glucose concentration is not controlled. You have **type 1 diabetes**.

Without insulin your blood glucose levels get very high after you eat. Eventually your kidneys excrete glucose in your urine. You produce lots of urine and feel thirsty all the time. Without insulin, glucose cannot get into the cells of your body, so you lack energy and feel tired. You break down fat and protein to use as fuel instead, so you lose weight. Type 1 diabetes is a disorder that usually starts in young children and teenagers. There seems to be a genetic element to the development of the disease.

Type 2 diabetes is another, very common type of diabetes. It gets more common as people get older and is often linked to obesity, lack of exercise, or both. There is also a strong genetic tendency to develop type 2 diabetes. In type 2 diabetes, the pancreas still makes insulin, although it may make less than your body needs. Most importantly, your body cells stop responding properly to the insulin you make. In countries such as the UK and the USA, levels of type 2 diabetes are rising rapidly as obesity becomes more common.

1 Define the terms:
   a insulin [1 mark] **b** diabetes [1 mark] **c** glycogen. [1 mark]

2 **a** Explain how the pancreas responds when blood glucose levels go above the ideal range. [3 marks]
   b 🅗 Explain how the pancreas responds when blood glucose levels go below the ideal range. [3 marks]
   c Why is it so important to control the level of glucose in your blood? [3 marks]

3 Explain the difference between type 1 and type 2 diabetes. 🖊 [6 marks]

4 🅗 Compare and contrast the roles of insulin and glucagon in controlling the body's blood glucose levels. 🖊 [6 marks]

Key points

- Your blood glucose concentration is monitored and controlled by your pancreas.
- The pancreas produces the hormone insulin, which allows glucose to move from the blood into the cells and to be stored as glycogen in the liver and muscles.
- 🅗 The pancreas also produces glucagon, which allows glycogen to be converted back into glucose and released into the blood.
- 🅗 Glucagon interacts with insulin in a negative feedback cycle to control glucose levels.
- In type 1 diabetes, the blood glucose may rise to fatally high levels because the pancreas does not secrete enough insulin.
- In type 2 diabetes, the body stops responding to its own insulin.

B11.3 Treating diabetes

Learning objective

After this topic, you should know:

- the differences in the way type 1 and type 2 diabetes are treated.

Before there was any treatment for diabetes, people would waste away. Eventually they would fall into a coma and die.

The treatment of diabetes has developed over the years and continues to improve today. There are now some very effective ways of treating people with diabetes. However, over the long term, even well-managed diabetes may cause problems with the circulatory system, kidneys, or eyesight.

Treating type 1 diabetes

If you have type 1 diabetes, you need replacement insulin before meals. Insulin is a protein that would be digested in your stomach, so it is usually given as an injection to get it into your blood.

This injected insulin allows glucose to be taken into your body cells and converted into glycogen in the liver. This stops the concentration of glucose in your blood from getting too high. Then, as the blood glucose levels fall, the glycogen is converted back to glucose. As a result, your blood glucose levels are kept as stable as possible.

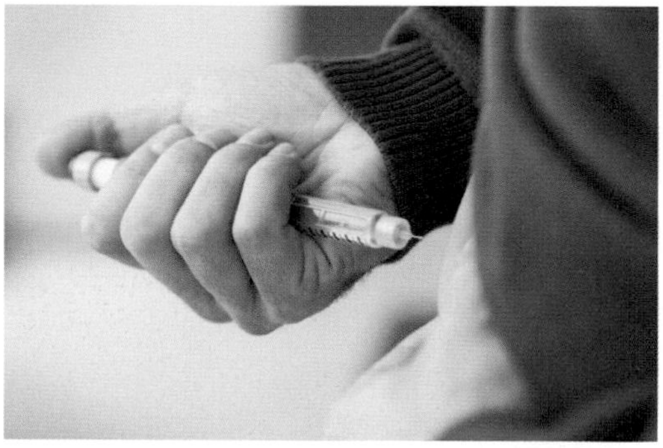

If you have type 1 diabetes, you also need to be careful about the levels of carbohydrate you eat. You need to have regular meals. Like everyone else, you need to exercise to keep your heart and blood vessels healthy. However, taking exercise needs careful planning to keep your blood glucose levels steady. Your cells need enough glucose to respire more rapidly to produce the energy required for your muscles to work.

Insulin injections treat diabetes successfully but they do not cure it. Until a cure is developed, someone with type 1 diabetes has to inject insulin every day of their life.

Curing type 1 diabetes

Scientists and doctors want to find a treatment that means people with diabetes never have to take insulin again.

- Doctors can transplant a pancreas successfully. However, the operations are difficult and risky. Only a few hundred pancreas transplants take place each year in the UK. There are 250 000 people in the UK with type 1 diabetes, but not enough pancreas donors are available. In addition, the patient exchanges one medicine (insulin) for another (immunosuppressants).

- Transplanting the pancreatic cells that make insulin from both dead and living donors has been tried, with limited success so far.

In 2005, scientists produced insulin-secreting cells from embryonic stem cells and used them to cure diabetes in mice. In 2008, UK scientists discovered a completely new technique. Using genetic engineering, they turned mouse pancreas cells that normally make enzymes into insulin-producing cells. Other groups are using adult stem cells from patients with diabetes to try the same technique.

Figure 1 *The treatment of type 1 diabetes involves regular blood glucose tests and insulin injections to keep the blood glucose levels constant*

Synoptic links

You learnt about stem cells and some of the issues surrounding their use in Topic B2.3 and Topic B2.4.

Synoptic link

For more information on the links between obesity and diseases such as type 2 diabetes, look back to Topic B7.4.

Scientists hope that eventually they will be able to genetically engineer faulty human pancreatic cells so that they work properly. Then they will be able to return them to the patient with no rejection issues. It still seems likely that the easiest cure will be to use stem cells from human embryos that have been specially created for the process. However, for some people, this is not ethically acceptable.

Much more research is needed. However, scientists hope that type 1 diabetes will soon be an illness they can cure rather than just manage.

Treating type 2 diabetes

Type 2 diabetes is linked to obesity, lack of exercise, and old age. If you develop the disease, your body cells no longer respond to any insulin made by the pancreas. This can often be treated without needing to inject insulin. Many people can restore their normal blood glucose balance by taking three simple steps:

- eating a balanced diet with carefully controlled amounts of carbohydrates
- losing weight
- doing regular exercise.

If this doesn't work, there are drugs that:

- help insulin work better on the body cells
- help your pancreas make more insulin
- reduce the amount of glucose you absorb from your gut.

If none of these treatments work, you will probably need insulin injections.

Type 2 diabetes usually affects older people. However, it is becoming more and more common in young people who are very overweight.

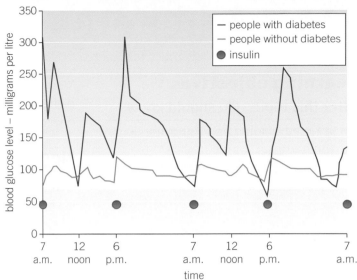

Figure 2 *The blood glucose levels of people with and without diabetes over two days. The yellow band shows normal blood glucose levels and the peaks show the effect of food intake. Insulin injections cannot mimic natural control, but enable people with diabetes to live active lives.*

Figure 3 *Losing weight and taking exercise are simple ways to help overcome type 2 diabetes*

1 State three differences between type 1 diabetes and type 2 diabetes. [3 marks]

2 It is a common misconception that diabetes is treated only by using insulin injections.
   a Explain why this is not always true for people with type 1 diabetes. [3 marks]
   b Explain why treatment with insulin injections is relatively uncommon for people with type 2 diabetes. [3 marks]

3 Transplanting a pancreas to replace natural insulin production seems to be the ideal treatment for type 1 diabetes. Compare this treatment with insulin injections and explain why it is not more widely used. [4 marks]

4 Explain the different methods used to treat type 1 diabetes and type 2 diabetes, linking these methods to how the types of diabetes are caused. ✏ [6 marks]

Key points

- Type 1 diabetes is normally controlled by injecting insulin to replace the hormone that is not made in the body.
- Type 2 diabetes is often treated by a carbohydrate-controlled diet and taking more exercise. If this doesn't work, drugs may be needed.

B11.4 The role of negative feedback

Learning objectives

After this topic, you should know:

- what adrenaline and thyroxine do in the body
- the importance of negative feedback systems.

Many hormones in your body are controlled as part of negative feedback systems. These involve the coordination of changes in the internal environment of your body with the amounts of hormone produced.

Negative feedback

Put simply, negative feedback systems work to maintain a steady state.

If a factor in the internal environment increases, changes take place to reduce it and restore the original level.

If a factor in the internal environment decreases, changes take place to increase it and restore the original level.

Whatever the initial change, in negative feedback the response causes the opposite (Figure 1). The principle is easier to understand when you see working examples. Many hormones are involved in negative feedback systems, including insulin and glucagon, most female sex hormones, and thyroxine (see below).

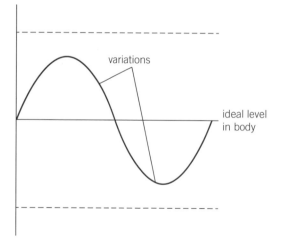

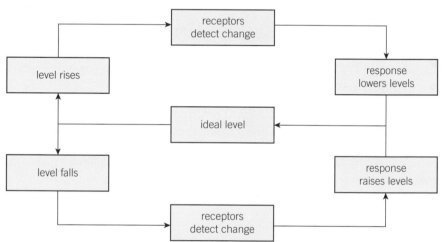

Figure 1 *A negative feedback loop means values will vary around a normal level within a limited range*

Thyroxine and negative feedback

The thyroid gland in your neck uses iodine from your diet to produce the hormone thyroxine. This controls the basal metabolic rate of your body – how quickly substances are broken down and built up, how much oxygen your tissues use, and how the brain of a growing child develops. Thyroxine plays an important part in growth and development. In adults the level of thyroxine in the blood usually remains relatively stable. This happens as a result of a negative feedback control involving the pituitary gland and the hormone it produces – thyroid stimulating hormone or TSH.

If levels of thyroxine in the blood begin to fall, it is detected by sensors in the brain. As a result, the amount of TSH released from the pituitary gland increases. This is a negative feedback system. TSH stimulates the production of thryoxine by the thyroid gland. As the level of thyroxine goes up, it is detected by the sensors and in turn the level of TSH released falls.

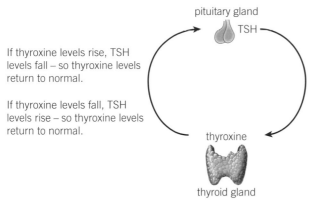

If thyroxine levels rise, TSH levels fall – so thyroxine levels return to normal.

If thyroxine levels fall, TSH levels rise – so thyroxine levels return to normal.

Figure 2 *The levels of thyroxine in your body are kept within narrow boundaries by a negative feedback loop*

Adrenaline

Not all hormones are involved in such clear-cut negative feedback systems. If you are stressed, angry, excited, or frightened your body needs to be ready for action. Your adrenal glands, located at the top of your kidneys, secrete lots of adrenaline that is carried rapidly around the body in your blood, affecting lots of organs. Adrenaline causes:

- your heart rate and breathing rate to increase
- stored glycogen in the liver to be converted to glucose for respiration
- the pupils of your eyes to dilate to let in more light
- your mental awareness to increase
- blood to be diverted away from your digestive system to the big muscles of the limbs.

Adrenaline boosts the delivery of oxygen and glucose to your brain and muscles, preparing your body for flight or fight. Once the danger is over, the raised levels of awareness are no longer needed. The adrenal glands stop releasing adrenaline and your systems return to their resting levels. This does not involve a negative feedback loop.

1 Describe how a negative feedback system works. [4 marks]

2 Compare the role of thyroxine in the body and the way it is controlled with the role and control system for adrenaline. [4 marks]

3 In Ethiopia up to 40% of the population, both adults and children, do not get enough iodine in their diet. Around 40% of the population also have non-communicable diseases caused by low thyroxine levels.
   a Explain why these two facts are linked. [3 marks]
   b Suggest a simple way of overcoming the problems of diseases linked to low thyroxine levels. [2 marks]

Key points

- Thyroxine from the thyroid gland stimulates the basal metabolic rate. It plays an important role in growth and development.
- Adrenaline is produced by the adrenal glands in times of fear or stress. It increases the heart rate and boosts the delivery of oxygen and glucose to the brain and muscles, preparing the body for 'fight or flight'.
- Thyroxine is controlled by negative feedback whereas adrenaline is not.

B11.5 Human reproduction

Learning objectives

After this topic, you should know:

- the main human reproductive hormones
- how hormones control the changes at puberty.

The big physical changes that make boys and girls look very different take place at the time of puberty. This is when the reproductive organs become active and the body takes on its adult form. Hormones play an important part in human reproduction at every stage.

Hormones and puberty

During puberty, the reproductive hormones control the development of the secondary sexual characteristics. The primary sexual characteristics are the ones you are born with. Primary sex characteristics are the ovaries in girls and the testes in boys. These reproductive organs produce the female and male sex hormones and the special sex cells or gametes that join together in reproduction. To understand the changes that take place in puberty it helps to understand the basic female and male anatomy (Figure 1 and Figure 2).

The timing of puberty and the order and rate of the changes varies but the basic changes are the same for everyone.

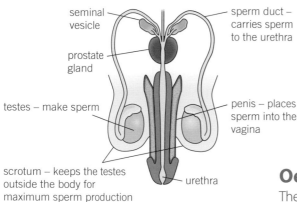

Figure 1 *Male reproductive organs*

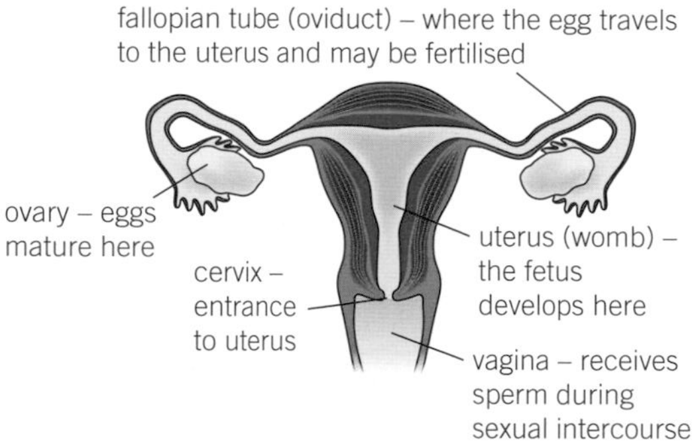

seminal vesicle

sperm duct – carries sperm to the urethra

prostate gland

testes – make sperm

penis – places sperm into the vagina

scrotum – keeps the testes outside the body for maximum sperm production

urethra

Oestrogen and puberty in females

The main female reproductive hormone is **oestrogen**, produced by the **ovaries**. Rising oestrogen levels trigger the development of the female secondary sexual characteristics usually between the ages of 8 and 14 years. The main changes include a growth spurt; the growth of hair under the arms and pubic hair; the breasts develop; the external genitals grow and the skin darkens; a female pattern of fat is deposited on the hips, buttocks and thighs; the brain changes and matures; mature ova start to form every month in the ovaries, the uterus grows and becomes active and menstruation begins.

The menstrual cycle

Once a girl has gone through puberty she will have a monthly menstrual cycle. Each month, eggs begin to mature in the ovary. At the same time the uterus produces a thickened lining ready for a pregnancy. Every 28 days a mature egg is released. This is called **ovulation**. If the egg is not fertilised, around 14 days later the lining of the uterus is shed along with the egg. This is the monthly period and this also takes place at approximately 28-day intervals. Several hormones are involved in controlling the menstrual cycle:

fallopian tube (oviduct) – where the egg travels to the uterus and may be fertilised

ovary – eggs mature here

cervix – entrance to uterus

uterus (womb) – the fetus develops here

vagina – receives sperm during sexual intercourse

Figure 2 *The female reproductive system*

- follicle stimulating hormone (FSH) causes the eggs in the ovary to mature (the eggs grow surrounded by cells called the follicle)
- luteinising hormone (LH) stimulates the release of the egg at ovulation
- oestrogen and progesterone stimulate the build-up and maintenance of the uterus lining.

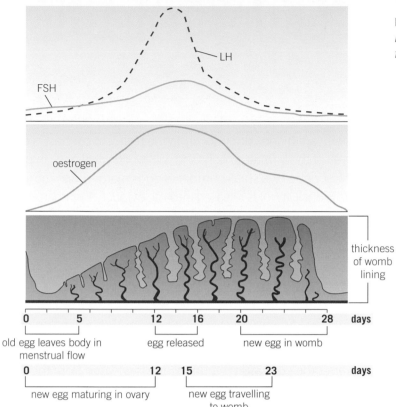

Figure 3 *The changing levels of the female sex hormones control the different stages of the menstrual cycle*

Female fertility

The ovaries of a baby girl contain all the eggs she will ever have. After puberty, eggs mature and are released every month, for an average of 35–40 years, except if she is pregnant. Eventually the supply of eggs runs out and the woman goes through the menopause – she can no longer have children. Approaching the menopause, a woman is less fertile and has a higher risk of having a baby with genetic problems.

Testosterone and puberty in males

The main male reproductive hormone is **testosterone**, produced by the testes. As levels of testosterone rise, all kinds of changes are triggered and the male secondary sexual characteristics develop. Boys usually go into the first stages of puberty slightly later than girls, between the ages of 9 and 15. The main changes include a growth spurt; pubic hair, underarm hair and facial hair grow; the larynx gets bigger and the voice breaks; the external genitalia and the skin darkens; the testes grow and become active, producing sperm throughout life; the shoulders and chest broaden as muscle develops; and the brain matures.

1 State three reasons why hormones are important in human reproduction. [3 marks]

2 Describe three similarities and three differences in puberty between boys and girls. [6 marks]

3 a Describe the role of hormones in the menstrual cycle. [4 marks]
  b State the differences between the production of mature eggs in women and mature sperm in men. ✐ [6 marks]

B11.6 Hormones and the menstrual cycle

Learning objectives

After this topic, you should know:

- the roles of hormones in human reproduction
- how hormones interact to control the menstrual cycle.

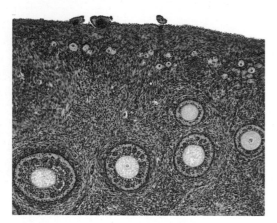

Figure 1 *You can see the ova (eggs) developing in the follicles in this light micrograph of a piece of ovarian tissue*

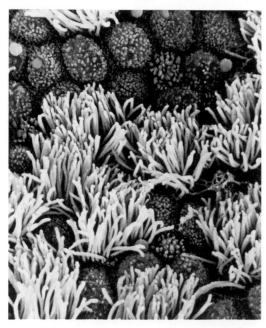

Figure 2 *The egg cannot move itself but the cilia of the fallopian tubes beat to move it along to the uterus (magnification: x2100)*

The levels of testosterone in a man's body remain relatively constant after puberty, and sperm are made continuously in the testes. In contrast, a baby girl is born with all of the immature eggs she will ever have already in place in her ovaries. After puberty, a small number of eggs mature each month until one (or occasionally two) is released into the oviduct. A woman's menstrual cycle is a good example of control by hormones. The levels of different hormones made in the pituitary gland and ovaries rise and fall in a regular pattern, affecting the way the body of the woman works.

The menstrual cycle

The average length of the menstrual cycle is about 28 days. Each month the lining of the uterus (womb) thickens ready to support a developing baby. At the same time, eggs start maturing in the follicles of the ovary.

About 14 days after the eggs start maturing, one is released from the ovary in ovulation. The lining of the uterus stays thick for several days after the egg has been released.

If the egg is fertilised by a sperm, then pregnancy may take place. The lining of the uterus provides protection and food for the developing embryo. If the egg is not fertilised, about 14 days after ovulation the lining of the uterus and the egg are shed from the body in the monthly period.

Control of the menstrual cycle

The complex events of the menstrual cycle are coordinated by the interactions of the four different hormones given in Topic B11.5. Once a month, a surge of hormones from the pituitary gland in the brain starts egg maturation in the ovaries. The hormones also stimulate the ovaries to produce the female sex hormone oestrogen.

Follicle stimulating hormone (FSH) is secreted by the pituitary gland. It makes eggs mature in their follicles in the ovaries. It also stimulates the ovaries to produce hormones including oestrogen.

Oestrogen is made and secreted by the ovaries in response to FSH. It stimulates the lining of the uterus to grow again after menstruation in preparation for pregnancy. High levels of oestrogen inhibit the production of more FSH and stimulate the release of LH.

Luteinising hormone (LH) from the pituitary gland stimulates the release of a mature egg from the ovary. Once ovulation has taken place, LH levels fall again.

Progesterone is secreted by the empty egg follicle in the ovary after ovulation. It is one of the hormones that helps to maintain a pregnancy if the egg is fertilised. Progesterone inhibits both FSH and LH and it maintains the lining of the uterus in the second half of the cycle, so it is ready to receive a developing embryo if the egg is fertilised.

The hormones produced by the pituitary gland and the ovary act together to control what happens in the menstrual cycle. As the oestrogen levels rise, they inhibit the production of FSH and encourage the production of LH by the pituitary gland. When LH levels reach a peak in the middle of the cycle, they stimulate the release of a mature egg.

FSH and LH are then suppressed and the body is kept ready for pregnancy until it becomes clear that the egg is not fertilised.

The levels of all the hormones then drop and the lining of the uterus pulls away and is lost from the body. At this stage a new cycle begins and the levels of FSH and oestrogen start to build up again.

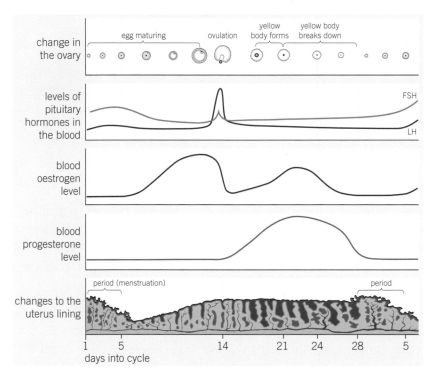

Figure 3 *The main events of the menstrual cycle*

1   a   State the four hormones that control the menstrual cycle.   [4 marks]
    b   State why the lining of the uterus builds up each month.   [1 mark]

2   Look at Figure 3 and answer the following questions.
    a   On which days is the woman having a menstrual period?   [1 mark]
    b   Explain what happens to the levels of FSH and LH during the cycle.   [4 marks]
    c   Name the hormone that controls the build up of the lining of the uterus.   [1 mark]

3   Explain the main events of the menstrual cycle, identifying the stage where a woman is most likely to become pregnant.   [6 marks]

B11.7 The artificial control of fertility

Learning objectives

After this topic, you should know:

- a number of different methods of hormonal and non-hormonal contraception.

To prevent pregnancy, you need to prevent the egg and sperm meeting or a fertilised egg implanting in the uterus. This is known as **contraception**. The different methods all have advantages and disadvantages.

Hormone-based contraception

Scientists have worked out a number of different ways to use the hormones of the menstrual cycle, or synthetic versions of them, to prevent pregnancy.

Oral contraceptives, often referred to as the contraceptive pill, use female hormones to prevent pregnancy. The mixed pill contains low doses of oestrogen along with some progesterone. The hormones inhibit the production and release of FSH by the pituitary gland, affecting the ovaries, so no eggs mature, preventing pregnancy. The pill hormones also stop the uterus lining developing, preventing implantation. They also make the mucus in the cervix thick to prevent sperm getting through. The contraceptive pill is easy to use but there is a slight risk of side effects, including raised blood pressure, thrombosis, and breast cancer.

Some contraceptive pills contain only progesterone. They have fewer side effects than the mixed pill. Women must take the pill very regularly, especially the progesterone-only pill. If they don't, the artificial hormone levels drop and their body's own hormones can take over very quickly. This can lead to the unexpected release of an egg – and an unexpected baby!

There are alternative ways of delivering hormones to prevent pregnancy. A contraceptive implant can last up to three years. A tiny tube is inserted under your skin by a doctor and slowly releases progesterone. This is 99.95% effective! Contraceptive injections also use progesterone but they only last about 12 weeks.

The contraceptive patch, like the mixed pill, contains a mixture of oestrogen and progesterone. You stick the patch to your skin, replacing it every 7 days, and the hormones are absorbed directly into your blood stream. All of these methods prevent an egg maturing and being released, as well as affecting the uterus lining and the mucus in the cervix.

Figure 1 *Different methods of hormonal contraception*

Chemical methods

Chemicals that kill or disable sperm are known as spermicides. They are readily available but are not very effective at preventing pregnancy.

Barrier methods

Barrier methods of contraception prevent the sperm reaching the egg. A condom is a thin latex sheath placed over the penis during intercourse to collect the semen and prevent the egg and sperm meeting. They have no side effects and do not need medical advice. Condoms offer some protection against sexually transmitted diseases such as syphilis and HIV/AIDS. They can, however, get damaged and let sperm through.

A diaphragm or cap is a thin rubber diaphragm placed over the cervix before sex to prevent the entry of sperm. Like condoms, they have no side effects but must be fitted by a doctor initially. If the cap is not positioned correctly, sperm may get past and reach the egg. Barrier methods work better combined with spermicide.

Intrauterine devices

Intrauterine devices are small structures inserted into the uterus by a doctor. They last for 3–5 years although they can be removed at any time if you want to get pregnant. Some intrauterine devices contain copper and prevent any early embryos implanting in the lining of the uterus. Others contain progesterone, releasing it slowly to prevent the build-up of the uterus lining and to thicken the mucus of the cervix. They are very effective but they may cause period problems or infections.

Abstinence

If people do not have sex they will not get pregnant. Some religious groups do not accept the use of artificial methods of contraception. Abstaining from intercourse around ovulation or when an egg is in the oviduct means sperm cannot fertilise the egg – this is known as the rhythm method. This method has no side effects but it is very unreliable. Ovulation indicators make it more effective.

Surgical methods

If people do not want any more children, they can be surgically sterilised. In men the sperm ducts are cut and tied, preventing sperm getting into the semen. This is called a vasectomy.

In women the oviducts are cut or tied to prevent the egg reaching the uterus and the sperm reaching the egg. Although this gives effective, permanent contraception with no risk of human error, women need a general anaesthetic for the surgery.

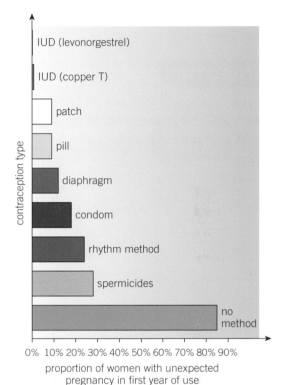

Figure 2 *Failure rates for different methods of birth control based on data from the American Academy of Paediatrics*

1 Define the term contraception. [1 mark]

2 The progesterone-only pill, the contraceptive implant, and the contraceptive patch are all forms of contraceptive.
  a State one similarity between each contraceptive. [1 mark]
  b For each contraceptive, state one way it is different. [3 marks]
  c Explain which of these three methods is likely to be the most effective contraceptive and why. [3 marks]

3 Compare the effectiveness of the three main types of contraception – hormone-based contraception, barrier methods, and surgical methods. [3 marks]

Key points

- Fertility can be controlled by a number of hormonal and non-hormonal methods of contraception.
- Contraceptive methods include oral contraceptives, hormonal injections, implants and patches, barrier methods such as condoms and diaphragms, intrauterine devices, spermicidal agents, abstinence, and surgical sterilisation.

B11.8 Infertility treatments

Learning objectives

After this topic, you should know:

- how hormones can be used to treat infertility.

Study tip

A combination of FSH and LH can be used in fertility treatments to cause eggs to mature in the ovary and to trigger ovulation.

In the UK as many as one couple in six have problems having a family when they want one. There are many possible reasons for this infertility. It may be linked to a lack of female hormones, to damaged oviducts, or to a lack of sperm in the semen. About a third of cases of infertility are due to problems with the female reproductive system, about a third due to the male system, and about a third are hard to explain, with both partners being a bit less fertile than normal. Common causes of infertility are obesity and eating disorders such as anorexia nervosa, but one of the most common causes of infertility is age. Increasingly, couples wait until they are in their late thirties to have children, and then find that they cannot conceive naturally.

Lack of ovulation

Some women want children but do not make enough FSH to stimulate the maturation of the eggs in their ovaries. Fortunately, artificial FSH can be used as a fertility drug. It stimulates the eggs in the ovary to mature and also triggers oestrogen production. An artificial form of LH can then be used to trigger ovulation. If a woman who is not ovulating as a result of a lack of her own FSH is treated in this way, she may be able to get pregnant naturally. In the early days of using fertility drugs there were big problems with the doses used. In 1971, an Italian doctor removed 15 four-month-old foetuses (ten girls and five boys) from the womb of a 35-year-old woman after treatment with fertility drugs. Not one of them survived. Now the doses are much more carefully controlled and most people using fertility drugs end up with just one or two babies.

In vitro fertilisation

Fertility drugs are also used in IVF (*in vitro* fertilisation). IVF is a form of fertility treatment used if the oviducts have been damaged or blocked by infection, if a donor egg has to be used, or if there is no obvious cause for long-term infertility.

In some cases a man produces very few sperm or the sperm do not mature properly. Individual sperm may be injected into an egg during the IVF process.

In all these cases it would not be possible to get pregnant naturally.

Fortunately, doctors can now help.

- They give the mother synthetic FSH to stimulate the maturation of a number of eggs at the same time, followed by LH to bring the eggs to the point of ovulation.

- They collect the eggs from the ovary of the mother and fertilise them with sperm from the father outside the body in the laboratory.

- The fertilised eggs are kept in special solutions in a warm environment to develop into tiny embryos.

- At the stage when they are minute balls of cells, one or two of the embryos are inserted back into the uterus of the mother. In this way they bypass the faulty tubes.

Table 1 *Data from 2010 showing the decreasing success rate of IVF as the mother gets older*

Age of mother	IVF % success rate
Under 35	32.2
35–37	27.7
38–39	20.8
40–42	13.6
43–44	5
Over 44	1.9

Modern infertility treatments such as these rely on advanced microscopy techniques. It takes a high level of manipulative skill and a high magnification to work on single eggs and sperm, or early embryos, without damaging them.

The advantages and disadvantages of fertility treatment

The use of hormones to help overcome infertility has been a major scientific breakthrough. It gives women and men who would otherwise be infertile the chance to have a baby of their own. But like most things there are advantages and disadvantages. Here are some points to think about:

IVF is expensive both for society, if it is provided by the National Health Service (NHS), and for individuals – many people end up paying thousands of pounds for repeated cycles of treatment.

It is not always successful. The older the parents, the less likely it is that they will have a baby (Table 1). Using donor eggs from younger women or donor sperm from younger men can help the success rate but then the baby is not biologically the parents' child.

The use of fertility drugs can have some health risks for the mother. The process of IVF is very emotionally and physically stressful.

IVF increases the chances of a multiple pregnancy. On average, 1 in 5 IVF pregnancies is a multiple pregnancy. The figure for natural pregnancies is 1 in 80. Multiple births increase the risks for both mothers and babies, and are more likely to lead to stillbirths and other problems including very premature births. It costs hospitals a lot of money to keep very small premature babies alive and if they survive many will have permanent and often severe disabilities.

The mature eggs produced by a woman using fertility drugs may be collected and stored, or fertilised and stored, until she wants to get pregnant later. But this raises ethical problems if the woman dies, the relationship breaks up, or one of the parents no longer wants the eggs or embryos.

1 Fertility drugs are used to make lots of eggs mature at the same time for collection.

2 The eggs are collected and placed in a special solution in a Petri dish.

3 A sample of semen is collected and the sperm and eggs are mixed in the Petri dish.

4 The eggs are checked to make sure they have been fertilised and the early embryos are developing properly.

5 When the fertilised eggs have formed tiny balls of cells, 1 or 2 of the tiny embryos are placed in the womb of the mother. Then, if all goes well, at least one baby will grow and develop successfully.

Figure 1 *New reproductive technology using hormones and IVF helps thousands of otherwise infertile couples to have babies each year*

1 What is IVF? [2 marks]

2 Explain how artificial female hormones can be used to:
   a help people overcome infertility and conceive naturally [3 marks]
   b help people overcome infertility and conceive through IVF. [3 marks]

3 a Draw a graph to show the effect of age on the chances of a woman having a baby successfully using IVF. [4 marks]
   b Some people think that IVF treatment should not be offered to people over the age of 40. Suggest arguments for and against this idea. [6 marks]

4 Suggest and explain some advantages and disadvantages of using artificial hormones to control female fertility? [6 marks]

Key points

- FSH and LH can be used as a fertility drug to stimulate ovulation in women with low FSH levels.
- In vitro fertilisation (IVF) uses FSH and LH to stimulate maturation of ova that are collected, fertilised, allowed to start development, and replaced in the uterus.
- IVF is emotionally and physically stressful, often unsuccessful, and can lead to risky multiple births.

B11 Hormonal coordination

Summary questions

1 Figure 1 shows the blood glucose levels of a person without diabetes and someone with type 1 diabetes managed with regular insulin injections. They both eat at the same times.

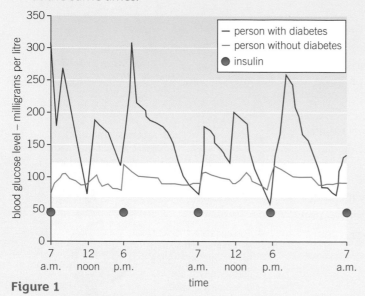

Figure 1

Use Figure 1 to help you answer these questions:
a What happens to the blood glucose levels of both individuals after eating? [1 mark]
b What is the range of blood glucose concentration of the person without diabetes? [1 mark]
c What is the range of blood glucose concentration of the person with diabetes? [1 mark]
d Figure 1 shows the effect of regular insulin injections on the blood glucose level of someone with diabetes. Why are the insulin injections important to their health and wellbeing? What limitations are suggested by the data? [4 marks]
e People with diabetes have to monitor the amount of carbohydrate in their diet. Explain why. [4 marks]

2 **a** Describe the main events of the menstrual cycle. [6 marks]
b 🄗 Explain the role of the following hormones in the menstrual cycle:
i FSH [2 marks]
ii oestrogen [3 marks]
iii LH [2 marks]
iv progesterone. [2 marks]

3 **a** What are secondary sexual characteristics? [2 marks]
b State three changes at puberty that are common to both boys and girls. [3 marks]
c State two changes that occur during puberty that are unique to girls. [2 marks]
d State two changes that occur during puberty that are unique to boys. [2 marks]
e Which hormone is important in the development of the secondary sexual characteristics in both females and males. Explain its role. [5 marks]
f 🄗 Explain how the hormones interact to bring about the main changes in the body during the menstrual cycle. [6 marks]

4 **a** Explain how artificial hormones can be used in pills to prevent people getting pregnant. [4 marks]
b Describe three other ways contraceptive hormones can be given and give one advantage for each method. [6 marks]
c 🄗 Explain how artificial hormones can be used to help treat infertility. [4 marks]
d 🄗 Compare the way artificial hormones are used in contraception and infertility treatment. [4 marks]

5 🄗 Many hormone systems in the body are controlled by negative feedback systems.
a Using the control of thyroxine levels in the body as an example, describe how a negative feedback system works. [5 marks]
b Explain why negative feedback control is so important in maintaining homeostasis. [5 marks]
c Not all hormones are controlled by negative feedback systems. Explain the role of adrenaline in the response of your body to stress. [6 marks]
d Discuss the importance of the lack of a negative feedback control system for adrenaline. [5 marks]

Practice questions

01 **Table 1** shows the changes in the blood glucose concentration of a person with type 1 diabetes after they ate a meal.

Table 1

Time in hours	Blood glucose concentration in units
0.0	8.0
0.5	15.0
1.0	18.0
1.5	19.4
2.0	17.6

01.1 Plot the data in **Table 1** as a line graph. Join the points with straight lines. [5 marks]

01.2 In a person who does not have diabetes, the blood glucose concentration would be between 3.5 and 5.5 units before meals and below 8 units two hours after eating a meal.
Use this information to sketch a line on your graph to predict the changes in blood glucose concentration for a person who ate the same meal, but who did not have diabetes. [3 marks]

01.3 Explain what causes the difference in the blood glucose concentrations of someone with type 1 diabetes compared to someone who does not have type 1 diabetes. [2 marks]

02 Fertility can be controlled by a variety of hormonal and non-hormonal methods of contraception.
Table 2 shows the pregnancy rate for some different methods of contraception. Pregnancy rate is the percentage of women who get pregnant in the first year of use.
Figures are given for perfect use, which is under medical study, and typical use, which is general everyday use.

Table 2

Contraception method	Pregnancy rate for perfect use	Pregnancy rate for typical use
no method	85.0	85.0
condom	2.0	17.4
hormone releasing intrauterine device (IUD)	0.1	0.1

Table 3 shows some information about condoms and a hormone-releasing intrauterine device (IUD).

Table 3

Condom	Hormone IUD
A condom is a rubber sheath that covers the erect penis. Condoms can be bought in supermarkets.	The IUD releases a hormone similar to progesterone. It will release the hormone for about five years and then has to be replaced. It is inserted and removed by a nurse or doctor. It is usually recommended for use by women who have had a child.

Evaluate the use of condoms and hormone releasing IUDs for contraception.
You should include how each method works as well as their advantages and disadvantages. [6 marks]

03 **Ⓗ** **Figure 1** shows the organs in the endocrine system.

Figure 1

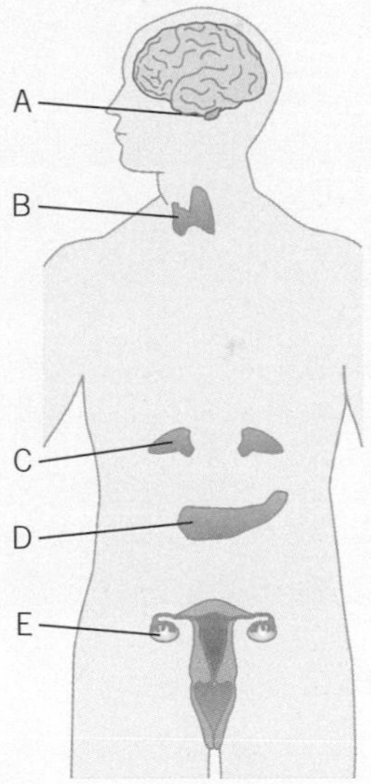

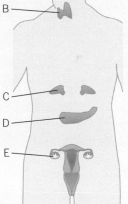

03.1 Thyroxine is produced by the thyroid gland. Which gland in **Figure 1** is the thyroid gland? [1 mark]

03.2 Describe **one** function of thyroxine in the body. [1 mark]

Figure 2 shows how the concentration of thyroxine in the blood is controlled by a negative feedback mechanism.

Figure 2

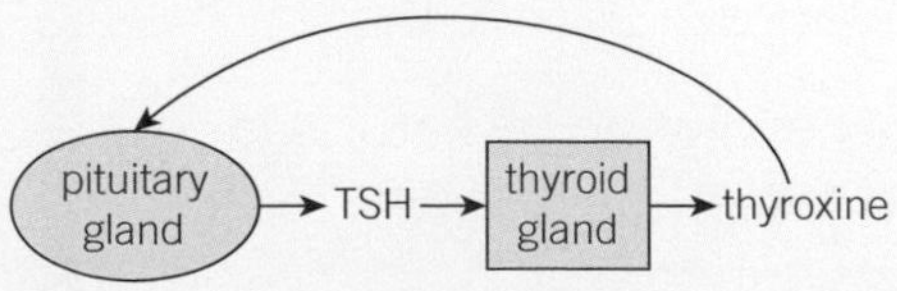

pituitary gland → TSH → thyroid gland → thyroxine

03.3 Use the information given in **Figure 2** to explain how the concentration of thyroxine in the blood is kept within a normal range. [6 marks]

4 Genetics and evolution

When the science of genetics was first developed, people hadn't even seen a chromosome. Now, in the 21st century, scientists can analyse the entire genome of an organism in a day or two. You will learn how the information in your genetic code controls the way the chemicals that make up your cells, tissues, and organs are built up. You will also consider some of the new gene technologies that scientists are using.

The organisms alive today have evolved from ancestral organisms over millions of years. You will explore how this has occurred and study examples of evolution in progress today. You will also see how our increasing knowledge of genomes allows scientists to classify organisms in a different way, allowing us to make sense of global biodiversity.

Key questions

- What is DNA, what is a genome, and why is it so important to be able to analyse the genome of an organism?

- How are characteristics passed from parents to their offspring?

- What is genetic engineering and what are the potential benefits and disadvantages of this technology?

- How does evolution by natural selection take place and why are mutations important?

Making connections

- You learnt about mitosis along with cell growth and differentiation in **B2 Cell division**.

- You learnt about antibiotics and their importance in treating bacterial diseases in **B6 Preventing and treating disease**.

- You discovered the importance of protein structure to protein function in **B3 Organisation and the digestive system**.

- You will learn the affect of mutations and natural selection on the adaptations of organisms to their environment in **B15 Adaptations, interdependence, and competition**.

- You will see how humans are changing their environment and driving extinction and changes in existing species in **B17 Biodiversity and ecosystems**.

I already know...

About the nucleus of the cell and the chromosomes it contains.

About mitosis and the cell cycle.

The process of reproduction.

How inheritance works.

How biological ideas develop.

About the characteristics of eukaryotic and prokaryotic cells, and the differences between animal, bacterial, and plant cells.

I will learn...

About the DNA that makes up the chromosomes, about the variants of the genes known as alleles, and how all the DNA of an organism can be analysed.

About meiosis in cell division and the formation of gametes.

How information is passed from one generation to another and how to use genetic diagrams, direct proportion, simple ratios, and probability to predict the outcome of a genetic cross.

About the importance of selective breeding in the development of plants and animals and the increasing use of genetic engineering to introduce desirable characteristics

How Charles Darwin built up the evidence for his theory of evolution by natural selection and some of the barriers to the acceptance of his ideas, as well as some of the modern evidence we have for evolution.

About new DNA-based systems for classifying organisms.

Learning objectives

After this topic, you should know:

- the main differences between asexual and sexual reproduction

Reproduction is essential to living things. It is during reproduction that genetic information in the chromosomes is passed on from parents to their offspring. There are two very different ways of reproducing – **asexual reproduction** and **sexual reproduction**.

Asexual reproduction

Asexual reproduction only involves one parent. The cells divide by mitosis. There is no joining (fusion) of special sex cells (gametes) and so there is no mixing of genetic information. As a result there is no variation in the offspring. Asexual reproduction gives rise to genetically identical offspring known as clones. Their genetic material is identical both to the parent and to each other. Only mitosis is involved in asexual reproduction.

Asexual reproduction is very common in the smallest animals and plants, and in fungi and bacteria. However, many larger plants, such as daffodils, strawberries, and brambles, can also reproduce asexually. The cells of your body reproduce asexually all the time. They divide into two identical cells for growth and to replace worn-out tissues.

Synoptic links

To remind yourself about chromosomes, genes and DNA and the role they play in your cells, as well as the way cells divide by mitosis, look back to Topic B2.1.

Figure 1 *Yeast are single-celled fungi that reproduce asexually. The cells divide by mitosis to form a mass of identical cells*

Sexual reproduction

Sexual reproduction involves a male sex cell and a female sex cell from two parents. These two special sex cells (gametes) fuse together to form a zygote, which goes on to develop into a new individual. Gametes are formed in a special form of cell division known as **meiosis**. The chromosome number of the original cell is halved, so that when gametes join together, the new cell has the right number of chromosomes. You will find out more about meiosis in Topic B12.2.

The offspring that result from sexual reproduction inherit genetic information from both parents. They will have some characteristics from both parents, but won't be identical to either of them. This introduces variation. The offspring of sexual reproduction show much more variation than the offspring from asexual reproduction.

● In plants, the gametes are the egg cells and pollen.

● In animals, the gametes are the egg cells (ova) and sperm.

Sexual reproduction is risky because it relies on two sex cells, often from two individuals, meeting and fusing. However, it also introduces variation, which is key to the long-term survival of a species. That's why sexual reproduction takes place in living things ranging from single-celled organisms to humans.

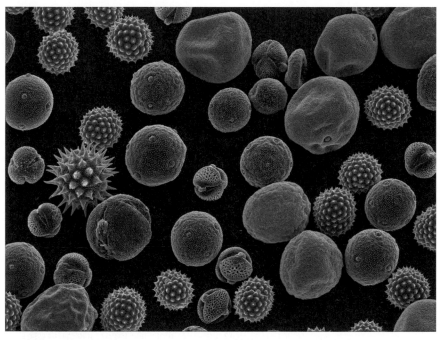

Figure 2 *Pollen, the male gametes in plants, are produced by meiosis. Each pollen grain contains half the genetic material of the parent cells*

1 Define the following:
   a asexual reproduction   [1 mark]
   b sexual reproduction   [1 mark]
   c gamete   [1 mark]
   d variation.   [1 mark]

2 Explain simply why the offspring of organisms that reproduce asexually are identical to their parents but the offspring of organisms that reproduce sexually are not.   [5 marks]

3 The conditions in many different environments are changing as a result of global warming. One theory is that this will affect organisms that reproduce asexually more than it affects organisms that reproduce sexually. Discuss this idea, using your knowledge of the differences between asexual and sexual reproduction. ⬤   [6 marks]

B12.2 Cell division in sexual reproduction

Learning objectives

After this topic, you should know:

- how cells divide by meiosis to form gametes
- how meiosis halves the number of chromosomes in gametes and fertilisation restores the full number
- how sexual reproduction gives rise to variation.

Mitosis takes place all the time, in tissues all over your body and whenever organisms reproduce asexually. There is, however, another type of cell division that takes place in the reproductive organs of animals and plants. In humans, these organs are the ovaries and the testes. **Meiosis** results in sex cells, called gametes, which have only half the original number of chromosomes.

Meiosis

In animals, the female gametes (egg cells or ova) are made in the ovaries. The male gametes (sperm) are made in the testes.

The gametes are formed by meiosis. In meiosis, the chromosome number is reduced by half. In a body cell there are two sets of each chromosome, one inherited from the mother and one from the father. When a cell divides to form gametes:

- The genetic information is copied so there are four sets of each chromosome instead of the normal two sets. Each chromosome forms a pair of chromatids. This is very similar to mitosis.

- The cell then divides twice in quick succession to form four gametes, each with a single set of chromosomes (Figure 1).

Each gamete that is produced is genetically different from all the others. Gametes contain random mixtures of the original chromosomes. This introduces variation.

The testes can produce around 400 million sperm by meiosis every 24 hours. Only one sperm is needed to fertilise an egg, but each sperm needs to travel 100 000 times its own length to reach the egg. Less than one sperm in a million actually make it.

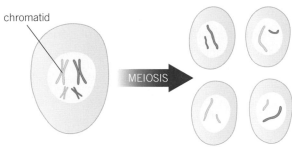

chromatid

MEIOSIS

Figure 1 *The formation of sex cells in the ovaries and testes involves meiosis to halve the chromosome number. The original cell is shown with only two pairs of chromosomes, to make it easier to follow what happens*

Fertilisation

More variation is added when fertilisation takes place. Each sex cell has a single set of chromosomes. When two sex cells join during fertilisation, the single new cell formed has a full set of chromosomes. In humans, the egg cell has 23 chromosomes and so does the sperm. When they join together, they produce a single new body cell with 46 chromosomes in 23 pairs – the correct number of chromosomes for human body cells.

The combination of genes on the chromosomes of every newly fertilised egg is unique. Once fertilisation is complete, the unique new cell begins to divide by mitosis to form a new individual. The number of cells increases rapidly. As the embryo develops, the cells differentiate to form different tissues, organs, and organ systems.

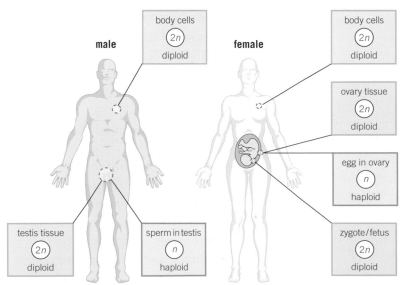

Figure 2 *Meiosis halves the number of chromosomes in the gametes, and fertilisation restores the full number of chromosomes in the body cells*

Variation

The differences between asexual and sexual reproduction are reflected in the different types of cell division involved.

In asexual reproduction, the offspring are produced as a result of mitosis from the parent cells. They contain exactly the same chromosomes and the same genes as their parents. There is no variation in the genetic material.

In sexual reproduction, the gametes are produced by meiosis in the sex organs of the parents. This introduces variation as each gamete is different. Then, when the gametes fuse, one of each pair of chromosomes, and so one of each pair of genes, comes from each parent, adding more variation. The combination of genes in the new pair of chromosomes will contain different forms of the same genes (alleles) from each parent. This also helps to produce variation in the characteristics of the offspring.

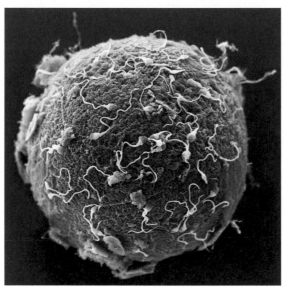

Figure 3 *At the moment of fertilisation, fusion of the gametes means the chromosomes are combined to give a new cell with a complete and unique set of chromosomes*

1 a State how many pairs of chromosomes there are in a normal human body cell. [1 mark]
b State how many chromosomes there are in a human sperm cell. [1 mark]
c State how many chromosomes there are in a fertilised human egg cell. [1 mark]

2 Sexual reproduction results in variation. Describe how this comes about. [4 marks]

3 a Name the special type of cell division that produces gametes from cells in the reproductive organs. Describe clearly what happens to the chromosomes in this process. [4 marks]
b State where in your body this type of cell division would take place. [1 mark]
c Explain why this type of cell division is so important in sexual reproduction. [2 marks]

Synoptic links

To remind yourself of the events of mitosis look at Topic B2.1.

Study tip

Learn to spell mitosis and meiosis.

Remember their meanings:

Mitosis – **m**aking **i**dentical **t**wo.

Meiosis – **m**aking **e**ggs (and sperm).

Key points

- Cells in the reproductive organs divide by meiosis to form the gametes (sex cells).
- Body cells have two sets of chromosomes, gametes have only one set.
- In meiosis, the genetic material is copied and then the cell divides twice to form four gametes, each with a single set of chromosomes.
- All gametes are genetically different from each other.
- Gametes join at fertilisation to restore the normal number of chromosomes. The new cell divides by mitosis. The number of cells increases and as the embryo develops, the cells differentiate.

B12.3 DNA and the genome

Learning objectives

After this topic, you should know:

- about DNA as the material of inheritance
- what a genome is
- some of the benefits of studying the human genome.

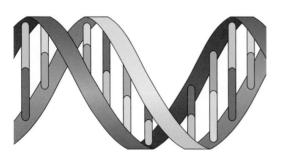

Figure 1 *The DNA double helix*

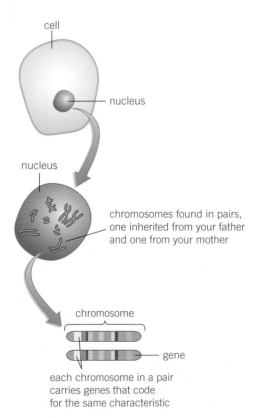

Figure 2 *The relationship between a cell, the nucleus, the chromosomes, and the genes*

Each different type of organism has a different number of chromosomes in their body cells. For example, humans have 46 chromosomes, potatoes have 48, and chickens have 78. In animals and flowering plants, chromosomes always come in pairs. You have 23 pairs of chromosomes in all your normal body cells. You inherit half your chromosomes from your mother and half from your father.

DNA – the molecule of inheritance

Inside the nuclei of all your cells, your chromosomes are made up of long molecules of a chemical known as deoxyribonucleic acid (DNA). DNA is a polymer, a long molecule made up of many repeating units. These very long strands of DNA twist and spiral to form a double helix structure (Figure 1).

Your genes are small sections of this DNA. This is where the genetic information – the coded information that determines inherited characteristics – is actually stored. Each of your chromosomes contains thousands of genes joined together. Each gene codes for a particular sequence of amino acids to make a specific protein. These proteins include the enzymes that control your cell chemistry. This is how the relationship between the genes and the whole organism builds up. The genes control the proteins, which control the make up of the different specialised cells that form tissues. These tissues then form organs and organ systems that make up the whole body.

In 2003, scientists announced that they had managed to sequence the human genome. Working in teams all around the world, the Human Genome Project finished two years early, and under budget. This was because the technology used to chop up the DNA and read all the base sequences had improved so fast during the life of the project. It was a scientific triumph – but why does it matter?

The human genome

The genome of an organism is the entire genetic material of the organism. That includes all of the chromosomes, and the genetic material found in the mitochondria as well. Mitochondria contain their own DNA. You always inherit your mitochondrial DNA from your mother because it comes from the mitochondria in the egg.

The human genome contains over 3 billion base pairs and almost 21 000 genes that code for proteins. That sounds a lot until you discover that rice has 36 000 coding genes! We are not simpler than rice. The human genome has the ability to make many different proteins from the same gene by using it in different ways, or by switching part of a gene on or off.

Since the initial human genome was read, scientists have carried on with the work. They went on to sequence the genomes of 1000 people, and now they are busy with the current 100 000 genomes project. The aim is to find out as much as possible about human DNA.

It isn't just human genomes that are sequenced. Scientists are sequencing the genomes of hundreds of different species of organisms. They use similarities and differences in the genomes to help them work out the relationships between different types of organisms. It is changing the way we classify living things. Sequencing the genomes of bacteria and viruses allows us to identify the causes of disease very rapidly and to choose the correct treatment.

Why does the genome matter?

Understanding the human genome has taken years of work and billions of pounds – and there is still a great deal to find out. There are many reasons why scientists feel that all of this effort is worthwhile.

Understanding the human genome helps us to understand inherited disorders such as cystic fibrosis and sickle cell disease. The more we can understand what goes wrong in these diseases, the more chance we have of overcoming them either through medicines or by repairing the faulty genes.

There are genes that are linked to an increased risk of developing many diseases, from heart disease to type 2 diabetes. Understanding the human genome is playing a massive part in the search for genes linked to different types of diseases. The more we understand about the genome, the more likely we are to predict the risk for each individual, so they can make lifestyle choices to help reduce the risks. This includes the changes that happen in the genome when a cancer develops. By analysing the genomes of cancer cells, scientists and doctors hope to become even better at choosing the best treatment for each individual.

Understanding the human genome helps us understand human evolution and history. People all over the world can be linked by patterns in their DNA, allowing scientists to trace human migration patterns from our ancient history. We can also be linked to early members of the human family tree. For example, most people have a small number of Neanderthal genes in their DNA, even though that branch of the human family died out around 40 000 years ago.

Figure 3 *Sequencing the first human genome took years. Machines like these mean it can now be done in days – soon it will take less than 24 hours*

Go further

Scientists are now analysing hundreds of thousands of human genomes from people around the world. They are searching for patterns that allow them to link particular genes or groups of genes with specific diseases. It seems likely that in the future everyone will have their genome sequenced. This will play an important part in keeping us healthy throughout our lives.

Key points

- The genome of an organism is the entire genetic material of that organism.
- The whole human genome has now been studied and this will have great importance for medicine in the future.
- The genetic material in the nucleus of a cell is composed of DNA. DNA is a polymer made up of two strands forming a double helix.
- A gene is a small section of DNA on a chromosome. Each gene codes for a particular sequence of amino acids, to make a specific protein.

1 a What is the genome of an organism. [1 mark]
  b The human genome is smaller than the genome of many other organisms. Explain how human beings can have more than 21 000 different chemicals within them. [3 marks]
2 a What was the Human Genome Project? [1 mark]
  b Give three ways in which the project was unusual. [3 marks]
  c Why do scientists continue to sequence:
    i more human genomes? [4 marks]
    ii the genomes of other organisms? [3 marks]
3 Discuss the use of the human genome in helping to trace human migration patterns from the past. [6 marks]

B12.4 Inheritance in action

Learning objectives

After this topic, you should know:

- that different forms of genes, called alleles, can be either dominant or recessive
- how to predict the results of genetic crosses when a characteristic is controlled by a single gene
- how to interpret Punnett square diagrams
- **H** how to construct Punnett square diagrams.

Most of your characteristics, such as your eye colour and nose shape, are controlled by several different genes interacting. However, some characteristics, such as black or brown fur colour in mice, are controlled by a single gene. Some human characteristics such as red–green colour blindness are also controlled by single genes. The way single gene features like these are passed from one generation to another follows some clear patterns. You can use them to predict what may be passed on. They are much easier to work with and understand than the majority of characteristics that result from multiple genes.

How inheritance works

The chromosomes you inherit carry your genetic information in the form of genes. Many of these genes have different forms, or **alleles** (sometimes called variants). Each allele codes for a different protein. The combination of alleles you inherit will determine your characteristics. We can make biological models that help us predict the outcome of any genetic cross.

Genetic terms

Some words are useful when you are working with biological models in genetics:

- **homozygote** – an individual with two identical alleles for a characteristic, for example, **BB** or **bb**
- **heterozygote** – an individual with different alleles for a characteristic, for example, **Bb**
- **genotype** – this describes the alleles present or genetic makeup of an individual regarding a particular characteristic, for example, **Bb** or **bb**
- **phenotype** – this describes the physical appearance of an individual regarding a particular characteristic, for example, black fur or brown fur in a mouse.

Figure 1 *You can tell the phenotype of these young mice by looking at them but it is more difficult to be sure about their genotype*

Picture a gene as a position on a chromosome. An allele is the particular form of information in that position on an individual chromosome. For example, the gene for coat colour in mice may have the black (B) or the brown (b) allele in place. Because the mouse inherits one allele from each parent, the mouse will have two alleles controlling whether it is black or brown. The alleles present in an individual, known as the genotype, work at the level of the DNA molecules to control the proteins made. These proteins result in characteristics – such as coat colour or the presence of dimples – that are expressed as the phenotype of the organism.

Some alleles are expressed in the phenotype even when they are only present on one of the chromosomes. The phenotype coded for by these alleles is **dominant**, so the alleles for black coats in mice and dangly earlobes in people are always expressed if they are present. Use a capital letter to represent the alleles for dominant phenotypes, for example, B for black coat in mice. So, if a mouse inherits BB or Bb from its parents, it will have a black coat.

Some alleles only control the development of a characteristic if they are present on both chromosomes – in other words, when no dominant allele is present. These phenotypes are **recessive**, such as brown coats in mice and attached earlobes. Use a lower case letter to represent recessive alleles, for example, b, and they are only expressed if the organism is homozygous. So a mouse would only have a brown coat if it inherited two recessive alleles, bb.

Genetic crosses

A genetic cross is when you consider the potential offspring that might result from two known parents. Remember that you need to look at both the possible genotypes and possible phenotypes of the offspring.

Using genetic diagrams

You can model genetic crosses using a genetic diagram such as a Punnett square to predict the outcome of different genetic crosses. A genetic diagram gives us:

- the alleles for a characteristic carried by the parents (the genotype of the parents)

- the possible gametes that can be formed from these

- how these may combine to form the characteristic in their offspring. The possible genotypes of the offspring allow you to work out the possible phenotypes too.

Inheriting different alleles can result in the development of quite different phenotypes. Genetic diagrams such as Punnett squares help to explain what is happening and predict what the possible offspring might be like. They give you the probability that a particular genotype or phenotype will be inherited in a given genetic cross.

1 a State what is meant by the term dominant allele. [1 mark]
  b State what is meant by the term recessive allele. [1 mark]
  c Describe the difference between being homozygous or heterozygous for a particular characteristic. [2 marks]

2 🄷 Draw a Punnett square to show the possible offspring from a cross between two people who both have dangly earlobes and the genotype Aa, when A is the dominant allele for dangly earlobes and a is the recessive allele for attached earlobes. [5 marks]

3 a Explain how real genetic crosses may be used to help work out the genotype of a black mouse. [4 marks]
  b 🄷 Draw a Punnett square diagram for:
    i a homozygous black mouse crossed with a heterozygous black mouse. [5 marks]
    ii heterozygous black mouse crossed with another heterozygous black mouse. [5 marks]

Phenotype: brown fur

Genotype: bb

Phenotype: black fur

Genotype: BB or Bb

Cross 1: bb × BB

Gametes	B	B
b	Bb	Bb
b	Bb	Bb

Offspring:

genotype: all Bb

phenotype: all black fur

Cross 2: bb × Bb

Gametes	B	B
b	Bb	bb
b	Bb	bb

Offspring:

genotype: 50% Bb, 50% bb

phenotype: 50% black fur, 50% brown fur

Figure 2 *Determining phenotype*

Key points

- Some characteristics are controlled by a single gene. Each gene may have different forms called alleles.
- The alleles present, or genotype, operate at a molecular level to develop characteristics that can be expressed as the phenotype.
- If the two alleles are the same, the individual is homozygous for that trait, but if the alleles are different they are heterozygous.
- A dominant allele is always expressed in the phenotype, even if only one copy is present. A recessive allele is only expressed if two copies are present.
- Most characteristics are the result of multiple genes interacting, rather than a single gene.

B12.5 More about genetics

Learning objectives

After this topic, you should know:

- how to use proportion and ratios to express the outcome of a genetic cross
- how sex is inherited
- how to use family trees.

Genetic diagrams such as Punnett squares show you the predicted ratios of the different phenotypes. They do not tell you the actual offspring because every time gametes meet they are carrying a unique and random mixture of genes. You only see the expected ratios of phenotypes if you carry out lots of genetic crosses. This is why plants and animals that breed fast and produce lots of offspring are widely used to study genetics.

Direct proportion and simple ratios

Using Punnet squares, you can work out the proportion of the offspring of a genetic cross expected to have a particular genotype or phenotype. You can also work out the ratios of one genotype or phenotype to another.

Figure 1 shows two Punnett squares. One is of a genetic cross between two heterozygous black mice. The other shows the cross you looked at in Topic B12.4 between a heterozygous black mouse and a homozygous recessive brown mouse.

In cross 1 you can look at both the genotype and the phenotype of the offspring. You can work out the proportions and the ratios between the different possible genetic combinations.

Genotypes: the proportions of the genotypes are:

$\frac{1}{4}$ or 25% homozygous dominant (BB)

$\frac{2}{4}$ or 50% heterozygous (Bb)

$\frac{1}{4}$ or 25% homozygous recessive (bb)

The possible genotypes appear in a ratio of 1:2:1 homozygous dominant : heterozygous : homozygous recessive.

Phenotypes: the proportions of the phenotypes are:

$\frac{3}{4}$ or 75% dominant (black)

$\frac{1}{4}$ or 25% recessive (brown)

The possible phenotypes appear in a ratio of 3:1 dominant : recessive.

The proportions and ratios of the possible offspring will be the same for every heterozygous cross you look at.

Look at cross 2 between a heterozygous black mouse and a homozygous recessive brown mouse. Work out the proportions and ratios of the genotypes and phenotypes for this cross.

Cross 1: Bb × Bb

Gametes	B	b
B	BB	Bb
b	Bb	bb

Cross 2: bb × Bb

Gametes	B	b
b	Bb	bb
b	Bb	bb

Figure 1 *Genetic crosses*

170

Sex determination

One feature of your phenotype is inherited not by a single gene or multiple genes but by a single pair of chromosomes. Humans have 23 pairs of chromosomes. In 22 cases, each chromosome in the pair is a similar shape. Each one has genes carrying information about the many different characteristics of your body. One pair of chromosomes is different – these are the **sex chromosomes** and they determine the sex of offspring.

● In human females the sex chromosomes are the same (XX).

● In human males the sex chromosomes are different (XY). The Y chromosome is very small and carries few genes other than those related to sexual characteristics.

When the cells undergo meiosis to form gametes, one sex chromosome goes into each gamete. This means that human egg cells contain an X chromosome. Half of the sperm also contain an X chromosome and the other half contain a Y chromosome. The inheritance of sex can be shown using a Punnett square (Figure 2).

XX × XY

Gametes	X	Y
X	XX	XY
X	XX	XY

Figure 2 *Using a Punnet square to determine sex inheritance*

Every pregnancy has a 50:50 chance of producing a boy and a 50:50 chance of producing a girl.

Family trees

You can trace genetic characteristics through a family by drawing a family tree (Figure 3). Family trees show males and females and can be useful for tracking inherited diseases, showing a family likeness, or showing the different alleles people have inherited. Family trees can be used to work out if an individual is likely to be homozygous or heterozygous for particular alleles.

1 State the sex chromosomes of:
   a human females   [1 mark]   b human males.   [1 mark]

2 Explain why we only get the expected ratios in a genetic cross if there are large numbers of offspring.   [2 marks]

3 A couple have three girls. They are expecting a fourth baby. Several people tell them they are sure to have a boy this time.
   a Explain why people might think this is the case.   [2 marks]
   b Explain why this statement is wrong.   [3 marks]

4 a Ⓗ Draw a Punnett square to show the genetic cross between female tiger **A** and the male tiger **B** at the top of the family tree in Figure 3. Use G for the dominant orange and g for the recessive white.   [5 marks]
   b Give the proportion of the different genotypes and phenotypes, and the ratios you would expect, and explain why they are not seen.   [4 marks]

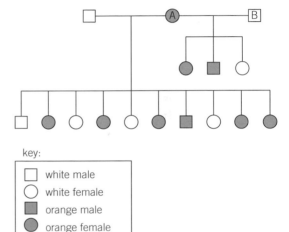

key:

□	white male
○	white female
■	orange male
●	orange female

Figure 3 *A family tree to show the inheritance of orange or white coat colour in tigers*

Key points

● Direct proportion and ratios can be used to express the outcome of a genetic cross.
● Use Punnett squares and family trees to understand genetic inheritance.
● Ⓗ Construct a Punnett square diagram to predict the outcome of a monohybrid cross.
● Ordinary human body cells contain 23 pairs of chromosomes; 22 control general body characteristics only, but the sex chromosomes carry the genes that determine sex.
● In human females the sex chromosomes are the same (XX) whilst in males the sex chromosomes are different (XY).

B12.6 Inherited disorders

Learning objectives

After this topic, you should know:

- how the human genetic disorders polydactyly and cystic fibrosis are inherited.

Some disorders are the result of a change in the bases or coding of our genes and can be passed on from parent to child. These types of diseases are known as inherited disorders. Some of them cause few problems, such as inherited colour blindness, but some inherited disorders are fatal.

Polydactyly

Sometimes babies are born with extra fingers or toes. This is called **polydactyly**. The most common form of polydactyly is caused by a dominant allele. It can be inherited from one parent who has the condition. People often have their extra digit removed, but some people live quite happily with them.

If you have polydactyly and are heterozygous, you have a 50% chance of passing on the disorder to any child you have. That's because half of your gametes will contain the faulty dominant allele (Figure 2). If you are homozygous, your children will definitely have the condition.

Some dominant genetic disorders have a much more widespread effect on the way the body works than polydactyly. For example, Huntington's disease is a dominant genetic disorder, in which symptoms develop in middle age. It affects the nervous system and eventually leads to death.

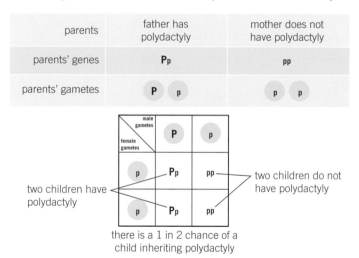

Figure 2 *A genetic diagram for polydactyly*

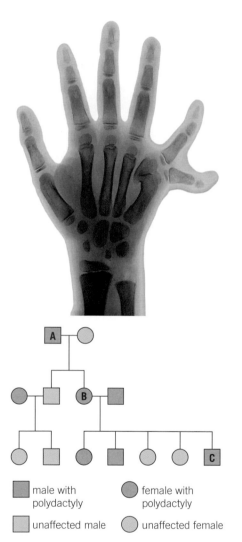

	male with polydactyly		female with polydactyly
	unaffected male		unaffected female

Figure 1 *Polydactyly is passed through a family tree by a dominant allele*

Cystic fibrosis

Cystic fibrosis is a genetic disorder that affects many organs of the body, particularly the lungs, the digestive system, and the reproductive system. Over 8500 people in the UK have cystic fibrosis.

Cystic fibrosis is a disorder of the cell membranes that prevents the movement of certain substances from one side to the other. As a result, the mucus made by cells in many areas of the body becomes very thick and sticky. Organs, especially the lungs, can become clogged up by the thick, sticky mucus, which stops them working properly. The pancreas cannot make and secrete enzymes properly because the tubes through which the enzymes are released into the small intestine are blocked with mucus. The reproductive system is also affected, so many people with cystic fibrosis are infertile.

Treatment for cystic fibrosis includes physiotherapy and antibiotics to help keep the lungs clear of mucus and infections. Enzymes are used to replace the ones the pancreas cannot produce and to thin the mucus. However, although treatments are getting better all the time, there is still no cure for the disorder.

Cystic fibrosis is caused by a recessive allele so it must be inherited from both parents. Children affected by cystic fibrosis are usually born to parents who do not suffer from the disorder. The parents have a dominant healthy allele, which means their bodies work normally. However, they also carry the recessive cystic fibrosis allele. Because it gives them no symptoms, they have no idea it is there. They are known as **carriers**.

In the UK, one person in 25 carries the cystic fibrosis allele. Most of them will never be aware of it. The only situation when it may become obvious is if they have children with a partner who also carries the allele. When looking at the possibility of inheriting genetic disorders, it is important to remember that every time an egg and a sperm fuse it is down to chance which alleles combine. So, if two parents who are heterozygous for the cystic fibrosis allele have four children, there is a 25% chance (one in four) that each child will have the disorder. In fact all four children could have cystic fibrosis, or none of them might be affected. They might all be carriers, or none of them might inherit the faulty alleles at all (Figure 3). It's all down to chance.

Curing genetic disorders

So far scientists have no way of curing genetic disorders, even those that are very serious and can shorten lives. Scientists hope that **genetic engineering** will be the answer, allowing them to replace faulty alleles with healthy ones. Currently scientists are working on gene replacement for cystic fibrosis and are beginning to make progress in halting the disease and improving lung function. Unfortunately, so far they have not managed to cure anyone with an inherited genetic disorder, but they remain very hopeful of this possibility.

1 a Define polydactyly. [1 mark]
   b State why just one parent with the allele for polydactyly can pass the condition on to their children. [2 marks]

2 a For each of the three people labelled A to C in Figure 1, give their possible alleles and explain your answers. [6 marks]
   b Explain why carriers of cystic fibrosis are not affected by the disorder themselves. [2 marks]
   c Explain why both of your parents must be carriers of the allele for cystic fibrosis before you can inherit the disease. [4 marks]

3 **H** A couple have a baby who has sickle cell anaemia – a genetic disorder caused by a recessive allele. Neither the mother or father have any signs of the disorder.
   Draw a genetic diagram to show the possible genotypes of the parents to explain how this could happen. Use S for the dominant healthy cell and s for the recessive sickle cell. [5 marks]

Synoptic links

Find out more about how scientists can change the genes of an organism in Topic B13.4.

C = dominant allele (normal metabolism)
c = recessive allele (cystic fibrosis)

Both parents are carriers, so (Cc)

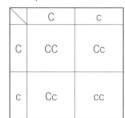

	C	c
C	CC	Cc
c	Cc	cc

Genotype of offspring:
25% normal (CC)
50% carriers (Cc)
25% affected by cystic fibrosis (cc)

Phenotype of offspring:

3/4, or 75% chance normal
1/4, or 25% chance cystic fibrosis

Figure 3 *A genetic diagram for cystic fibrosis*

Key points

- Some disorders are inherited.
- Polydactyly is a dominant phenotype caused by a dominant allele which can be inherited from either or both parents.
- Cystic fibrosis is a recessive phenotype and is caused by recessive alleles which must be inherited from both parents.

B12.7 Screening for genetic disorders

Learning objectives

After this topic, you should know:

- that embryos can be screened for some of the alleles that cause genetic disorders
- some of the concerns and issues associated with these screening processes.

chorionic villus sampling – transcervical method

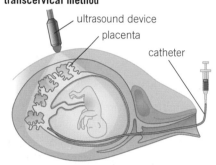

chorionic villus sampling – transabdominal method

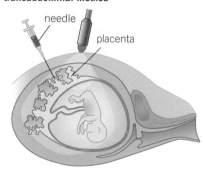

amniocentesis

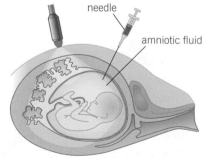

Figure 1 *Amniocentesis and chorionic villus sampling enable us to take cells from a developing fetus. The cells can then be screened for genetic diseases*

Genetic tests are now available that can show people if they carry a faulty allele. This allows them to make choices such as whether or not to have a family. It is also possible to screen embryos and fetuses during pregnancy for the alleles that cause inherited disorders. Embryos can also be screened before they are implanted in the mother during *in vitro* fertilisation (IVF) treatment. These tests are very useful, but they do raise ethical issues.

Screening embryos

To screen an embryo or fetus, you first need to harvest some cells from the developing individual. There are two main methods by which this is done once pregnancy is underway – amniocentesis and chorionic villus sampling.

- Amniocentesis is carried out at around 15–16 weeks of pregnancy. It involves taking some of the fluid from around the developing fetus. This fluid contains fetal cells, which can then be used for genetic screening.

- Chorionic villus sampling of embryonic cells is done at an earlier stage of pregnancy – between 10 and 12 weeks – by taking a small sample of tissue from the developing placenta. This again provides fetal cells to screen.

Both of these tests have an associated risk of causing a miscarriage, but are currently the main methods used to obtain fetal cells for screening. New methods that depend on analysing fetal cells found in the blood of the mother offer the promise of much less invasive testing in future.

- Another alternative taken by some couples with an inherited disorder in the family is for embryos produced by IVF to be tested before they are implanted in the mother, so only babies without that disorder are born.

Carrying out the screening

Once cells have been collected from an embryo or fetus – either before implantation or by techniques such as amniocentesis and chorionic villus sampling – they need to be screened. Whatever the potential genetic problem, the screening process is similar. DNA is isolated from the embryo cells and tested for specific disorders.

If the screening shows that a fetus is affected, the parents have a choice. They may decide to keep the baby, knowing that it will have a genetic disorder when it is born. On the other hand, they may decide to have an abortion or not proceed with implantation. This prevents the birth of a child with serious problems. Then the parents can try again to have a healthy baby. They may choose to have pre-implantation embryo screening using IVF to avoid having another affected pregnancy.

Concerns about embryo screening

Today, scientists not only understand the causes of many genetic disorders, they can also test for them. However, being able to test for a genetic disorder doesn't necessarily mean parents should always do it. There are several concerns linked to the process. The processes used to collect cells from a developing fetus increase the risk of miscarriage. So, in some cases, a healthy fetus will be miscarried as a result of a test to see whether it has a genetic abnormality, which is obviously very distressing for the parents.

The screening procedures are becoming more reliable and accurate all the time, but sometimes they can still give a false positive or a false negative result. This can, very occasionally, lead to the termination of a healthy pregnancy or the unexpected birth of a child with a genetic disorder.

Embryo screening also means that people have to make decisions about whether or not to terminate a pregnancy. This is never an easy decision. It is one of the questions to which science cannot provide an answer. Each couple, faced with the knowledge that their embryo is affected by a genetic disorder, makes their decision based not only on their scientific understanding of the situation explained by their doctors, but also on their emotions and their own ethical framework and religious beliefs.

There are economic considerations too. Screening is expensive. At the moment it is usually offered to people with family histories of a genetic disorder and older parents who are more at risk of having a child with genetic problems. Some people think everyone should have the tests regardless of the cost. If a couple have a child affected by a genetic disorder it can be very costly for society to provide health care and support for the family.

Some people prefer not to have the screening tests so they do not have to make decisions such as these. Others are concerned that genetic screening could give rise to a demand for 'designer babies'. This would occur if parents used genetic screening to choose children with 'desirable' characteristics, such as a particular sex, good looks, or intelligence.

At the moment, embryo screening can tell parents that an embryo or fetus has a genetic disorder, but there are no cures to offer. In future, using genetic engineering techniques, doctors and scientists may be able to repair damaged genes so that the child is born unaffected by the original inherited disorder.

Synoptic links

You will find out more about gene therapy and the ethical issues which arise in Topic B13.4 and Topic B13.5.

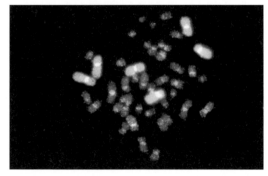

Figure 2 *A fluorescent gene probe showing up entire faulty chromosomes – the blue ones are normal*

1 Describe the main methods used to extract cells from a fetus in the uterus for use in genetic screening. [4 marks]

2 Describe the steps involved in carrying out genetic screening of:
 a a fetus at 15 weeks of pregnancy [2 marks]
 b embryos produced by IVF. [2 marks]

3 Discuss the advantages and disadvantages of offering embryo screening for genetic disorders to all pregnant couples. 🖊
 [6 marks]

Key points

● Cells from embryos and fetuses can be screened for the alleles that cause many genetic disorders.
● Embryo and fetal cells are used to identify genetic disorders but screening raises economic, social, and ethical issues.

B12 Reproduction

Summary questions

1 **a** What is meiosis and where does it take place? [2 marks]

   b Why is meiosis so important? [3 marks]

   c Explain, using labelled diagrams, what takes place when a cell divides by meiosis. [3 marks]

2 **a** What is a gamete? [1 mark]

   b How do the chromosomes in a gamete vary from the chromosomes in a normal body cell? [1 mark]

3 **a** In peas the green phenotype (allele G) is dominant to the yellow phenotype (allele g). Explain what this means. [4 marks]

   b Copy and complete this Punnett square for a cross between two heterozygous green pea plants. [2 marks]

	G	g
G		
g		

   c What is the probability of getting a plant that produces green peas from this cross? Explain your answer. [4 marks]

   d What is the probability of getting a homozygous recessive plant from this cross? Explain your answer. [3 marks]

4 Many human features are the result of different genes interacting, but there are some that are the result of single gene inheritance.

   a What is meant by the terms dominant phenotype and recessive phenotype? [2 marks]

   b Describe two characteristics inherited by single gene inheritance. [2 marks]

   c In cystic fibrosis, explain why both parents must have the allele for the disease to pass it on to their children. [3 marks]

   d Explain why people who are carriers for cystic fibrosis are usually unaffected by the disease themselves. [3 marks]

   e Ⓗ Draw a genetic diagram to show how two healthy parents could have a child with cystic fibrosis, and give the ratio of the different phenotypes and genotypes that might be expected. [5 marks]

f Huntington's disease is a serious human genetic disorder carried on a dominant allele. The problems it causes do not show up until the person is middle-aged. Explain how an affected individual could pass this disease to their offspring. [5 marks]

5 The allele for a straight thumb, S, is dominant to the curved alleles. Use this information to help you answer these questions.
Josh has straight thumbs but Sami has curved thumbs. They are expecting a baby.

   a We know exactly what Sami's thumb alleles are. What are they and how do you know? [2 marks]

   b i If the baby has curved thumbs, what does this tell you about Josh's thumb alleles? [2 marks]

   ii Draw a Punnett square to show the genetics of your explanation. [3 marks]

6 Amjid grew some purple-flowering pea plants from seeds he had bought at the garden centre. He planted them in his garden.
Here are his results:

Total seeds planted	247	White-flowered plants	1
Purple-flowered plants	242	Seeds not growing	4

   a Suggest the origin of the white-flowered plant. [3 marks]

   b Amjid was interested in these plants, so he collected the seed from some of the purple-flowered plants and used them in the garden the following year. He made a careful note of what happened.
Here are his results:

Total seeds planted	406	White-flowered plants	105
Purple-flowered plants	295	Seeds not growing	6

Amjid was slightly surprised. He did not expect to find that a third of his flowers would be white.

   i Ⓗ The purple allele, P, is for the dominant phenotype and the allele for white flowers, p, is recessive. Draw a genetic diagram that explains Amjid's numbers of purple and white flowers. [5 marks]

   ii Compare the expected ratio of phenotypes suggested by the genetic diagram in **b i** with Amjid's actual results. [3 marks]

   c i Suggest another genetic cross that would confirm the genotype of the purple plants. [2 marks]

   ii Ⓗ What results would you expect from this cross? [3 marks]

Practice questions

01 Sickle cell anaemia is an inherited disorder that causes red blood cells to develop abnormally. **Figure 1** shows a normal red blood cell and a sickle red blood cell.

Figure 1

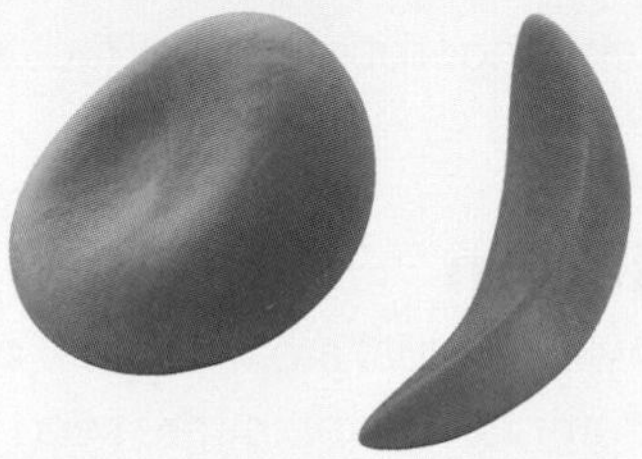

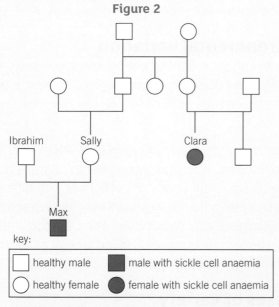

Sickle red blood cells cannot carry out their function properly.

01.1 Suggest **one** symptom of sickle cell anaemia. [1 mark]

01.2 **Figure 2** shows a family tree. Some members of this family have sickle cell anaemia.

Figure 2

key:

☐ healthy male   ■ male with sickle cell anaemia
○ healthy female ● female with sickle cell anaemia

Sickle cell anaemia is caused by a recessive allele. What evidence for this is there in **Figure 2**? [1 mark]

01.3 **h** can be used to represent the recessive allele for sickle cell anaemia.
H can be used to represent the dominant allele for healthy red blood cells.
Give the genotypes for Ibrahim and Sally. [2 marks]

01.4 ⒽConstruct a genetic cross diagram to show how Max inherited the disorder. [4 marks]

01.5 ⒽSarah would like another child.
Use your genetic cross diagram in **02.4** to predict the probability of another child having sickle cell anaemia. [1 mark]

02 Strawberry plants can reproduce both sexually and asexually.

02.1 Name the two gametes involved in sexual reproduction in plants. [2 marks]

02.2 The seeds produced in sexual reproduction will grow into new plants that are all different from each other. Explain why. [2 marks]

02.3 How do strawberry plants reproduce asexually? [1 mark]

02.4 Scientists investigated the effect of X-rays on mutations in strawberry plants.
What is a mutation? [1 mark]
Figure 3 shows the scientists' results.

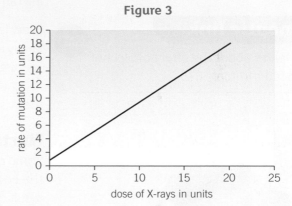

Figure 3

02.5 Give **two** conclusions that can be made from **Figure 3** [2 marks]

03 There are **two** types of cell division in humans: mitosis and meiosis.
Copy and complete the table to compare the two types of cell division in humans. [6 marks]

Table 1

	Mitosis	Meiosis
One organ where it occurs		
Number of daughter cells produced		
Number of chromosomes in daughter cells		
Daughter cells all identical (YES / NO)		

13.1 Variation

Synoptic link

You learnt about the effect of alcohol on a fetus in Topic B7.5.

Figure 1 *However much this Chihuahua eats, it will never be as big as the Great Dane – it just isn't in its genes*

No one else in the world will have exactly the same fingerprints as you. Even identical twins have different fingerprints. What factors make you so different from other people?

Nature – genetic variation

The basic characteristics of every individual are the result of the genes they have inherited from their parents. An apple tree seed will never grow into an oak tree.

Every individual looks different, but there is usually less variation between family members than between members of the general population. Features such as eye colour, nose shape, your sex, and dimples are the result of genetic information inherited from your parents. The variation in them is due to genetic causes – but your genes are only part of the story.

Nurture – environmental variation

Some differences between you and other people are due entirely to the environment you live in. For example, you may have a scar as a result of an accident or an operation. Such variation is environmental, not genetic.

Genes play a major part in deciding how an organism will look but the conditions in which it develops are important too. Genetically identical plants can be grown under different conditions of light or with different mineral ions. The resulting plants do not look identical. Plants deprived of light, carbon dioxide, or mineral ions do not make as much food as plants with everything. The deprived plants are smaller and weaker as they have not been able to fulfil their 'genetic potential'.

Combined causes of variety

Many of the differences between individuals of the same species are the result of both their genes and the environment. For example, you inherit your hair and skin colour from your parents. However, whatever your inherited skin colour, it will darken if you live in a sunny environment.

Your height and weight are also affected by both your genes and your environment. You may have a genetic tendency to be overweight, but if you never have enough to eat, you will be thin.

Investigating variety

It is quite easy to produce genetically identical plants to investigate variety. You can then put them in different situations to see how the environment affects their appearance. Scientists also use groups of animals that are genetically very similar to investigate variety.

The only genetically identical humans are identical twins who come from the same fertilised egg. Scientists are very interested in finding out how similar identical twins are as adults.

It would be unethical to separate identical twins just to investigate environmental effects. However, there are cases of identical twins who have been adopted by different families. Some scientists have researched these separated identical twins.

Often, identical twins look and act in a remarkably similar way. Scientists have measured features such as height, weight, and IQ (a measure of intelligence). The evidence shows that human beings are just like other organisms. Some of the differences between us are mainly due to genetics and some are largely due to our environment.

In one study, scientists compared four groups of adults:

- identical twins brought up together
- separated identical twins
- non-identical twins brought up together
- same sex, non-twin siblings brought up together.

The differences between the pairs were measured. A small difference means that the individuals in a pair are very alike. If there was a big difference between the identical twins, the scientists could see that their environment had more effect than their genes (Table 1).

Figure 2 *Whether identical twins are brought up together or apart, their behaviours are often very similar but never exactly identical as adults. This is because their different environments will have made some subtle differences*

Table 1 *Differences in pairs of adults*

Measured difference in:	Identical twins brought up together	Identical twins brought up apart	Non-identical twins brought up together	Non-twin siblings brought up together
Height in cm	1.7	1.8	4.4	4.5
Mass in kg	1.9	4.5	4.6	4.7
IQ	5.9	8.2	9.9	9.8

1 Both genes and the environment affect the appearance of individuals. Give an example of:
   a how genes affect a person's appearance [1 mark]
   b how the environment affects a person's appearance. [1 mark]
2 a Suggest why identical twins that were reared together and identical twins that were reared separately were studied. [3 marks]
   b Using the data from Table 1, explain which human characteristic seems to be mostly controlled by genes and which seems to be most affected by the environment. [4 marks]
3 You are given 20 pots containing identical cloned seedlings, all the same height and colour. Suggest how you might investigate the effect of temperature on the growth of these seedlings compared to the impact of their genes. 🖊 [6 marks]

B13.2 Evolution by natural selection

Learning objectives

After this topic, you should know:

- how natural selection works
- how evolution occurs via natural selection.

Figure 1 *From this mass of floating seeds, a tiny number will survive, germinate, and grow into adults. These will be the seeds with a combination of genes that gives them an edge over all the others*

Synoptic links

You learnt about meiosis and sexual reproduction in Topic B12.2.

For more information on Darwin's theory of evolution and how it was developed, see Chapter B14; for more about the competition between plants and animals in the natural world, look at Topic B15.4, and Topic B15.5.

Figure 2 *The natural world is often brutal. Only the best-adapted predators capture prey and survive to breed – and only the best-adapted prey animals escape to breed as well*

Animals and plants are always in competition with other members of their own species. Organisms that gain an advantage are more likely to survive and breed. This is **natural selection**. By looking at what is happening at the level of the genes, you can understand how the process works.

Mutation and genetic variation

The individual organisms in any species may show a wide range of variation. This is partly because of differences in the genes they inherit that arise through meiosis and sexual reproduction. New variants – that is changes in the genes themselves – arise as a result of **mutation**, a change in the DNA code.

Mutations can take place whenever cells divide. Those that take place when the gametes are formed may affect the phenotype of the offspring and introduce new variants into the genes of a species. In terms of survival, this is very important. Many mutations have no effect on the phenotype of an organism, and some mutations are so harmful that the organism does not survive. Very rarely a mutation produces an adaptation that makes an organism better suited to its environment, or gives it an advantage if there is an environmental change.

Survival of the fittest and evolution

The theory of evolution by natural selection states that all species of living things have evolved from simple life forms that first developed more than 3 billion years ago. Evolution through natural selection produces changes in the inherited characteristics of a population over time that result in organisms that are well suited to their environment. It may result in the formation of new species. The process can be summarised as follows:

- individual organisms within a particular species may show a wide range of phenotype and genetic variation
- individuals with characteristics most suited to the environment are more likely to survive to breed successfully
- the alleles (variants) that have enabled these individuals to survive are then passed on to the next generation.

If two populations of one species become so different they can no longer interbreed to produce fertile offspring, they have formed two new species.

Natural selection in action

When new variants arise from a mutation, there may be a relatively rapid change in a species. This is particularly true if the environment changes. If the mutation gives the organism an advantage in the changed environment, making it more likely to survive and breed in the new conditions, the new allele will become common quite quickly.

Oyster problems

In 1915, the oyster fishermen in Malpeque Bay, Canada noticed a few diseased oysters among their healthy catch. By 1922, the oyster beds were almost empty. The oysters had been wiped out by a destructive new disease.

Fortunately, a few of the oysters had a mutation that made them resistant to the disease. These were the only ones to survive and breed. The oyster beds filled up again, and by 1940 they were producing more oysters than ever.

A new population of oysters had evolved. As a result of natural selection, almost every oyster in Malpeque Bay now carries an allele that makes them resistant to Malpeque disease. The disease is no longer a problem.

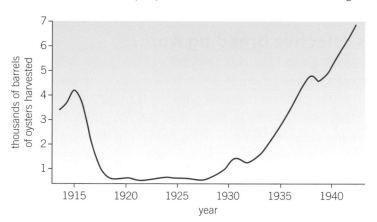

Figure 3 *The effect of disease and natural selection on oyster harvests*

Natural selection can bring about change very quickly. In bacteria, the genetic make-up of a population can change in days. The Malpeque Bay oysters took about 20 years. However, to produce an entire new species rather than a different population usually takes much longer. It has taken millions of years for the organisms present on Earth in the 21st century to evolve. There have been many different species that no longer exist, which lived on Earth many millions of years ago. The descendants of some of those species are the animals and plants you see around you.

1 Explain what is meant by survival of the fittest. [3 marks]
2 a Explain what is meant by genetic variation. [2 marks]
  b The environments that organisms live in can change. Explain how genetic mutations can be beneficial when this occurs. [3 marks]
3 Suggest how the following characteristics of animals and plants may have resulted from evolution by natural selection.
  a Male red deer have large sets of antlers. [3 marks]
  b Seals have thick fur and a thick layer of fat (blubber). [3 marks]
  c Buff tip moths look exactly like small broken silver birch twigs. [3 marks]

B13.3 Selective breeding

Learning objectives

After this topic, you should know:

- what selective breeding is
- how selective breeding works
- what the benefits and risks of selective breeding are.

Figure 1 *Ancient images, like this one of agricultural work in Egypt, give us evidence that plants and animals have changed dramatically in appearance and nature from their wild ancestors as a result of selective breeding*

For centuries people have attempted to speed up evolution to get the characteristics of animals and plants they want. This happened long before any scientific idea of evolution was developed, and also long before the mechanisms of genetics were understood. From the earliest times, farmers bred from the plants that produced the biggest grain and the animals that produced the most milk. This resulted in plants that all had bigger grains and cows that all produced a lot of milk. This is called **selective breeding**.

How does selective breeding work?

You can change animals and plants by artificially selecting which members of a group you want to breed. Farmers and breeders select animals and plants from a mixed population that have particularly useful or desirable characteristics. They use these organisms as their breeding stock. They then select from the offspring and only breed again from the ones that show the desired characteristic. This continues over many generations until all of the offspring show the desired characteristic (Figure 2).

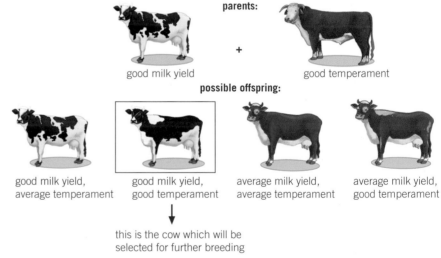

parents:

good milk yield + good temperament

possible offspring:

good milk yield, average temperament | good milk yield, good temperament | average milk yield, average temperament | average milk yield, good temperament

this is the cow which will be selected for further breeding

Figure 2 *Sometimes an animal or plant with one desirable trait will be cross-bred with organisms showing another desirable trait. Only the offspring showing both of the favoured features will be used for further breeding*

The process can be used to select for a whole range of features. Examples include:

- disease resistance in food crops or garden plants
- animals that produce more meat or milk
- domestic dogs and farm animals with a gentle nature
- large, unusual, brightly coloured or heavily scented flowers.

The results of centuries of selective breeding have been dramatic. Our placid dairy cows that produce litres and litres of milk each day are very different to their aggressive wild ancestors that produced enough milk for their single calf and little more. Fields of wheat with their large and heavy heads of grain show little resemblance to the wild grasses that were their ancestors. Genetic manipulation by selective breeding has resulted in animals and plants with strange combinations of genes that would probably never have occurred naturally. However, organisms that are either useful or simply enjoyable have been produced as a result.

Figure 3 *The dog in part **b** is a long way from his ancestor, the wolf. Most pets are the result of years of selective breeding for largely cosmetic reasons*

Limitations of selective breeding

Selective breeding has been responsible for much of the agricultural progress that has been made over the centuries, but there is a major problem with it. Selective breeding greatly reduces the number of alleles in the population – because only individuals with the chosen alleles are allowed to breed. This not only reduces the variation between individuals, but also the variation in the alleles for a given characteristic. This is not a problem when conditions are stable. However, as soon as there is a problem – the climate changes or a new disease emerges – the lack of variation can mean that none of the animals or plants in the population can cope with this change. This can result in the population dying out. For example, bananas are all genetically very similar. The banana industry is at risk of being wiped out as a result of new aggressive diseases, because none of the plants are resistant to the pathogens.

There is also the problem of 'inbreeding'. Some breeding populations have been so closely bred to achieve a particular appearance that animals are mated with close relatives. This results in very little variation in the population. Consequently, some breeds are particularly prone to certain diseases or inherited defects. For example, boxer dogs are at high risk of epilepsy, and King Charles spaniels have brains that are too big for their small skulls. Scientists at Imperial College, London discovered that the 10 000 pugs in the UK are very inbred. Not only do they struggle to breathe but their genetic variation is the equivalent of 50 healthy dogs. Persian cats have similar problems with breathing difficulties and watering eyes due to their very flat skulls. Some breeds of cows have problems giving birth because their calves are so big.

Figure 4 *Selective breeding for high milk yields, and a lot of inbreeding, means many dairy cattle have problems with their feet*

1. **a** Define selective breeding. [3 marks]
   b Explain why people have bred animals and plants selectively through the centuries. [4 marks]

2. **a** Explain why selective breeding reduces variation in the alleles of a breed of animals or plants. [3 marks]
   b Explain why variation is useful in a population. [3 marks]
   c Discuss the dangers of reducing the genetic variation in a population. [3 marks]

3. Explain, using an example, why inbreeding has caused health problems in some dog breeds. ⬤ [6 marks]

Key points

- Selective breeding is a process where humans breed plants and animals for desired characteristics.
- Desired characteristics include: disease resistance, increased food production in animals and plants, domestic dogs with a gentle nature, and heavily scented flowers.
- Problems can occur with selective breeding including defects in some animals due to lack of variation.

B13.4 Genetic engineering

Learning objectives

After this topic, you should know:

- **Ⓗ** how genes are transferred from one organism to another in genetic engineering to obtain a desired characteristic
- the potential benefits and problems associated with genetic engineering in agriculture and medicine.

Synoptic links

You learnt about use of insulin to treat diabetes, in Topic B11.2 and Topic B11.3, and about inherited problems such as cystic fibrosis in Topic B12.6.

What is genetic engineering?

Genetic engineering involves modifying (changing) the genetic material of an organism. The gene for a desirable characteristic is cut out of one organism and transferred to the genetic material in the cells of another organism. This gives the genetically engineered organism a new, desirable characteristic. For example, plant crops have been genetically engineered to be resistant to certain diseases, or to produce bigger, better fruits.

Principles of genetic engineering

Genetic engineering involves changing the genetic material of an organism using the following process (Figure 1):

- Enzymes are used to isolate and 'cut out' the required gene from an organism, for example, a person.
- The gene is then inserted into a vector using more enzymes. The vector is usually a bacterial plasmid or a virus.
- The vector is then used to insert the gene into the required cells, which may be bacteria, animal, fungi, or plants (Figure 1).
- Genes are transferred to the cells at an early stage of their development (in animals, the egg, or very early embryo). As the organism grows, it develops with the new desired characteristics from the other organism. In plants, the desired genes are often inserted into meristem cells which are then used to produce identical clones of the genetically modified plant.

Higher

Transferring genes to animal and plant cells

Genetically engineered bacteria and fungi can be cultured on a large scale to make huge quantities of protein from other organisms, for example, human insulin and human growth hormone. There is, however, a limit to the proteins that bacteria can make. Scientists have now found that genes can be transferred to animal and plant cells as well as bacteria and fungi. New techniques are making genetic modification of a wide range of organisms easier all the time. For example, genes from jellyfish have been used to produce crops that glow in the dark when lacking water. This shows where irrigation is needed. Animals have been genetically engineered in several ways, for example, some have

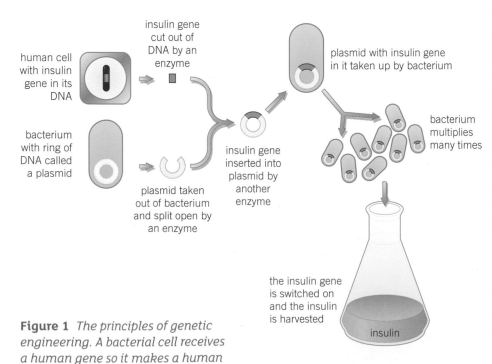

Figure 1 *The principles of genetic engineering. A bacterial cell receives a human gene so it makes a human protein – in this case, the hormone insulin*

Labels in Figure 1:
- human cell with insulin gene in its DNA
- insulin gene cut out of DNA by an enzyme
- plasmid with insulin gene in it taken up by bacterium
- bacterium multiplies many times
- bacterium with ring of DNA called a plasmid
- plasmid taken out of bacterium and split open by an enzyme
- insulin gene inserted into plasmid by another enzyme
- the insulin gene is switched on and the insulin is harvested
- insulin

been engineered to make complex human proteins in their milk. Others, particularly mice, have been modified to model human diseases including Alzheimer's, cancer, and diabetes. This allows scientists both to learn more about the diseases and to see how new treatments might work. However, it is with plants that most progress has been made.

Genetically modified crops

Crops that have had their genes modified by genetic engineering techniques are known as genetically modified crops (GM crops). Genetically modified crops often show increased yields. For example, genetically modified crops include plants that are resistant to attack by insects because they have been modified to make their own pesticide. This means that more of the crops survive to provide food for people.

GM plants that are more resistant than usual to herbicides mean farmers can spray and kill weeds more effectively without damaging their crops. Again, this increases the crop yield. Sometimes the genetic modification directly increases the size of the fruit or the nutritional value of the crop.

Increasing crop yields is extremely important in providing food security for the world's human population, which is growing all the time. For example:

- potatoes have been modified to make more starch and to be more resistant to several common pests

- soybeans have been modified to produce a healthier balance of fatty acids

- rice plants have been modified to withstand being completely covered in water for up to three weeks and still produce a good crop. Globally, 3.3 billion people rely on rice for their main food, and severe flooding in many rice growing countries is becoming more common so this genetic modification could save millions of lives

- some GM grasses can absorb and break down explosive residues in the soil

- GM crops resistant to common diseases such as mosaic viruses and blights are being produced.

Sometimes GM crops contain genes from a completely different species, such as the jellyfish genes added to crop plants described earlier. Sometimes genetic modification simply speeds up normal selective breeding, by taking a gene from a closely related plant and inserting it into the genome, for example, flood-resistant rice. Many plant scientists hope that GM technologies will help feed the world population. Perhaps GM plants can also remove more carbon dioxide from the atmosphere and solve the problems of global warming. As genetic engineering may also help wipe out some human disorders, it is no wonder that scientists are excited by the technology.

1 Explain what is meant by genetic engineering. [4 marks]

2 Give three examples of ways in which food crops have been genetically modified, and explain the advantage of each modification. [6 marks]

3 ⓗ Draw a flowchart that explains the stages of transferring a gene for a shorter stem from one plant to another using a bacterial plasmid as a vector. [6 marks]

Figure 2 *In many parts of the world, global warming is causing crop fields, such as those used for rice production, to flood. Flood-resistant GM rice offers hope to millions of people*

Key points

- Crops that have had their genes modified are known as genetically modified (GM) crops. GM crops often have improved resistance to insect attack or herbicides and generally produce a higher yield.

- ⓗ Genes can be transferred to the cells of animals and plants at an early stage of their development so they develop desired characteristics. This is genetic engineering.

- ⓗ In genetic engineering, genes from the chromosomes of humans and other organisms can be 'cut out' using enzymes and transferred to the cells of bacteria and other organisms using a vector, which is usually a bacterial plasmid or a virus.

B13.5 Ethics of genetic technologies

Learning objectives

After this topic, you should know:

- some of the concerns and uncertainties about the new genetic technologies, such as cloning and genetic engineering.

Figure 1 *Yellow beta carotene is needed to make vitamin A in the body. The amount of beta carotene in golden rice and golden rice 2 is reflected in the depth of colour of the rice*

One huge potential benefit of genetic engineering yet to be fully realised is its potential to cure inherited human disorders. Modern medical research is exploring ways of putting 'healthy' genes into affected cells using genetic modification, so the cells work properly. Perhaps in the future, the cells of an early embryo could be engineered so that the individual would develop into a healthy person unaffected by their inherited disorder. If these treatments become possible, many people would have new hope of a normal life for themselves or their children.

Gene therapy in humans has a long way to go. However, recently, scientists have had some very promising results while trying to find cures for diseases such as cystic fibrosis and a condition called macular degeneration that often results in blindness in older people. The potential benefits of genetic engineering in agriculture and medicine are becoming more apparent all the time. There are, however, some concerns about using these new technologies before scientists fully understand any long-term impact they may have on individuals or the environment.

Benefits of genetic engineering

People are already seeing many benefits from genetic engineering in medicine. Genetically engineered microorganisms can make the proteins humans need in large quantities and in a very pure form. For example, pure human insulin and human growth hormones are mass-produced using genetically engineered bacteria and fungi. Also, scientists can genetically modify mice so they mimic human diseases. These mice are very useful in developing cures for conditions ranging from cancer to diabetes. You can remind yourself about genetic engineering in Topic B13.4.

There are many advantages of genetic engineering in agriculture. These include:

- improved growth rates of plants and animals
- increased food value of crops, as genetically modified (GM) crops usually have much bigger yields than ordinary crops
- crops can be designed to grow well in dry, hot, or cold parts of the world
- crops can be engineered to produce plants that make their own pesticide or are resistant to the herbicides used to control weeds.

Short-stemmed GM crops, flood- and drought-resistant GM crops, and high yielding, high-nutrition GM crops are already helping to solve the problems of world hunger, for example, almost 60% of the soybeans grown globally are GM strains.

Ever since genetically modified foods were first introduced, there has been controversy and discussion about them. For example, varieties of GM rice known as 'golden rice' and 'golden rice 2' have been developed.

These varieties produce large amounts of vitamin A. Up to 500 000 children go blind each year as a result of lack of vitamin A in their diets. In theory, golden rice offers a solution to this problem but some people object to the way trials of the rice were run and the cost of the product. As a result of this controversy, no golden rice is yet being grown in countries affected by vitamin A blindness.

Concerns about genetic engineering

Genetic engineering is still a very new science, and no one can be sure what the long-term effects might be. For example, insects may become pesticide-resistant if they eat a constant diet of pesticide-forming plants.

Some people are concerned about the effect of eating GM food on human health. However, people eat a wide range of organisms with many different types of DNA every day as part of a normal diet. Our bodies have the enzymes needed to break down all sorts of DNA as part of normal digestion, so they should be able to deal with any extra genes.

Another concern is that genes from GM plants and animals might spread into the wildlife of the countryside. Some people are very anxious about the effect that these GM organisms might have on populations of wild flowers and insects. GM crops were originally often made infertile, which meant farmers in poor countries had to buy new seed each year. Many people were unhappy with this practice. If these infertility genes spread into wild populations, it could cause major problems in the environment – although scientists are working hard to prevent this and there is little evidence so far of problems arising.

The majority of plant scientists believe that GM crops are the way forward and probably the only way to solve the problem of feeding the world's expanding population and to cope with global warming.

Some of the main ethical objections to genetic engineering involve fears about human engineering. People may want to manipulate the genes of their future children to make sure they are born healthy. There are also concerns that people might want to use this process to have 'designer' children with particular characteristics such as high intelligence or good looks. Genetic engineering raises issues for everyone to think about.

Synoptic link

Find out more about curing genetic diseases in Topic 12.6, and about screening for genetic diseases and associated ethical concerns in Topic 12.7.

Go further

A new technique called gene editing has been developed. It makes it possible for scientists to make very precise changes in DNA itself. In 2016, UK researchers looked at the effect of using gene editing on very early human embryos. They were working on the cells that produce the placenta.

1 State three advantages of genetic engineering in agriculture.
[3 marks]

2 Genetic engineering may one day be used to cure human genetic disorders. Suggest how. [4 marks]

3 Discuss the ethical concerns surrounding genetic engineering. [6 marks]

Key points

- Modern medical research is exploring the possibility of genetic modification to overcome some inherited disorders.
- There are benefits and risks associated with genetic engineering in agriculture and medicine.
- Some people have ethical objections to genetic engineering.

B13 Variation and evolution

Summary questions

1 In a class experiment a teacher gave every student a small packet of seeds, a petri dish, and some cotton wool. The students were told to take the seeds home and grow them for three weeks, measuring the mean height of the seedlings every day for a class investigation into genetic variation in plants.

   a What is variation? [2 marks]

   b Give two factors that will affect variation in a population of plants. [2 marks]

   c Give three factors that could have affected the investigation and the validity of the students' results on the genetic variation in the plants. For each point state why it would have affected the results. [6 marks]

   d Suggest two ways to make the investigation more scientific, and explain why your suggested method is an improvement. [4 marks]

2 Look at the birds in Figure 1. They are known as Darwin's finches. They live on the Galapagos Islands. Each one has a slightly different beak and eats a different type of food.
   Explain how natural selection can result in so many different beak shapes from one original type of finch. [6 marks]

Figure 1 *Darwin's finches – more evidence for evolution*

3 Human growth is usually controlled by growth hormones produced by the pituitary gland in the brain. If you don't make enough hormones, you don't grow properly and remain very small. This condition affects 1 in every 5000 children. Until recently, the only way to get growth hormone was from the pituitary glands of dead bodies. Genetically engineered bacteria can now make plenty of pure growth hormone.

   a Ⓗ Draw and label a diagram to explain how a healthy human gene for making growth hormone can be taken from a human chromosome and put into a working bacterial cell. [6 marks]

   b What are the advantages of using genetic engineering to produce substances such as growth hormone? [4 marks]

4 The original wild sheep were very similar to Soay sheep. Soay sheep are small and light. They are very wild and can be hard to handle. They usually produce only one lamb each year.

   a By what process have modern sheep been obtained from the original wild types of sheep? [1 mark]

   b Suggest three features that farmers have bred for in modern sheep to make them more useful to us. Explain why each feature you suggested is an advantage. [3 marks]

   c Explain how the process of improving a breed works. [3 marks]

   d Discuss some of the problems that result from breeding animals or plants for characteristics chosen by people. [6 marks]

5 **a** What is meant by the term GM crops? [2 marks]

   b Explain the main concerns of people about the use of GM crops around the world. [3 marks]

   c Most plant scientists believe GM technology will be the key to producing enough food to feed the world population. How can it be used to do that? [6 marks]

   d One concern people have about GM crops is that they might cross-pollinate with wild plants. Scientists need to research how far pollen from a GM crop can travel to be able to answer these concerns. Describe how a trial to investigate this might be set up. [6 marks]

Practice questions

01 The characteristics of an individual are affected by their genes and environmental factors.
A boy has brown eyes, a scar on his right arm, five toes, and is 182 cm tall.
For each characteristic state whether it is affected by:
a genes only
b environment only
c both genes and environment
d neither genes nor environment. [4 marks]

02 Some student nurses carried out a survey at a blood donor clinic. They recorded the blood group of 40 blood donors.
Table 1 shows the results of their survey.

Table 1

Blood group	Number of people
A	16
B	4
AB	2
O	18

02.1 Plot the results in a bar chart. [4 marks]
02.2 Which blood group was the most common? [1 mark]
02.3 Calculate the percentage of the blood donors who had this blood group. [1 mark]
02.4 One of the students said that the results of their survey would represent the proportions for each blood group in the UK.
Suggest why this statement may not be correct. [1 mark]

03 The World Canine Association states that there are 339 different breeds of dog in the world.
03.1 How could you show that different breeds of dog are of the same species? [2 marks]
03.2 🅗 A new breed of dog with an extremely good sense of smell was produced in Russia. The new breed is used as a 'sniffer dog' to search for drugs and explosives at airports.
The new breed was selectively bred from huskies and jackals. Huskies have a good sense of smell and are very obedient and easy to train. Jackals, which are a wild type of dog, have an even better sense of smell but are very hard to train. For this reason the new breed is only a quarter jackal.

Suggest how the new breed was produced to have a very good sense of smell and be an obedient working dog. [6 marks]

04 Some babies are allergic to a protein found in cow's milk. Scientists have created a genetically engineered cow called Bella. Bella produces milk that does not contain this protein.
The scientists identified the gene that blocks the manufacture of the milk protein. They inserted this gene into the DNA of a skin cell from Cow **X**.
The genetically engineered skin cell and an egg cell from Cow **Y** were used to produce an embryo. The embryo was inserted into the uterus of Cow **Z**.
The embryo developed to form Bella.

Figure 1

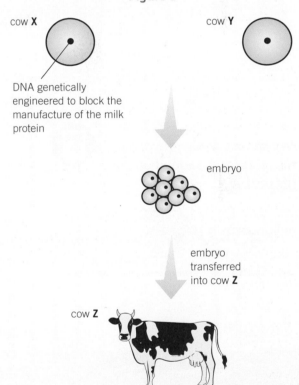

04.1 Describe in detail how the skin cell and the egg cell would be used to create the embryo. [4 marks]
04.2 Which cow would Bella be identical to?
Choose the correct answer from the box.

cow **X**   cow **Y**   cow **Z**   none of the other cows

Give a reason for your answer. [2 marks]

Learning objectives

After this topic, you should know:

- the evidence for the origins of life on Earth
- how fossils are formed
- what we can learn from fossils.

There is no record of the origins of life on Earth. It is a puzzle that can never be completely solved – no one was there to see it! Scientists don't even know exactly when life on Earth began. However, most think it was somewhere between 3 and 4 billion years ago.

Darwin's theory of evolution by natural selection is now widely accepted as the best explanation of the living world. The understanding of genetics and how characteristics are passed to offspring in the genes gives a clear mechanism for the process and helps supply the supporting evidence needed. There are several other strong strands of evidence that also support the theory of evolution by natural selection.

What can you learn from fossils?

Some of the best evidence scientists have about the history of life on Earth comes from fossils. Fossils are the remains of organisms from millions of years ago that are found preserved in rocks, ice, and other places. For example, fossils have revealed the world of dinosaurs. These giant reptiles dominated the Earth at one stage and died out many millions of years before the evolution of the first human beings.

You have probably seen a fossil in a museum or on TV, or maybe even found one yourself. Fossils can be formed in a number of ways.

- When an animal or plant does not decay after it has died. This happens when one or more of the conditions needed for decay are not there. This may be because there is little or no oxygen present. It could be because poisonous gases kill off the bacteria that cause decay. Sometimes the temperature is too low for decay to take place. Then the animals and plants are preserved almost intact, for example, in ice (Figure 1) or peat. These fossils are rare, but they give a clear insight into what an animal looked like. They can also tell us what an animal had been eating or the colour of a long-extinct flower. Scientists can even extract the DNA and compare it to the DNA of modern organisms.

- Many fossils are formed when harder parts of the animal or plant are replaced by minerals as they decay and become part of the rock. This takes place over long periods. Mould fossils are formed when an impression of an organism is made in mud and then becomes fossilised, while cast fossils are made when a mould is filled in. Rock fossils are the most common form of fossils (Figure 2). One of the biggest herbivores found so far is *Argentinosaurus huinculensis*. It lived around 70 million years ago, was nearly 40 metres long and probably weighed about 80–100 tonnes! Among the largest carnivores was *Giganotosaurus*. It was about 14 metres long, with a brain the size of a banana and 20 cm-long serrated teeth. For comparison, the biggest living carnivorous lizard, the Komodo dragon, is about 3 metres long and weighs around 140 kg.

Using timescales

Timescales for the evolution of life are big:

a thousand years is 10^3 years

a million years is 10^6 years

a billion years is 10^9 years.

Figure 1 *This baby mammoth was preserved in ice for at least 10 000 years. Examining this kind of evidence helps scientists to check the accuracy of ideas, based on fossil skeletons alone*

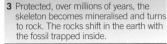

1 The reptile dies and falls to the ground.

2 The flesh rots, leaving the skeleton to be covered in sand or soil and clay before it is damaged.

3 Protected, over millions of years, the skeleton becomes mineralised and turns to rock. The rocks shift in the earth with the fossil trapped inside.

4 Eventually, the fossil emerges as the rocks move and erosion takes place.

Figure 2 *It takes a very long time for fossils to form, but they provide us with invaluable evidence of how life on Earth has developed*

Figure 3 *These footprints were found fossilised in volcanic ash. They were made by early humans and you can see clearly the adult prints and those of a child who walked along 3.6 million years ago*

- Some of the fossils found are not of actual animals or plants, but are preserved traces they have left behind. Fossil footprints (Figure 3), burrows, rootlet traces (evidence of roots), and droppings are all formed. These help us to build up a picture of life on Earth long ago.

An incomplete record

The fossil record is not complete for several reasons.

- Many of the very earliest forms of life were soft-bodied organisms. This means they have left little fossil trace. The majority of any fossils that were formed in the earliest days will have been destroyed by geological activity including the formation of mountain ranges, continental movements, erosion, volcanoes, and earthquakes. This is why scientists cannot be absolutely certain about how life on Earth began.

- Most organisms that died did not become fossilised – the right conditions for fossil formation were rare.

- Finally, there are many fossils that are still to be found.

In spite of all these limitations, the fossils that have been found can still give you a 'snapshot' of life millions of years ago.

1 There are several theories about how life on Earth began.
  a Explain why it is impossible to know for sure how life on Earth began. [2 marks]
  b Explain why fossils are such important evidence for the way life has developed. [3 marks]

2 **a** What is the most common type of fossil? [1 mark]
  b How long ago were many of these fossils formed? [1 mark]
  c Summarise the main ways in which fossils are formed. 🖊 [6 marks]

3 **a** Describe how ice fossils are formed. [3 marks]
  b Explain why ice fossils are so valuable to scientists. [4 marks]

Key points

- Fossils are the remains of organisms from millions of years ago that can be found in rocks, ice, and other places.
- Fossils may be formed in different ways including the absence of decay, parts replaced by minerals as they decay, and as preserved traces of organisms.
- Fossils give us information about organisms that lived millions of years ago.
- It is very difficult for scientists to know exactly how life on Earth began because there is little valid evidence. Early forms of life were soft bodied so left few traces behind and many traces of early life have been destroyed by geological activity.

191

B14.2 Fossils and extinction

Learning objectives

After this topic, you should know:

- what fossils can reveal about how organisms have changed over time
- how organisms can become extinct.

Go further

In 2015, a team of paleontologists in South Africa found hundreds of fossilised early human bones deep in a cave. The fossils are thought to be anything from 3 million to 100 000 years old. Scientists are still working on them - and arguing about where they fit on the human evolutionary tree.

Using the fossil record

The fossil record helps scientists to understand how much organisms have changed since life developed on Earth. However, this understanding is often limited. Only small pieces of skeletons or little bits of shells have been found. Luckily, there is a very complete fossil record for a few animals, including the horse. These relatively complete fossil records can show you how some organisms have changed and developed over time. They can also help you to reconstruct the ecology, climate, and environment of millions of years ago.

Fossils also show you that not all animals have changed very much. For example, fossil sharks from millions of years ago look very like modern sharks. They evolved early into a form that was almost perfectly adapted for their environment and their way of life. Their environment has not changed much for millions of years, so sharks have also remained the same.

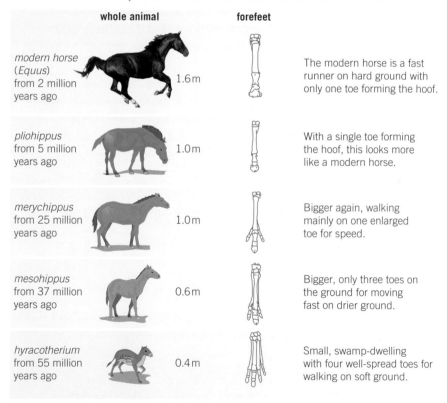

whole animal		forefeet	
modern horse (*Equus*) from 2 million years ago	1.6 m		The modern horse is a fast runner on hard ground with only one toe forming the hoof.
pliohippus from 5 million years ago	1.0 m		With a single toe forming the hoof, this looks more like a modern horse.
merychippus from 25 million years ago	1.0 m		Bigger again, walking mainly on one enlarged toe for speed.
mesohippus from 37 million years ago	0.6 m		Bigger, only three toes on the ground for moving fast on drier ground.
hyracotherium from 55 million years ago	0.4 m		Small, swamp-dwelling with four well-spread toes for walking on soft ground.

Figure 1 *The evolutionary history of the horse based on the fossil record*

Extinction

Throughout the history of life on Earth, scientists estimate that about 4 billion different species have existed. Yet scientists currently estimate that there are only 8.7 million species of organisms are alive today. The rest have become extinct. **Extinction** is the permanent loss of all the members of a species.

As conditions change, new species evolve that are better suited to survive the new conditions. The older species that cannot cope with the changes gradually die out because they are not able to compete so well for food and other resources.

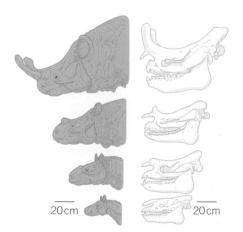

| 20 cm | 20 cm |

Figure 2 *A sequence of fossil skulls of a group of animals known as the perissodactyls. Their modern living relatives are rhinos*

This is how evolution takes place by natural selection and the number of species on Earth slowly changes. Some of the species that have become extinct are lost forever or only exist in the fossil record, while other species have descendants living on Earth today.

There are many different causes of extinction, but they always involve a change in the environment of the organism.

This could be a change in temperature, new predators, new diseases, or new, more successful competitors. It could also be changes to the environment over geological time or single catastrophic events, such as volcanic eruptions or collisions with asteroids.

Organisms that cause extinction

Living organisms can change an environment and cause extinction in several different ways:

- New predators can wipe out unsuspecting prey animals very quickly if the prey animals do not have adaptations to avoid them. New predators may evolve, or an existing species might simply move into new territory. Sometimes this can be due to human intervention. Hedgehogs were introduced to the Scottish island of North Uist in 1974 to combat the problem of garden slugs. Unfortunately, the hedgehogs bred rapidly and now eat the eggs and chicks of the many rare sea birds that breed on the island. Now people are trying to kill or remove the hedgehogs from the island to save the birds.

- New diseases can bring a species to the point of extinction. They are most likely to cause extinctions on islands, where the whole population of an animal or plant is close together. The Australian Tasmanian devil is one example of this. These rare animals are dying from a new form of communicable cancer that attacks and kills them very quickly.

- Finally, one species can cause another to become extinct by successful competition. New mutations can give one type of organism a real advantage over another. Sometimes new species are introduced into an environment by mistake. This means that a new, more successful competitor can take over from the original animal or plant and cause it to become extinct. For example, in Australia, the introduction of rabbits has caused severe problems. They eat so much and breed so fast that the other native Australian animals are dying out because they cannot compete with the rabbits for food.

1 Look at the evolution of the horse shown in Figure 1. Explain how the fossil evidence of the legs helps us to understand how hyracotherium and mesohippus lived and what they were like. [4 marks]

2 How does fossil evidence help us to understand just how much organisms have changed – or not changed – over time? [4 marks]

3 Explain how the following can cause extinction:
   a new predators [3 marks]
   b new diseases [3 marks]
   c successful competition. [3 marks]

Figure 3 *Hedgehogs love to eat the eggs of ground nesting birds such as lapwings – this may drive some birds to extinction*

Study tip

Always mention a *change* when you give possible reasons for extinction of a species.

Key points

- You can learn from fossils how much or how little organisms have changed as life has developed on Earth.
- Extinction may be caused by a number of factors including new predators, new diseases or new, more successful competitors.

B14.3 More about extinction

Learning objectives

After this topic, you should know:

- how environmental change can cause extinction
- how single catastrophic events can cause extinction on a massive scale.

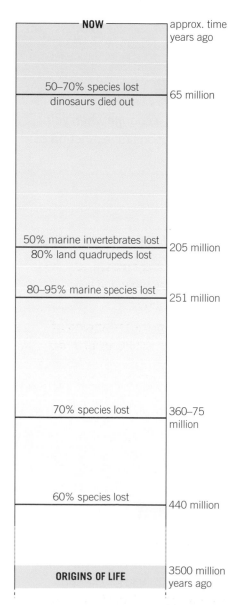

Figure 2 *The five main extinction events so far in the evolutionary history of the Earth*

It isn't just changes in living organisms that bring about extinctions. The biggest influences on survival are changes in the environment.

Environmental changes

Throughout history, the climate and environment of the Earth has been changing. At times, the Earth has been very hot. At other times, temperatures have fallen and the Earth has been in the grip of an Ice Age. These changes take place over millions and even billions of years – this is known as geological time.

Organisms that do well in the temperature of a tropical climate wouldn't do well in the freezing conditions of an Ice Age. Many of them would become extinct through lack of food or being too cold to breed. However, species that cope well in cold climates would evolve and thrive by natural selection.

Changes to the climate or the environment have been the main cause of extinction throughout history. There have been five occasions during the history of the Earth when significant climate change has led to extinction on an enormous scale (Figure 2).

Figure 1 *Dinosaurs ruled the Earth for millions of years, but when the whole environment changed, they could not adapt and most of them died out. Mammals, which could control their own body temperature, had an advantage and became dominant*

Extinction on a large scale

Fossil evidence shows that at times there have been mass extinctions on a global scale. During these events, many (or even most) of the species on Earth died out. This usually happens over a relatively short time period of several million years. Huge numbers of species disappear from the fossil record.

The evidence suggests that a single catastrophic event is often the cause of these mass extinctions. This could be a colossal volcanic eruption or the collision of giant asteroids with the surface of the Earth. These events also affect climate over a long period, which in turn has an impact on extinction rates.

What destroyed the dinosaurs?

The most recent mass extinction was when the dinosaurs became extinct around 65 million years ago. In 2010, an international team of scientists published a review of all the evidence put together over the past 20 years. They agreed that around 65 million years ago a giant asteroid collided with the Earth in Chicxulub in Mexico.

There is a huge crater (180 km in diameter) there. Scientists have identified a layer of rock formed from crater debris in countries across the world. The further you move away from the crater, the thinner the layer of crater debris in the rock. Also, deep below the crater, scientists found lots of iridium, a mineral that is only formed when rock is hit with a massive force such as an asteroid strike.

The asteroid impact would have caused huge fires, earthquakes, landslides, and tsunamis. Enormous amounts of material would have been blasted into the atmosphere. The accepted theory is that the dust in the atmosphere made everywhere almost dark. Plants struggled to survive and the drop in temperatures caused a global winter. Between 50 and 70% of all living species, including dinosaurs, became extinct.

No sooner had this work been published than a group of UK scientists published different ideas and evidence. They suggest that the extinction of the dinosaurs started sooner (137 million years ago) and that it was much slower than previously thought.

Their theory is that the melting of the sea ice (caused by global warming) flooded the seas and oceans with very cold water. A drop in the sea temperature of about 9 °C triggered the mass extinction. Their evidence is based on an unexpected change in fossils and minerals that they found in areas of Norway.

As you can see, building up a valid, evidence-based history of events that happened so long ago is not easy to do. Events can always be interpreted in different ways.

Perhaps both of the theories above played their part in the extinction of dinosaurs. What is certain is that it will be a very long time before there is enough evidence for scientists to be completely sure.

Figure 3 *This dark layer of debris from the asteroid crater appears in rocks that are 65 million years old – the time that dinosaurs died out*

Study tip

Remember that the timescales in forming new species and mass extinctions are huge.

Try to develop an understanding of time in millions and billions of years.

1   **a**   Give four causes of extinction in species of living organisms.
                                                              [4 marks]

      b   Give two possible causes of mass extinction events.   [2 marks]

2   Suggest why extinction is an important part of evolution.   [3 marks]

3   **a**   Summarise the evidence for a giant asteroid impact as the cause of the dinosaur mass extinction event.   [4 marks]

      b   **i**   Explain why scientists think that low light levels and low temperatures would have followed a massive asteroid strike.   [3 marks]

           ii   Why would these have caused mass extinctions?   [4 marks]

Key point

- Extinction can be caused by a variety of factors including changes to the environment over geological time and single catastrophic events, such as massive volcanic eruptions or collisions with asteroids.

B14.4 Antibiotic resistant bacteria

Synoptic link

To remind yourself about bacteria, their role in communicable diseases, and the importance of antibiotics in treating bacterial diseases, look back to Chapter B5.

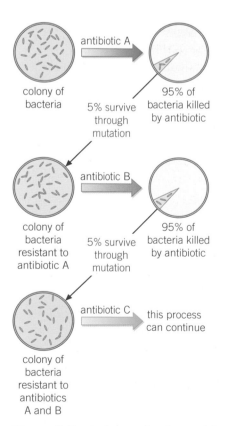

Figure 1 *Bacteria can develop resistance to many different antibiotics in a process of natural selection, as this simple model shows*

If you are given an antibiotic and use it properly, the bacteria that have made you ill are killed off. However, some bacteria develop resistance to antibiotics. They have a natural mutation (change in their genetic material) that means they are not affected by the antibiotic. These mutations happen by chance and they produce new strains of bacteria by natural selection. Bacteria can evolve rapidly because they reproduce at a fast rate.

Antibiotic resistant bacteria

Normally, an antibiotic kills the bacteria of a non-resistant strain. However, individual resistant bacteria survive and reproduce, so the population of resistant bacteria increases. Antibiotics are no longer active against this new resistant strain of the pathogen. As a result, the new strain will spread rapidly because no one is immune to it and there is no effective treatment. This is what has happened with bacteria such as MRSA (see later). More types of bacteria are becoming resistant to more antibiotics, so bacterial diseases are becoming more difficult to treat.

To prevent more resistant strains of bacteria appearing:

- It is important not to overuse antibiotics. For this reason, doctors no longer use antibiotics to treat non-serious infections such as mild throat or ear infections. Also, since antibiotics don't affect viruses, people should not request antibiotics to treat an illness that their doctor believes is caused by a virus.

- It is also important that patients finish their course of medicine every time. This is to make sure that all bacteria are killed by the antibiotic, so none survive to mutate and form resistant strains.

- It is important to restrict the agricultural use of antibiotics. This is to prevent the spread of antibiotic resistance from animal to human pathogens. In the UK we already have many restrictions in place. Other countries are not so careful. In the USA, 70% of human antibiotics are also used on farm animals, often to speed up growth or prevent rather than cure infections.

Hopefully, these measures will slow the development of resistant strains.

The MRSA story

Hospitals use a lot of antibiotics to treat infections. As a result of natural selection, some of the bacteria in hospitals are resistant to many antibiotics. This is what has happened with MRSA, the bacterium methicillin-resistant *Staphylococcus aureus*. As doctors and nurses move from patient to patient, these antibiotic resistant bacteria are spread easily.

At its peak in the early 21st century, MRSA alone caused or contributed to over 1000 deaths every year in UK hospitals and care homes. A number of simple measures to reduce the spread of microorganisms such as MRSA have been widely implemented in England and Wales. The proof of their effectiveness can be seen in the data (Figure 2).

- Antibiotics should only be used when they are really needed.
- Specific bacteria should be treated with specific antibiotics.
- Medical staff should wash their hands with soap and water or use alcohol gel between patient visits and wear disposable clothing or clothing that is regularly sterilised.
- Hospitals should have high standards of hygiene so that they are really clean.
- Patients who become infected with antibiotic resistant bacteria should be looked after in isolation from other patients.
- Visitors to hospitals and care homes should wash their hands as they enter and leave.

Simple commonsense measures such as these can have a considerable effect on reducing deaths from antibiotic resistant bacteria.

Medicines for the future

In recent years, doctors have found strains of bacteria that are resistant to even the strongest antibiotics. When that happens, there is nothing more that antibiotics can do for a patient and he or she may well die. Scientists are constantly looking for new antibiotics. However, it isn't easy to find chemicals that kill bacteria without damaging human cells.

Penicillin and several other antibiotics are made from microorganisms. Scientists are exploring sources from crocodile blood and fish slime, to soil samples and the ocean depths to produce a new antibiotic to kill antibiotic resistant bacteria. The development of new antibiotics is both expensive and slow. It seems unlikely that the appearance of new antibiotics will keep up with the development of new antibiotic resistant strains of bacteria.

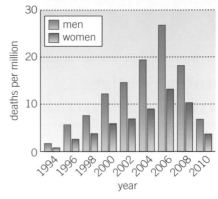

Figure 2 *The number of deaths in England and Wales in which MRSA played a part, 1994–2010 (Source: Office for National Statistics)*

1 **a** Is MRSA a bacterium or a virus? [1 mark]
  b How does MRSA illustrate the importance of not using antibiotics too frequently or when they are not really necessary? [3 marks]

2 Make a flowchart to show how bacteria develop resistance to antibiotics. [6 marks]

3 Use Figure 2 to help you answer these questions.
  a Suggest what caused the increase in deaths linked to MRSA from 1994 to 2006. [1 mark]
  b Deaths from MRSA and other hospital-acquired infections are falling. Explain this continuing reduction in death rates. [4 marks]

Key points

- Bacteria can evolve rapidly because they reproduce at a fast rate.
- Mutations of bacterial pathogens produce new strains.
- Some strains might be resistant to antibiotics and so are not killed. They survive and reproduce, so the population of the resistant strain increases by natural selection. The resistant strain will then spread because people are not immune to it and there is no effective treatment.
- MRSA is resistant to antibiotics.
- To reduce the rate of development of antibiotic resistant strains, it is important that doctors do not prescribe antibiotics inappropriately, patients use the correct antibiotics prescribed, and patients complete each course of antibiotics.
- The development of new antibiotics is costly and slow and is unlikely to keep up with the emergence of new resistant strains.

B14.5 Classification

Learning objectives

After this topic, you should know:

- the basic principles of classification and the system developed by Linnaeus
- the binomial naming system of genus and species
- how new technologies have changed classification.

Table 1 *Examples of the scientific names of some common organisms*

Common name	Scientific name
Human being	*Homo sapiens*
Domestic cat	*Felis domesticus*
Blackthorn (sloes)	*Prunus spinosa*
Malarial parasite	*Plasmodium falciparum*

Classification is the organisation of living things into groups according to their similarities. Biologists classify organisms to make it easier to study them. Classification allows us to make sense of the living world. It also helps us to understand how the different groups of living things are related to each other. Perhaps most importantly, it enables us to recognise the biodiversity present in the world and gives scientists a common language in which to talk about it.

How are organisms classified?

Living things are classified by studying their similarities and differences. Classification as it's known today really began in the 18th century with Carl Linnaeus, a Swedish botanist (plant biologist). He grouped organisms together depending on their structure and characteristics. Linnaeus made many careful observations of living things and used a hierarchical structure to classify them. The way we classify organisms today is still based on the Linnaean system, which is also known as the natural classification system. Linnaeus classified organisms into the following groups – kingdom, phylum, class, order, family, genus, and species.

Kingdoms

When Linnaeus first devised his classification system, the number of known types of living things was much smaller than it is today and so he suggested just two kingdoms, the animals and the plants. Kingdoms contain lots of organisms with many differences but a few important similarities. For example, all animals move their whole bodies about during at least part of their life cycle, and their cells do not have cellulose cell walls. On the other hand, plants do not move their whole bodies about and their cells have cellulose cell walls. Also, some plant cells contain chloroplasts full of chlorophyll for photosynthesis.

Now scientists know of many more organisms and they also know much more about them. Developments in microscopes have enabled scientists to compare the internal structures of cells. They also know a great deal more about the biochemistry of different organisms. You can even compare the genomes of organisms. As a result, classification systems have changed. At one stage scientists used five kingdoms to classify organisms – the prokaryotes, protista, fungi, plants, and animals. However, later research has shown even that model to be too simple. Now most scientists accept a three domain, six kingdom model (Topic B14.10). To give the six kingdoms, the prokaryotes were divided into the **archaea**, ancient bacteria-like organisms, and the eubacteria (the normal bacteria that we might grow in the school laboratory).

Species

The smallest group of clearly identified living organisms in Linnaeus's classification system is a **species**. Members of the same species are very similar. A species is a group of organisms that can breed together and produce fertile offspring. Orang-utans, dandelions, and brown trout are all examples of separate species of living organisms.

Naming living things

The huge variety of living organisms and the number of different languages spoken means that the same organism can have many different names around the world. This makes it impossible for one biologist to know what organism another is talking about!

The problem is solved because every organism has a scientific name given using a binomial system again first put forward by Carl Linnaeus. Binomial means two names. The two names of an organism are in Latin and they give the genus and the species of the organism. Even in the time of Linnaeus, Latin was no longer spoken but was the language of scholars everywhere. This meant no-one was offended because their language was not chosen – yet most people could understand the names.

Simple rules for writing scientific names

● The first name is the name of the genus to which the organism belongs. It is written with a capital letter.

● The second name is the name of a species to which the organism belongs. It is written with a lower case letter.

● The two names are underlined when hand written or are in italics when printed.

Vanessa atalanta   *Plebejus argus*

Figure 1 *Organisms are identified by the differences between them. These two animals are butterflies, but they belong to different species*

1 Define classification. [1 mark]

2 **a** Name the main kingdoms of the living world in:
   i the original Linnaean classification system [2 marks]
   ii the modern classification system. [6 marks]
   b Explain why there is such a difference between the two systems. [3 marks]

3 **a** Define a species. [2 marks]
   b Give five examples of species of living organisms, including at least one plant species and *not* including those given in Table 1. [5 marks]

4 **a** Define the binomial naming system. [2 marks]
   b Explain the importance of the binomial system for scientists around the world. *ⓘ* [3 marks]

B14.6 New systems of classification

Learning objectives

After this topic, you should know:

- more about the ways in which new technology has changed how scientists classify organisms
- how scientists use evolutionary trees.

Synoptic link

To remind yourself about prokaryotic and eukaryotic cells look back to Topic B1.3.

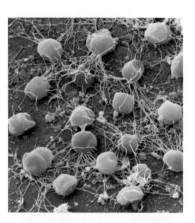

Figure 1 *The three domains of the living world - the archaea, the bacteria, and the eukaryota*

Classification became very unfashionable amongst biologists at one stage. New technologies and new threats to global biodiversity mean it is now a high-profile area of biology again.

The three-domain system

Under the Linnaean system of classification, the largest group is a kingdom. In recent years, our understanding of the internal processes and biochemical makeup of cells has increased rapidly. Many more species are known and recognised. One of the results of this new knowledge is the work of Carl Woese and others. In the 1970s they introduced the idea of a new, higher level of classification above kingdoms called a **domain**.

The evidence for this idea comes particularly from a detailed analysis of the biochemistry of cell ribosomes and the way the different cells reproduce. Prokaryotes are by far the most numerous organisms on Earth, although we often forget them because they are so small. It seems they come in two distinct forms, fitting two different domains. The eukaryotes, including humans, are different again. Using these divisions, Woese proposed three domains, in turn divided into six kingdoms:

- Archaea: primitive forms of bacteria that include the extremophiles, organisms that can live in extreme conditions. This domain contains one kingdom, the archaebacteria.
- Bacteria: these are the true bacteria and the cyanobacteria, bacteria-like organisms that can photosynthesise. This domain contains one kingdom, the eubacteria.
- Eukaryota: these organisms all have cells that contain a nucleus enclosing the genetic material. There are four eukaryotic kingdoms – the protista, the fungi, the plants, and the animals.

Classification and evolutionary trees

In the past, scientists relied on careful observation of organisms to decide which species they belonged to. Out in the field, this is still the main way to identify organisms. However, microscopy and biochemical analysis produce different models of the relationships between organisms.

Since Darwin's time, scientists have used classification to show how they think different organisms are related. These models are called **evolutionary trees**. They are built up by looking at the similarities and differences between different groups of organisms.

However, observation may not tell you the whole story. Some organisms look very different but are closely related. Others look very similar but come from very different groups. Now scientists are using DNA evidence to decide what species an organism belongs to. They look for differences as well as similarities in the DNA. This allows them to work out the evolutionary relationships between organisms. To build up evolutionary trees for living organisms, scientists use all the current types of classification data, including DNA analysis. To build up evolutionary trees for extinct organisms, they use fossil data.

Pandas and red pandas – an evolutionary tree

Giant pandas have a wrist thumb to grip bamboo. The only other animals with a similar 'wrist thumb' are red pandas. Both red pandas and giant pandas eat bamboo. Based on their feeding relationships, it looks as though they are closely related in evolution.

Combining traditional observation, modern classification data based on DNA, and fossil evidence, scientists have produced a rather different evolutionary tree. It seems the two species had a common ancestor a very long time ago, but the special 'wrist thumb' evolved separately to solve two different ecological problems (Figure 3).

Figure 3 *Scientists think giant pandas evolved a wrist-thumb for bamboo eating, but that red pandas evolved one to help them escape into trees carrying food stolen from sabre-toothed cats*

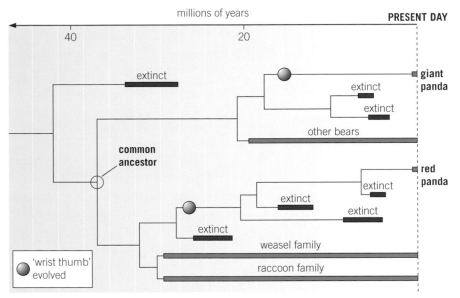

Figure 2 *Evolutionary trees such as this show us the best model of the evolutionary relationships between organisms*

1 a State the three domains. [3 marks]

b Describe the characteristics of the domains you listed in part **a**. [6 marks]

c Explain the difference between domains and kingdoms. [3 marks]

d Suggest why Woese put forward the idea of a new classification group and why you think it has been widely accepted. [3 marks]

2 a Describe what observations can be made to compare living organisms, and how modern technology has affected the way organisms are classified. [4 marks]

b Explain how classification helps us to build up models of evolutionary relationships. [3 marks]

3 Explain how evolutionary trees are useful to us. [5 marks]

Key points

● Studying the similarities and differences between organisms allows us to classify them into archaea, bacteria, and eukaryota.

● Classification also helps us to understand evolutionary and ecological relationships.

● Models such as evolutionary trees allow us to suggest relationships between organisms.

B14 Genetics and evolution

Summary questions

1 a What is a fossil? [2 marks]

Figure 1 *fossil A*

Figure 1 *fossil B*

b Explain how fossil A was formed. [6 marks]

c Explain how fossil B was formed. [4 marks]

d How can fossils like these be used as evidence for the development of life on earth? [5 marks]

e Explain **three** of the limitations of fossils as evidence for the evolution of life on earth. [3 marks]

2 a i How many million years ago did the first living animals appear on earth? [1 mark]

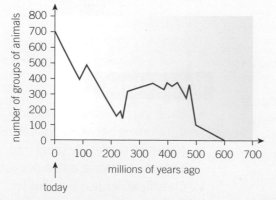

Figure 2

ii How many million years did it take for the number of animal groups to rise to 400? [1 mark]

b i Calculate the proportion of groups that disappeared between 100 million years ago and 80 million years ago. Show your working. [3 marks]

ii Give **two** reasons why some groups of animals have disappeared during Earth's history. [2 marks]

3 The skulls in Figure 3 come from the fossil record of a group of animals known as perissodactyls.

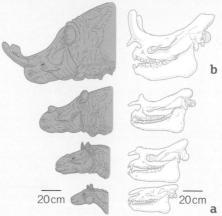

Figure 3

a Suggest a possible living relative of these animals. [1 mark]

b Skull A is based on the oldest fossil, skull B is relatively recent. How do you think these organisms changed as they evolved, based on the evidence in Figure 3? [4 marks]

c What are the limitations of this type of evidence? [3 marks[

d Discuss the other fossil remains you would want to see to understand more about the lives of these extinct organisms. [6 marks]

4 a What is extinction? [1 mark]

b How does mass extinction differ from species extinction? [3 marks]

c What is the evidence for the occurrence of mass extinctions throughout the history of life on Earth? [2 marks]

d Suggest two theories about the possible causes of mass extinctions and explain the sort of evidence that is used to support these ideas. [5 marks]

5 a Define classification. [1 mark]

b Explain two alternative ways of deciding how to classify an organism. [3 marks]

c Discuss how ideas of classification have developed over time. [6 marks]

Practice questions

01 Match each scientist with the area of science they developed.

Scientist	Theory
Darwin	The three domain system of classification
Linnaeus	The theory of evolution by natural selection
Woese	The binomial system of naming organisms

[3 marks]

02 Gregor Mendel carried out breeding experiments on pea plants in the mid-19th century.
He crossed pure-breeding tall plants with pure-breeding short plants. All the seeds produced from this cross grew into tall plants.
Use the symbol **T** to represent the allele for a tall plant and the symbol **t** to represent the allele for a short plant.

02.1 What alleles (genotype) did the seeds from this cross have? [1 mark]

02.2 Which term best describes the genotype of the offspring you gave in **02.1**?
A heterozygous
B homozygous recessive
C homozygous dominant [1 mark]

02.3 Mendel grew hundreds of plants from the seeds of the offspring.
He crossed these plants with each other and produced 2480 tall plants and 620 short plants.
Draw a genetic diagram to explain these results.
Use the symbol **T** to represent the allele for a tall plant and the symbol **t** to represent the allele for a short plant. [4 marks]

02.4 From his many experiments Mendel concluded that the inheritance of each characteristic is determined by 'units' that are passed on to descendants unchanged.
What do we call these units of inheritance today? [1 mark]

03 Fossils provide evidence for evolution.
Figure 1 shows a fossil of an organism called a trilobite.

Figure 1

03.1 What is a fossil? [1 mark]
03.2 How are fossils formed? [5 marks]
03.3 Give **one** other source of evidence for evolution. [1 mark]

04 **Figure 2** shows the evolutionary tree for two types of panda.

Figure 2

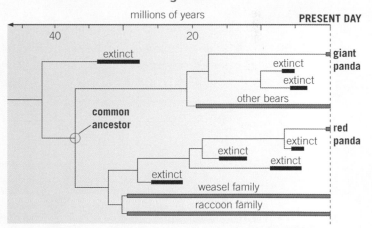

04.1 Which animals are the closest surviving relatives to the Giant panda? [1 mark]
04.2 How long ago did the common ancestor of the red panda and giant panda exist? [1 mark]
04.3 **Figure 2** shows that many animals have become extinct.
Suggest **three** reasons why a species may become extinct. [3 marks]

5 Ecology

Ecology is the study of organisms and their relationships with the living and non-living environment in which they live. We are discovering the importance of the balance of nature in maintaining the health of Earth. You will be looking at the way organisms are adapted to their environment and how they compete for mates and resources such as light and food.

Life on Earth as we know it depends on the sugar made by plants using energy from the Sun. You will be learning about the feeding relationships and material cycles that maintain the diversity of life. You will also consider the impact of the human population on the Earth.

Key questions

- What is adaptation and why is it so important?

- Why is the cycling of materials in nature so vital to life on Earth?

- What is global warming and why does it matter?

- How can we reduce the negative impact people have on ecosystems and biodiversity?

Making connections

- You learnt about natural selection and survival of the fittest in **B14 Genetics and evolution**.

- You learned about communicable diseases and how they can affect organisms in an environment in **B5 Communicable diseases**.

- You looked at how plants make food by photosynthesis in **B8 Photosynthesis**.

- You found how the biodiversity of life is classified in **B14 Genetics and evolution**.

I already know...

That plants and animals have different requirements from their environments.

Darwin's theory and about natural selection.

That plants need mineral ions and water from the soil, carbon dioxide from the air, and light to make the chemicals they need.

How organisms are interdependent in an ecosystem.

How people reproduce.

The importance of biodiversity.

I will learn...

How to investigate and measure the distribution and abundance of species in a system.

About the competition between organisms for scarce resources, and about the adaptations of organisms that result from natural selection and enable them to compete successfully in specific environments.

About the material cycles in nature that return chemicals from the bodies of organisms to the soil, water, and air.

About the levels of organisation within an ecosystem, including the cyclical relationships between predators and their prey.

The reasons for the human population explosion and its impact in terms of pollution of the land, water, and air.

Some of the ways people interact with their environment, and how these ways can have negative or positive effects on biodiversity.

Required Practicals

Practical		Topic
7	Investigate the population size of a common species in a habitat	B15.3

B 15 Adaptations, interdependence, and competition

15.1 The importance of communities

Learning objectives

After this topic, you should know:

- what is meant by a stable community
- how organisms are adapted to the conditions in which they live
- the relationship between communities and ecosystems.

Synoptic links

You will learn more about the role of microorganisms such as bacteria, protista, and fungi in Topics B16.2 and B16.3

Figure 2 *Animals are often involved in seed dispersal – for example, animals will eat these berries and spread the seeds they contain*

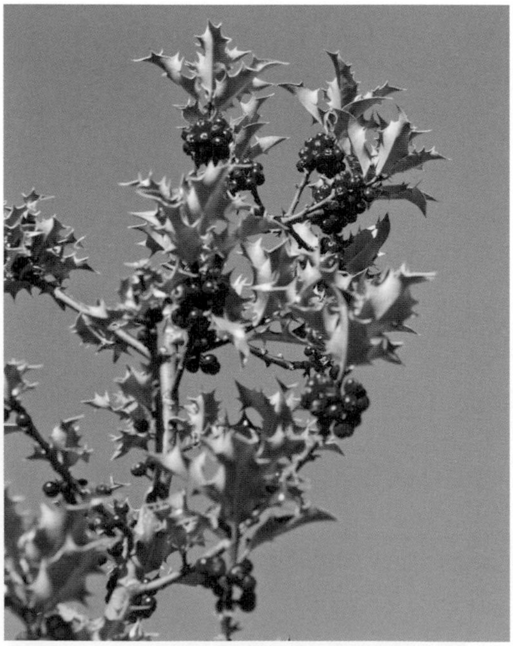

Organisms do not live in isolation. Any individual – even if one of a species that lives a solitary existence most of the time, such as a polar bear or a tiger – will be part of a population of organisms of the same species. Populations do not exist in isolation either – they live in complex **communities**. A community is made up of the populations of different species of animals and plants, protista, fungi, bacteria, and archaea that are all interdependent in a habitat. Within a community each species depends on other species for things including food, shelter, pollination, and seed dispersal.

Communities and ecosystems

An ecosystem is made up of a community of organisms interacting with the non-living or abiotic elements of their environment. The interactions of the living things make up the biotic elements of the ecosystem. All species live in ecosystems composed of complex communities of animals, plants, and other organisms that are dependent on each other and that are adapted to their particular conditions. The Sun is the source of the energy that is transferred through ecosystems within the chemical bonds that make up the organisms. Materials – including carbon, nitrogen, and water – are constantly being recycled through the living world in processes where microorganisms play a major part.

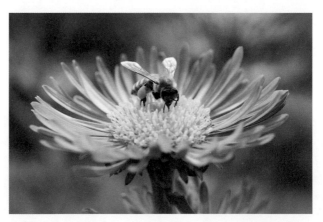

Figure 1 *Without the insects, birds, and mammals, many plants would not be able to reproduce*

Within any community the different animals and plants are often interdependent:

- plants produce food by photosynthesis
- animals eat plants
- animals pollinate plants

- animals eat other animals
- animals use plant and animal materials to build nests and shelters
- plants need the nutrients from animal droppings and decay.

Different animals and plants compete for various resources both within each species and with other species. Different communities exist close to each other and may overlap. It is important that you understand the relationships within and between communities. If one species is removed, or becomes very numerous, it can affect the whole community. This is called **interdependence**.

Stable communities

In some communities the environmental factors are relatively constant – they may change, but if they do it is in a regular pattern such as the seasons of the year in the UK. In these stable environments, the species of living organisms may also be in balance. The number of species remains relatively constant, as does the population sizes of the different species, although they will vary slightly. These stable communities are very important. Examples include tropical rainforests, ancient oak woodlands, and mature coral reefs. These communities include a wide range of species – a single mature oak tree can house up to 1000 other species. Within limits, change can be tolerated and absorbed. For example, a falling tree allows light into the forest floor, so new seedlings can grow up. But when a large, stable community is lost, it cannot easily be replaced.

Figure 3 *Stable communities such as this tropical rainforest in Costa Rica take a very long time to evolve – and cannot easily be replaced*

1 **a** Describe a community. [2 marks]
   b Explain how an ecosystem differs from a community. [3 marks]
2 Give five examples of how animals and plants can interact in an ecosystem. [5 marks]
3 Describe an example of a stable community, and state why it is so important that it is stable. Your answer should include examples of at least one plant and animal that live in this community. [6 marks]

Figure 4 *Stable communities like this ancient oak woodland in the New Forest can be centuries old*

Key points

- An ecosystem is the interaction of a community of living organisms with the non-living (abiotic) parts of their environment.
- Organisms require materials from their surroundings and other living organisms to survive and reproduce.
- Within a community, each species depends on other species for food, shelter, pollination, seed dispersal, etc. If one species is removed it can affect the whole community. This is called interdependence.
- A stable community is one where all the species and environmental factors are in balance so that population sizes remain fairly constant.

B15.2 Organisms in their environment

Learning objectives

After this topic, you should know:

- some of the factors that affect communities.

Figure 1 *Snow leopards are one of the rarest big cats. They live in cold, high-altitude environments where there is very little prey for them to hunt*

Figure 2 *The flowering of the desert after rain*

To survive and breed successfully, organisms need to be well adapted to the environment in which they live. For example, reindeer live in cold environments where most of the plants are small because low temperatures and light levels limit growth. They eat grass, moss, and lichen. Reindeer travel thousands of miles as they feed, because they cannot get enough food to survive in just one area. They travel in herds as protection against predators.

Abiotic factors affecting communities

Non-living factors that affect living organisms and so affect communities include the following:

- Light intensity: light limits photosynthesis, so light intensity also affects the distribution of plants and animals. Some plants are adapted to living in low light levels, for example, they may have more chlorophyll or bigger leaves. Nettles growing in the shade of other bushes have leaves with a much bigger surface area than nettles growing in the open. However, most plants need plenty of light to grow well. The breeding cycles of many animal and plant species are linked to day length and light intensity.

- Temperature: temperature is a limiting factor on photosynthesis and therefore growth in plants. In cold climates, temperature is always limiting. For example, the low Arctic temperatures mean the plants are all small. This in turn affects the numbers of herbivores that can survive and so the number of carnivores in the community.

- Moisture levels: if there is no water, there will be little or no life. As a rule, plants and animals are relatively rare in a desert as the availability of water is limited. However, after it rains, many plants grow, flower, and make seeds very quickly while the water is available. These plants are eaten by many animals that move into the area to take advantage of them.

- Soil pH and mineral content: the level of mineral ions, for example, nitrate ions, has a considerable impact on the distribution of plants. Carnivorous plants such as sundews thrive where nitrate levels are very low because they can trap and digest animal prey. The nitrates they need are provided when they break down the animal protein. Most other plants struggle to grow in areas with low levels of mineral ions. The pH of the soil also has a major effect on what can grow in it and on the rate of decay and therefore on the release of mineral ions back into the soil. A low (acidic) pH inhibits decay.

- Wind intensity and direction: in areas with strong prevailing winds, the shape of the trees and the whole landscape is affected by the wind. It also means that plants transpire fast.

- Availability of oxygen: the availability of oxygen has a huge impact on water-living organisms. Some invertebrates can survive in water with very low oxygen levels. However, most fish need a high level of dissolved oxygen. The proportion of oxygen in the air varies very little.

- Availability of carbon dioxide: the level of carbon dioxide acts as a limiting factor for photosynthesis and plant growth. It can also affect the distribution of organisms. For example, mosquitoes are attracted to their food animals by high carbon dioxide levels.

The abiotic factors that affect communities of organisms do not work in isolation. They interact to create unique environments where different animals and plants can live.

Figure 3 *The distribution of plants such as this carnivorous pitcher plant depends heavily on soil pH and nutrient levels*

Biotic factors affecting communities

It isn't only abiotic factors that affect communities – biotic factors are very important too. The main biotic factors that affect living organisms and so affect communities are:

- Availability of food: when there is plenty of food, organisms breed successfully. When food is in short supply, animals struggle to survive and often do not breed.

- New pathogens or parasites: when a new pathogen or parasite emerges, organisms have no resistance to the disease. A new pathogen can damage and even wipe out populations in a community.

- New predators arriving: organisms that have no defences against new predators may quickly be wiped out.

- Interspecific competition (competition between species): a new species may outcompete another to the point where numbers become too low for successful breeding. The grey squirrels that were introduced to Britain and outcompeted the native red squirrels are a good example. Another example is Japanese knotweed, which has become a very invasive plant pest.

Figure 4 *In the UK, red squirrels now only survive in areas such as Brownsea Island, northern England, and Scotland where grey squirrels have not reached*

Animals, plants, fungi, protista, bacteria, and archaea are all involved in constant struggles between members of the same species and between members of different species in their community for its resources . Their success in this competition makes the difference between life, including reproduction – and death.

1 State the abiotic factors most likely to affect living organisms.
[6 marks]

2 Explain how carnivorous plants survive in areas with very low levels of nitrate ions. [2 marks]

3 **a** Light intensity is a non-living factor that affects the distribution of living organisms. State another factor that also does this. [1 mark]
   b Explain how light intensity and your chosen factor affect the distribution of living organisms. [4 marks]

4 Explain how a new predator can change the balance of organisms in a community, and ultimately the balance of living organisms in an entire habitat. [5 marks]

Key points

- Abiotic factors that may affect communities of organisms include:
  - light intensity
  - temperature
  - moisture levels
  - soil pH and mineral content
  - wind intensity and direction
  - the carbon dioxide levels for plants
  - the availability of oxygen for aquatic animals.
- Biotic factors that may affect communities of organisms include:
  - availability of food
  - new predators arriving
  - new pathogens
  - new competitors.

B15.3 Distribution and abundance

Learning objectives

After this topic, you should know:

- how to measure the distribution of living things in their natural environment
- how finding the mean, median and mode can help you understand your data.

Figure 1 *Measuring the number of plants in a particular quadrat*

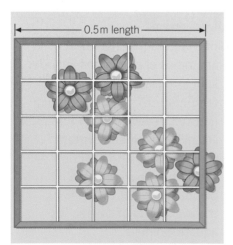

Figure 2 *It doesn't matter if organisms partly covered by a quadrat are counted as in or out, as long as you decide and do the same each time. In this diagram of a quadrat, you have six or seven plants per 0.25 m² (that's 24 or 28 plants per square metre), depending on the way you count*

Ecologists study the make-up of biological communities and ecosystems. They look at how living and non-living factors affect the **abundance** and **distribution** of organisms. They also investigate the effect of changes in the environment on the organisms in a particular ecosystem. To do this, they must be able to measure how many organisms there are and how those organisms are distributed in the first place.

Quadrats

The simplest way to count the number of organisms is to use a sample area called a **quadrat**. We often use a square frame laid on the ground to outline our sample area. People refer to these frames as quadrats too.

A quadrat with sides 0.5 m long gives you a 0.25 m² sample area. Quadrats are used to investigate the size of a population of plants. They can also be used for animals that move very slowly, such as snails or sea anemones.

You use the same size quadrat every time, and sample as many areas as you can. This makes your results as valid as possible. **Sample size** is very important. You must choose your sample areas *at random*. This ensures that your results reflect the true distribution of the organisms and that any conclusions you make will be valid.

There are a number of ways to make sure that the samples you take are random. For example, the person with the quadrat closes their eyes, spins round, opens their eyes, and walks 10 paces before dropping the quadrat. A random number generator is a more scientific way of deciding where to drop your quadrat.

You need to take several random readings and then find the **mean** number of organisms per m². This technique is known as **quantitative sampling**. You can use quantitative sampling to compare the distribution of the same organism in different habitats. You can also use it to compare the variety of organisms in several different habitats.

Finding the range, the mean, the median, and the mode
A student takes 10 random 1 m² quadrat readings looking at the number of snails in a garden. The results are:

3	4	3	4	5	2	6	7	3	3

The **range** of the data is the range between the minimum and maximum values – in this case from **2–7 snails per m².**

To find the **mean** distribution of snails in the garden, add all the readings together and divide by 10 (the number of readings):

3 + 4 + 3 + 4 + 5 + 2 + 6 + 7 + 3 + 3 ÷ 10 = 40 ÷ 10 = **4 snails per m²**

The **median** is the middle value when the numbers are put in order – in this case, the median is **3 snails per m².**

The **mode** is the most frequently occurring value – in this case, **3 snails per m².**

Sampling is also used to measure changes in the distribution of organisms over time. You do this by repeating your measurements at regular time intervals and calculating the mean. Finding the **range** of distribution and the **median** and **mode** of your data can also give you useful information.

Counting along a transect

Sampling along a **transect** is another useful way of measuring the distribution of organisms. There are different types of transect. A line transect is most commonly used.

Transects are not random. You stretch a tape between two points, for example up a rocky shore, across a pathway, or down a hillside. This is often done where you suspect a change is linked to a particular abiotic factor. You sample the organisms along that line at regular intervals using a quadrat. This shows you how the distribution of organisms changes along that line. You can also measure some of the physical factors, such as light levels and soil pH, that might affect the growth of the plants along the transect.

Measure the population size of a common species in a habitat and use sampling techniques to investigate the effect of a factor on the distribution of this species.

Once you have decided on the habitat you are going to investigate you need to make a few more decisions before you start collecting your data:

- Which species are you going to study? It has to be common but may not be the most common – for example, your results on a school field might be more interesting if you look at the population size of daisies or dandelions rather than grass.

- Which factor will you investigate? This could be an easily measured abiotic factor such as light levels or soil pH, or a biotic factor such as trampling, grazing, or competing organisms.

- What will you do? You need to measure the population of your chosen organism throughout the habitat. Then take a transect across your habitat. Take regular quadrats along your transect to investigate any changes in both your environmental factor and the populations of your chosen organism.

Safety: Follow health and safety instructions.

1 State the function of a quadrat. [1 mark]
2 a Describe how a quadrat is used to gain quantitative data. [4 marks]
  b Explain why it is important for samples to be random. [2 marks]
  c In a series of 10 random 1 m² quadrats, a class found the following numbers of dandelions: 6, 3, 7, 8, 4, 6, 5, 7, 9, 8. Determine the mean density of dandelions per m² on the school field, the median value, and the mode from the data. [4 marks]
3 Explain the ways in which the information you get from quadrats and transects is similar and how it differs. [4 marks]

Figure 3 *Carrying out a transect of a rocky shore*

B15.4 Competition in animals

Learning objectives

After this topic, you should know:

- why animals compete
- the factors that organisms are competing for in a habitat
- how organisms are adapted to the environment they live in
- what makes an animal a successful competitor.

Figure 1 *Pandas only eat bamboo, so they are vulnerable to competition from other animals that eat bamboo as well as anything that damages bamboo*

Figure 2 *The dramatic colours of this Costa Rican poison dart frog are a clear warning to predators to keep well away*

Each animal or plant lives with many other organisms. Some will be from the same species while others will be completely different. In any area there is only a limited number of resources. As a result, living organisms have to compete for the things they need.

The best-adapted organisms are those most likely to win the **competition** for resources. They will be most likely to survive and produce healthy offspring. There is competition between members of different species for the same resources and competition between members of the same species. Animals often avoid direct competition with members of other species when they can. It is the competition between members of the same species that is most intense.

What do animals compete for?

Animals compete for many things, including:

- food
- territory
- mates.

Competition for food

Competition for food is very common. Herbivores sometimes feed on many types of plant, and sometimes on only one or two different sorts. Many different species of herbivores will all eat the same plants. Just think how many types of animals eat grass! The animals that eat a wide range of plants are most likely to be successful. If you are a picky eater, you risk dying out if anything happens to your only food source (Figure 1).

Competition is also common among carnivores. They compete for prey. Small mammals such as mice are eaten by animals such as foxes, owls, hawks, and domestic cats. The animal best adapted to finding and catching mice will be most successful. Carnivores have to compete with their own species for their prey as well as with different species. Some successful predators are adapted to have long legs for running fast and sharp eyes to spot prey. These features will be passed on to their offspring.

Prey animals compete with each other too – to be the one that *isn't* caught! Their adaptations help prevent them becoming a meal for a predator. Some animals contain poisons that make anything that eats them sick. Very often these animals also have warning colours so that predators learn which animals to avoid (Figure 2).

The introduction of a new herbivore can drastically reduce the amount of plant material available for other animals. For example, the introduction of rabbits into Australia led to the extinction of a number of common species that simply could not compete with the grass eating and breeding abilities of the rabbits.

Competition for territory

For many animals, setting up and defending a territory is vital. A territory may simply be a place to build a nest, or it could be all the space needed for an animal to find food and reproduce. Most animals cannot reproduce successfully if they have no territory, so they will compete for the best spaces.

This helps to make sure they will be able to find enough food for themselves and for their young. For example, for many small birds such as tits, the number of territories found in an area varies with the amount of food available. Many animals use urine or faeces to mark the boundaries of their territories.

Competition for a mate

Competition for mates can be fierce. In many species, the male animals put a lot of effort into impressing the females. The males compete in different ways to win the privilege of mating with a female.

In some species – such as deer, lions, and elephant seals – the males fight between themselves. The winner then gets to mate with several females.

Many male animals display themselves to females to get their attention. Some birds have spectacular adaptations to help them stand out. Male peacocks display extravagant tail feathers to warn off other males and attract females. Male lizards often display bright colours too.

What makes a successful competitor?

A successful competitor is an animal that is adapted to be better at finding food or a mate than the other members of its own species. It also needs to be better at finding food than the members of other local species. It must also be able to breed successfully.

Many animals are successful because they avoid competition with other species as much as possible. They feed in a way that no other local animals do, or they eat a type of food that other animals avoid. For example, many different animals can feed on one plant without direct competition. While caterpillars eat the leaves, greenfly drink the sap, butterflies suck the nectar from the flowers, and beetles feed on the pollen.

Figure 3 *The striking blue markings and dramatic crests of the male green basilisk lizard (Basiliscus plumifrons) in part **a** are adaptations to attract a mate, as are the extravagant tail feathers of the male peacock in part **b**.*

1 Animals that rely on a single type of food can easily become extinct. Explain why. [2 marks]

2 **a** Give two ways in which animals compete for mates. [2 marks]
b Suggest the disadvantages of the methods chosen in part **a**. 🖊 [4 marks]

3 Suggest two adaptations you would expect to find in each of the following organisms and give one advantage the adaptation would give.
a an animal that hunts small mammals such as mice [4 marks]
b an animal that eats grass [4 marks]
c an animal that is hunted by many different predators [4 marks]
d an animal that feeds on the tender leaves at the top of trees. [4 marks]

B15.5 Competition in plants

Learning objectives

After this topic, you should know:
- what plants compete for
- how plants compete
- adaptations that plants have to make them successful competitors.

Plants compete fiercely with each other. They compete for:

- light for photosynthesis, to make food
- water for photosynthesis and for keeping their tissues rigid and supported
- nutrients (minerals) from the soil, to make all the chemicals they need in their cells
- space to grow, allowing their roots to take in water and nutrients and their leaves to capture light.

Why do plants compete?

As with animals, plants are in competition both with other species of plants and with their own species. Big, tall plants such as trees take up a lot of water and nutrients from the soil. They also reduce the amount of light reaching the plants beneath them. The plants around them need adaptations to help them survive.

When a plant sheds its seeds they might land nearby. In this case, the parent plant will be in direct competition with its own seedlings. As the parent plant is large and settled, it will take most of the water, mineral ions, and light. So the plant will deprive its own offspring of everything they need to grow successfully. The roots of some desert plants even produce a chemical that stops seeds from germinating, killing the competition before it even begins to grow!

Coping with competition

Plants that grow close to other species often have adaptations to help them avoid competition. Small plants found in woodlands often grow and flower very early in the year. This is when plenty of light gets through the bare branches of the trees. The dormant trees take very little water out of the soil. The leaves that shed the previous autumn have rotted down to provide mineral ions in the soil. Plants such as snowdrops, anemones, and bluebells are all adapted to take advantage of these things. They flower, make seeds, and die back again before the trees are in full leaf.

Another way plants successfully avoid competition is by having different types of roots. Some plants have shallow roots taking water and nutrients from near the surface of the soil, while other plants have long, deep roots that go far underground.

If one plant is growing in the shade of another, it may grow taller to reach the light. It may also grow leaves with a bigger surface area to take advantage of all the light it does get. Some plants have adaptations such as tendrils or suckers that allow them to climb up artificial structures or large trees to reach the light.

Figure 1 *Small woodland plants like these wild daffodils grow and flower early in the year, before their competitors even have leaves*

Investigating competition in plants

Carry out an investigation to look at the effect of competition on plants. Set up two trays of seeds – one crowded and one spread out. Then monitor the height and wet mass (mass after watering) of the plants. Keep all of the conditions – light level, the amount of water and mineral ions available, and the temperature – exactly the same for both sets of plants. The differences in their growth will be the result of overcrowding and competition for resources in one of the groups.

This data shows the growth of tree seedlings. You can get results in days rather than months by using cress seeds.

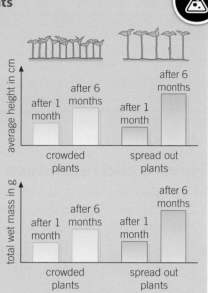

Figure 2 *The light seeds and fluffy parachutes of dandelions mean they are spread widely and compete very successfully*

Spreading the seeds

To reproduce successfully, a plant has to avoid competition with its own seedlings for light, space, water, and mineral ions. Many plants use the wind to help them spread their seeds as far as possible. They produce fruits or seeds with special adaptations for flight to carry their seeds away (Figure 2). Plants also use explosive seed pods, animals, or even water to carry their seeds as far away as possible (Figure 3).

Figure 3 *Coconuts will float for weeks or even months on ocean currents, which can carry them hundreds of miles from competition with their parents – or any other coconuts*

1 **a** Suggest three ways in which plants can overcome the problems of growing in the shade of another plant. [3 marks]
b Explain how snowdrops and bluebells grow and flower successfully in spite of living under large trees in woodlands. [2 marks]

2 **a** Describe why so many plants have adaptations to make sure that their seeds are spread successfully. [2 marks]
b Give three successful adaptations for spreading seeds. [3 marks]

3 The dandelion is a successful weed. Explain why its adaptations make it a better competitor than many other plants on a school field or a garden lawn. [6 marks]

Key points

● Plants often compete with each other for light, space, water, and mineral ions from the soil.
● Plants have many adaptations that make them good competitors.

B15.6 Adapt and survive

Learning objectives

After this topic, you should know:

- what organisms need in order to survive
- how organisms are adapted to survive in many different conditions.

Figure 1 *Flowers that have evolved to be pollinated by bats are just one of the adaptations of the amazing saguaro cacti of the Sonoran desert in North America*

The variety of conditions on the surface of the Earth is huge. It ranges from hot, dry deserts to permanent ice and snow. There are deep saltwater oceans and tiny freshwater pools. Whatever the conditions, almost everywhere on Earth you will find living organisms capable of surviving and reproducing.

Survive and reproduce

Living organisms need a supply of materials from their surroundings and from other living organisms in order to survive and reproduce successfully. What they need depends on the type of organism.

- Plants need light, carbon dioxide, water, oxygen, and mineral ions to produce glucose to give them the energy they need to survive.
- Animals need food from other living organisms, water, and oxygen.
- Microorganisms need a range of things. Some are similar to plants, while some are similar to animals, and some don't need oxygen or light to survive.

Living organisms have special features known as **adaptations**. These features make it possible for them to survive in their particular habitat, even when the conditions are very extreme.

Plant adaptations

Plants need to photosynthesise to produce the glucose needed for energy and growth. They also need enough water to maintain their cells and tissues. They have adaptations that enable them to live in many different places. For example, most plants get water and mineral ions from the soil through their roots.

Epiphytes are found in rainforests. They have adaptations which allow them to live high above the ground attached to other plants. They collect water and nutrients from the air and in specially adapted leaves.

Some plant adaptations are about reproduction. For example, the saguaro cactus (*Carnegiea gigantea*) is one of a small number of plants that rely on bats to pollinate their flowers. Their flowers open at night, have a strong perfume, and produce lots of nectar. The flowers stand on top of the cactus making it easy for the bats to feed from them. At the same time pollen is transferred from one flower to another on the bat's fur.

Animal adaptations

Animals cannot make their own food – they have to eat plants or other animals. Many of the adaptations of animals help them to get the food they need. This means that you can tell what a mammal eats by looking at its teeth.

- Herbivores have teeth for grinding up plant cells.
- Carnivores have teeth adapted for tearing flesh or crushing bones.

Animals also often have adaptations to help them find and attract a mate.

Study tip

Practise recognising how plant and animal adaptations are related to where they live.

Adapting to the environment

Some of the adaptations seen in animals and plants help them to survive in a particular environment. For example:

● Some sea birds get rid of all the extra salt they take in from the sea water by 'crying' very salty tears from a special salt gland. Plants such as mangroves get rid of excess salt in a similar way.

● Animals and plants that survive extreme winter temperatures often produce a chemical in their cells that acts as an antifreeze. It stops the water in the cells from freezing and destroying the cell.

● Plants such as water lilies have lots of big air spaces in their leaves. They float on top of the water to photosynthesise.

Living in extreme environments

Organisms that survive and reproduce in the most difficult conditions are known as **extremophiles** (Figure 2). Many extremophiles are microorganisms, especially archaea. Microorganisms have a range of adaptations that mean they can live in many different environments.

Some extremophiles live at very high temperatures. Bacteria known as thermophiles can survive at temperatures of over 45 °C and often up to 80 °C or higher. These extremophiles have specially adapted enzymes that do not denature at these high temperatures (Figure 3).
In fact, many thermophiles cannot survive and reproduce at lower temperatures. Other bacteria and archaea live at very low temperatures, down to −15 °C. They are found in ice packs and glaciers around the world.

Most living organisms struggle to survive in a very salty environment because of the problems it causes with water balance. However, there are species of extremophile bacteria that can only live in extremely salty environments, such as the Dead Sea and salt flats. These bacteria have adaptations to their cytoplasm so that water does not move out of their cells into their salty environment by osmosis. However, in ordinary sea water they would swell up and burst!

Not all extremophiles are microorganisms. Other species including specialised worms and fish have adaptations for survival in extreme conditions (Figure 2).

1 a Define the term extremophile. [3 marks]
  b Give two examples of adaptations found in different extremophiles. [4 marks]
2 State what the following organisms need to survive and reproduce:
  a plants [5 marks]
  b animals. [3 marks]
3 a Explain what is meant by an adaptation. [2 marks]
  b Give three examples of adaptations in either animals or plants to a particular environment or way of life. [3 marks]
  c Explain how these adaptations help the organism survive in its environment. [6 marks]

Figure 2 *Extremophiles from the deep oceans such as this angler fish are adapted to cope with enormous pressure, no light, and very cold, salty water. If they are brought to the surface too quickly, they explode because of the rapid change in pressure*

Figure 3 *Black smoker bacteria live in deep ocean vents, 2500 m down, at temperatures of well over 100 °C, with enormous pressure, no light, and an acid pH of about 2.8. They have adaptations to cope with some of the most extreme conditions on Earth*

Key points

● To survive and reproduce, organisms need a supply of materials from their surroundings and from the other living organisms in their habitat.
● Organisms, including microorganisms, have features (adaptations) that enable them to survive in the conditions in which they normally live.
● Extremophiles have adaptations that enable them to live in environments with extreme conditions of salt, temperature, or pressure.

B15.7 Adaptation in animals

Learning objectives

After this topic, you should know:

- some of the ways in which animals are adapted in order to survive.

Figure 1 *This moth in an English woodland needs to avoid the birds which might eat it, so camouflage is an important adaptation for survival*

Synoptic link

Remind yourself about surface area : volume ratios by looking back at Topic B1.10.

Animals have adaptations that help them to get the food and mates they need to survive and reproduce. They also have adaptations for survival in the conditions where they normally live. These include:

- structural adaptations, for example the shape or colour of the organism or part of the organism

- behavioural adaptations, such as migration to move to a better climate for the summer or winter, basking to absorb energy from the sun and to warm up, and tool-using to obtain food

- functional adaptations related to processes such as reproduction and metabolism, for example, delayed implantation of embryos or antifreeze in cells.

Animals in cold climates

To survive in a cold environment, you must be able to keep warm. The surface area to volume ratio is very important when you look at the adaptations of animals that live in cold climates. The smaller the surface area to volume ratio the easier it is to reduce the transfer of energy to the environment and minimise cooling. This explains why so many Arctic mammals, such as seals, walruses, whales, and polar bears, are relatively large.

Animals in very cold climates often have other adaptations too. The surface area of the thin-skinned areas of their bodies, such as their ears, is often small, reducing cooling through energy transfers to the environment. Many mammals in cold environments have plenty of insulation. Inside they have blubber, a thick layer of fat that builds up under the skin. On the outside they have a thick fur coat providing very effective insulation.

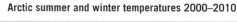

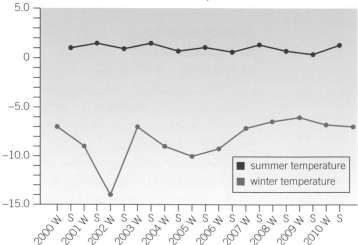

Figure 2 *The Arctic woolly bear moth caterpillar is adapted to survive up to 14 years of freezing and thawing before it becomes an adult moth. Adaptations include antifreeze in the cells, and dark, long hairs that absorb energy and provide insulation*

These adaptations reduce cooling by minimising energy transfers to the surroundings from the skin.

Camouflage

Camouflage is a form of structural adaptation that is important both to **predators** (so their prey doesn't see them coming) and to prey (so they can't be seen).

The colours that would camouflage an Arctic animal in summer against plants would stand out against the snow in winter. Many Arctic animals, for example, the Arctic fox, have grey or brown summer coats that change to pure white in the winter.

The colour of the coat of a lioness is another example of effective camouflage. The sandy brown colour matches perfectly with the dried grasses of the African savannah.

Surviving in dry climates

Dry climates are often also hot climates – like deserts. Deserts are very difficult places for animals to live, because very high temperatures during the day are often followed by very low temperatures at night. Water is also in short supply. However, very cold climates can also be dry because the water is frozen.

The biggest challenges if you live in a desert are:

● coping with the lack of water

● stopping your body temperature from getting too high or too low.

Many desert animals have functional adaptations in their kidneys so they can produce very concentrated urine and need little or nothing to drink. They get the water they need from the food they eat. Animals that live in hot conditions often adapt their behaviour to keep cool. They are often most active in the early morning and evening, when it is cooler. During the cold nights and the hottest times of the day, they rest in burrows or shady areas. Many desert animals are relatively small, with relatively large surface area to volume ratios. They often have large, thin ears, to help them transfer energy to the surroundings through their skin, cooling them down.

Figure 3 *Bactrian camels have to survive extremes of both high temperature (38 °C) and low temperature (-29 °C) in the rocky deserts of East and Central Asia. Adaptations which help them survive include: a thick winter coat; very little sweating; tissues that can cope with big changes in their core temperature; a large store of fat in the humps; the ability to take water from their food; drinking large amounts of water (around 135 litres) at a time when it becomes available*

1 **a** List the main problems that face animals living in cold conditions such as the Arctic. [2 marks]

b List the main problems that face animals living in deserts. [2 marks]

2 Describe three of the adaptations seen in Arctic animals and explain how they work. 🖊 [6 marks]

3 **a** Describe three visible adaptations of an elephant that enable it to keep cool in hot conditions and explain how they work. 🖊 [6 marks]

b Suggest three ways in which animals might be adapted to survive in hot, dry conditions. [3 marks]

c Describe and explain at least two adaptations which enable marine mammals, such as whales and seals, to survive. 🖊 [6 marks]

Key points

● Organisms, including animals, have features (adaptations) that enable them to survive in the conditions in which they normally live. These adaptations may be structural, behavioural, or functional.

B15.8 Adaptations in plants

Learning objectives

After this topic, you should know:

● some of the ways in which plants are adapted to survive.

Synoptic links

To find out more about transport in plants and transpiration, look back to Topics B4.8 and B4.9.

Study tip

Remember that plants need their stomata open to exchange gases for photosynthesis and respiration. However, this leads to loss of water by evaporation, so desert plants have adaptations to conserve water.

Figure 1 *The 'leaves' of butcher's broom are really stems, not leaves. This adaptation reduces water loss so the plant can survive in the dry, shady conditions under big woodland trees*

Plants need light, water, space, and mineral ions to survive. There are some places where plants cannot grow. In deep oceans, no light penetrates and so plants cannot photosynthesise. In the icy wastes of the Antarctic, it is simply too cold for plants to grow.

Almost everywhere else, including the hot, dry areas of the world, you can find plants growing. Without them there would be no food for animals. However, plants need water for photosynthesis and to keep their tissues supported. If a plant does not get the water it needs, it wilts and eventually dies.

Plants take in water from the soil through their roots. It moves up through the plant and into the leaves. There are small openings called stomata in the leaves of a plant. These open to allow gases in and out for photosynthesis and respiration. At the same time, water vapour is lost through the stomata by diffusion after it evaporates from the surface of the cells into the air spaces in the leaves.

The rate at which a plant loses water is linked to the conditions in which it grows. When a plant grows in hot and dry conditions, photosynthesis and respiration take place quickly and as a result, plants lose water vapour very quickly. Plants that live in very hot, dry conditions therefore need special adaptations to survive. Most plants either reduce their surface area so they lose less water or store water in their tissues, and some plants do both!

Changing surface area

When it comes to stopping water loss through the leaves, the surface area to volume ratio is very important to plants. A few desert plants have broad leaves with a large surface area. These leaves collect the dew that forms in the cold evenings. They then funnel the water towards their shallow roots.

Some plants in dry environments have curled leaves to reduce the surface area of the leaf. This also traps a layer of moist air around the leaf to reduce the amount of water the plant loses by evaporation. Most plants that live in dry conditions have leaves with a very small surface area. This adaptation cuts down the area from which water can be lost. Some desert plants have small fleshy leaves with a thick cuticle to keep water loss down. The cuticle is a waxy covering on the leaf that stops water evaporating.

Some examples of plants with specialised adaptations are:

● Marram grass grows on sand dunes. It has tightly curled leaves to reduce the surface area for water loss so it can survive the dry conditions.

● Butcher' broom lives in shady, dry conditions under woodland trees and in hedgerows. To reduce water loss its 'eaves' are really flattened leaf-like bits of stem. The flowers and berries grow straight out from these stems. Stems have far fewer stomata than true leaves, and so the butcher' broom loses very little water. This ensures that it can survive and reproduce in conditions where there is little competition from other species.

The best-known desert plants are the cacti. Their leaves have been reduced to spines with a very small surface area. This means that cacti only lose a tiny amount of water. Not only that, their sharp spines also discourage animals from eating them.

Collecting water

Many plants that live in very dry conditions have specially adapted root systems. They may have extensive root systems that spread over a very wide area, roots that go down a very long way, or both. These adaptations allow the plant to take up as much water as possible from the soil. For example, the mesquite tree has roots that grow as far as 50 m down into the soil.

Storing water

Some plants cope with dry conditions by storing water in their tissues. When there is plenty of water after a period of rain, the plant stores it. Some plants use their fleshy leaves to store water, while other plants use their stems or roots.

For example, cacti don't just rely on their spiny leaves to help them survive in dry conditions. The fat green body of a cactus is its stem, which is full of water-storing tissue. These adaptations make cacti the most successful plants in a hot, dry climate. After a storm, a large saguaro cactus in the desert can take in 1 tonne of water in a single day. Its adaptations for water conservation mean that it normally loses less than one glass of water a day even in the desert heat. A UK apple tree can lose a whole bath's worth of water in the same amount of time!

Synoptic links

For information on surface area to volume ratio, look back at Topic B1.10.

Figure 2 *Plants such as these cacti and the UK apple tree live in very different environments and have very different adaptations for maintaining their water balance*

1 **a** State why plants need water. [4 marks]
 b Describe how plants get the water they need. [2 marks]

2 **a** Describe how plants lose water from their leaves. [3 marks]
 b Explain why this makes living in a dry place such a problem. [2 marks]

3 **a** Plants living in dry conditions have adaptations to reduce water loss from their leaves. Give three of these and explain how they work. 🕖 [6 marks]
 b Describe and explain three other adaptations seen in plants to help them survive in dry conditions. 🕖 [6 marks]

Key points

● Organisms, including plants, have features (adaptations) that enable them to survive in the conditions in which they normally live. These adaptations may be structural, behavioural, or functional.

B15 Adaptations, interdependence, and competition

Summary questions

1 a What is a community of organisms? [2 marks]
   b Organisms within a community are interdependent. In the three communities listed below, explain one way in which some of the organisms are interdependent:
      i an ancient oak woodland [2 marks]
      ii a desert [2 marks]
      iii a pond. [2 marks]
   c Organisms within a community often compete with each other.
      i Describe three ways in which organisms of the same species might compete against each other. [3 marks]
      ii Describe three ways in which organisms of different species might compete against each other. [3 marks]

2 Suggest how each of the following factors affects the distribution of living organisms in an environment. In each case, try to give a plant or animal example.
   a Availability of nesting sites or sheltered areas [3 marks]
   b Availability of nutrients [3 marks]
   c Temperature [3 marks]
   d Amount of light [3 marks]

3 Students carried out an investigation into the distribution of worm casts in different areas of the school grounds – heavily trampled areas of a path across the games field and a well-composted flower bed. They took nine 0.25 m² quadrat readings at random in each area. The results are in Table 1:

Table 1

Trampled area	Flower bed
4	6
3	7
7	5
4	8
5	9
2	9
2	6
0	9
4	4

a Why is it important that the quadrat samples are random? [2 marks]
b Describe a method students could use to ensure the quadrat samples are random. [2 marks]
c For a data set, state what is meant by:
   i the mean [1 mark]
   ii the median [1 mark]
   iii the mode. [1 mark]
d Determine the mean, the median, and the mode of the results from the flower bed. [4 marks]
e What do the results suggest about the distribution of worms in an environment? [1 mark]

4 a Some students carried out samples along a line transect of a rocky shore. What is a line transect? [1 mark]
   b What factors might affect the distribution of organisms along this transect? [4 marks]
   c Describe how you might measure how the distribution of organisms changes along this transect. [3 marks]
   d Explain how the distribution of organisms on a shore will differ from the distribution of organisms in a field. [4 marks]

5 Animals such as amphibians and reptiles do not control their own body temperature internally. They absorb energy from their surroundings and cannot move until they are warm.
   a Why do you think that there are no frogs or snakes in the Arctic? [2 marks]
   b What problems do you think reptiles face in desert conditions and what adaptations could they have to cope with them? [6 marks]
   c Most desert animals are quite small. Explain how this adaptation helps them survive in the harsh conditions. [2 marks]

6 a Explain why competition between animals of the same species is so much more intense than competition between different species. [1 mark]
   b How does marking out and defending a territory help an animal to compete successfully? [2 marks]
   c What are the advantages and disadvantages for males of having an elaborate courtship ritual and colouration compared with fighting over females? [5 marks]

Practice questions

01   Match each ecological term to its definition.

Ecological term	Definition
community	How different species rely on each other for survival
ecosystem	The place where an organism lives
	All the organisms that live in a habitat
habitat	A sampling square
interdependence	The interaction of a community with the non-living parts of their environment

[4 marks]

02   **Figure 1** shows part of a gorse plant.

Figure 1

02.1   Suggest how each of the labelled features helps the gorse plant to survive.   [3 marks]

02.2   Organisms compete in order to survive.
Give **three** resources that plants compete for.
[3 marks]

02.3   Bluebells are wild plants that do not compete well with other plants. Bluebells grow in woods.
The leaves of bluebells appear early in February. Large, dark green leaves quickly grow. Bluebells flower in April and May, but the plants die back by July and become dormant until the next spring.
The trees in the wood are in full leaf by July.
Explain how the habitat and life cycle of bluebells helps them to survive.   [6 marks]

03   The table gives some information about the African elephant and the woolly mammoth.

African elephant	Woolly mammoth
Mass of male: 6000 kg	Mass of male: 8000 kg
Habitat: near the equator	Habitat: northern Europe
An endangered species	Extinct

Use information from the table to help you to answer the following questions.

03.1   The diagrams show that both animals have tusks.
Tusks help animals to compete.
Suggest **two** things animals may compete for.
[2 marks]

03.2   The woolly mammoth was adapted to survive during the ice age.
Use information from the table to suggest two ways the woolly mammoth was adapted to survive in the cold.   [4 marks]

03.3   Darwin's theory of evolution says that elephants developed a trunk because animals with a longer nose had an advantage over animals with a shorter nose.
The elephants with a longer nose survived to breed and pass on the gene for a longer nose to their offspring.
Name the process by which evolution happens.
[1 mark]

03.4   Describe how Lamarck's theory would explain how elephants developed a trunk.   [1 mark]

AQA, 2013

04   **Figure 2** shows a wood.

Figure 2

There is a wide variety of life in the wood.
The different organisms depend on each other for their survival.
Describe how the organisms interact with each other for survival.   [5 marks]

Learning objectives

After this topic, you should know:

- the importance of photosynthesis in feeding relationships
- the main feeding relationships within a community
- how the numbers of predators and prey in a community are related.

Figure 1 *Food chains can be used to represent feeding relationships within a community*

Light from the Sun falls continually on to the surface of the Earth. It is the source of energy for most communities of living organisms. Green plants and algae absorb a small amount of this light for photosynthesis. During photosynthesis, glucose and oxygen are made. The glucose is then used to produce the range of chemicals that make up the cells of the plants and algae. This new material adds to the **biomass** of the organisms. Plants and algae are called **producers** because they produce most of the biomass for life on Earth.

Feeding relationships

The organisms within a community are connected by feeding relationships, and these relationships can be represented simply by food chains. Producers are at the beginning of all food chains because they produce glucose by photosynthesis. On land, the producers are almost always green plants. In the oceans, the main producers are the algae and tiny photosynthetic organisms called phytoplankton.

The animals that eat the producers are known as **primary consumers**. On land these are the herbivores, the plant-eating animals. They include animals ranging from sheep, hippos, and rabbits, to caterpillars, aphids, and an array of birds from parrots to hummingbirds. In the oceans, the primary consumers are often zooplankton, shrimps, crabs, sea urchins, and small fish.

Primary consumers are eaten by animals known as **secondary consumers**. These include carnivores such as lions, foxes, blue tits, eagles, and chameleons. In the oceans the secondary consumers include larger fish, turtles, and seals. Going up the food chain, secondary consumers may themselves be eaten by tertiary consumers – usually large carnivores such as polar bears, birds of prey, or tigers. Food chains are very simple models of the feeding relationships in a community – for example, most organisms eat a variety of food, not just a single species – but they are nevertheless useful in helping us understand how communities work.

producer	$\rightarrow$	primary consumer	$\rightarrow$	secondary consumer	$\rightarrow$	tertiary consumer
phytoplankton	$\rightarrow$	fish	$\rightarrow$	seal	$\rightarrow$	killer whale

Predators and prey

Primary consumers eat plants or algae. This has its problems, because cellulose is very difficult to digest. Herbivores have to use a variety of methods to break down plant cell walls to get at the contents of the cells. Primary consumers have to find and eat enough plant material to provide them with the nutrients they need. The big advantage of eating plants is that they don't move around.

Secondary and tertiary consumers have a different problem. Their food is other animals, so it is high in protein and fat and relatively easy to digest. But animals move about. Before you can eat them you have to catch them. Consumers that eat other animals are known as predators. Consumers that are eaten are known as **prey**. Prey animals are often primary consumers but they may be secondary or even tertiary consumers in different food chains. Predators are always secondary consumers or above. In a stable community, the numbers of predators and prey rise and fall in linked cycles.

- If there is plenty of food available, the prey animals grow and reproduce successfully, so numbers increase.

- As prey animal numbers go up, there is plenty of food available for the predators, so predators can reproduce successfully and predator numbers increase.

- The high number of predators eat a larger proportion of the prey animals, so prey numbers fall.

- With fewer prey animals, there is less food for the predators, so they are less successful and predator numbers fall.

- With the reduction in predators, and the good food supply that results from fewer animals, prey numbers go up again and the cycle repeats itself.

Synoptic links

You learnt about photosynthesis and how the products of photosynthesis are used to make the chemicals needed in cells in Topic B8.1 and Topic B8.3.

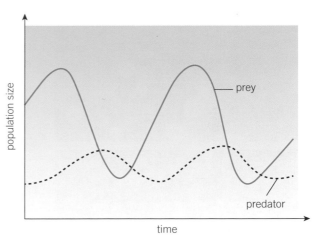

Figure 2 *The numbers of many different predator–prey populations show typical rises and falls*

Key points

- Photosynthetic organisms are the producers of biomass for life on Earth.
- Feeding relationships within a community can be represented by food chains. All food chains begin with a producer which synthesises new molecules. On land this is usually a green plant that makes glucose by photosynthesis.
- Producers are eaten by primary consumers, which in turn may be eaten by secondary consumers and then tertiary consumers.
- Consumers that eat other animals are often predators and those that are eaten are prey. In a stable community the numbers of predators and prey rise and fall in cycles.

1 **a** What is a producer? [1 mark]
   b State why producers are the first organism in a food chain. [1 mark]

2 **a** Describe how food chains are a useful way to model feeding relationships in a community. [2 marks]
   b Suggest one limitation of food chains as a model of what is happening in a community. [2 marks]

3 Use the graphs in Figure 2 to help you answer the following questions:
   a Explain the way predator and prey numbers rise and fall in cycles in a stable community. [3 marks]
   b The numbers in predator–prey populations do not always follow the expected pattern. Suggest reasons why the cycles might differ from the expected model. ✔ [5 marks]

B16.2 Materials cycling

Learning objectives

After this topic, you should know:

- how materials are recycled in a stable community
- the importance of decay.

Figure 1 *This mole was broken down by decomposers over a period of several weeks and the material recycled into the environment*

Synoptic links

You learnt the structure of the main chemicals that make up cells in Topic B1.3. You learnt about the chemistry of respiration in Topic B9.1 and about transpiration in Topic B4.8 and Topic B4.9.

Living organisms remove materials from the environment for growth and other processes. For example, plants take mineral ions from the soil all the time. These nutrients are passed on into animals through feeding relationships. If this were a one-way process, the resources of the Earth would have been used up long ago.

Fortunately, all the materials taken from the environment by plants are returned to the environment and recycled to provide the building blocks for future organisms. For example, many trees shed their leaves each year, and most animals produce droppings at least once a day. Animals and plants eventually die as well. A group of organisms known as the **decomposers** break down the waste and the dead animals and plants. The chemicals that make up living organisms are made up mainly of carbon, oxygen, hydrogen, and nitrogen. These are the elements that need to be recycled to provide the building materials for all new life on Earth. The decay process means that the same material is recycled over and over again and often leads to very stable communities of organisms.

Microorganisms and the recycling of materials

The decomposers are a group of microorganisms that include bacteria and fungi. They feed on waste droppings and dead organisms.

Detritus feeders, or detritivores, such as maggots and some types of worms and beetles, often start the process of decay. They eat dead animals and produce waste material. The bacteria and fungi then digest everything – dead animals, plants, and detritus feeders plus their waste. They use some of the nutrients to grow and reproduce and release carbon dioxide, water, and mineral ions as waste products.

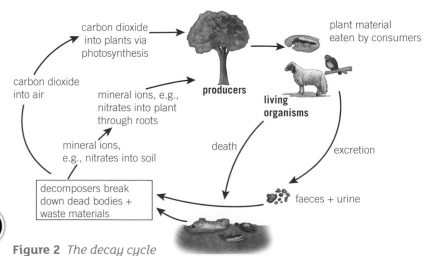

Figure 2 *The decay cycle*

Decomposers return mineral ions, including nitrates, to the soil. Plants take them up through their roots and use them to make proteins and other chemicals in their cells. Decay returns carbon to the atmosphere as carbon dioxide that can be used by producers in photosynthesis. The decomposers also 'clean up' the environment, removing the bodies of all the dead organisms.

The water cycle

Not all of the recycling in an ecosystem relies on the action of microorganisms. Water is vital for life. The water cycle provides fresh water for animals and plants on land before draining into the seas and oceans (Figure 3). Water evaporates constantly from the surface of the land and the rivers, lakes, and oceans of the world. It condenses as it rises into the cooler air and forms clouds where it is then precipitated onto the surface of the Earth as rain, snow, hail, or sleet. Water passes through the bodies of animals and plants, released during respiration in their lifetime, as well as when organisms decay. Animals also release water in urine, faeces, and sweat (in mammals), whilst plants release water into the atmosphere during transpiration. Every drop of water you drink has been through the bodies of many living organisms before you.

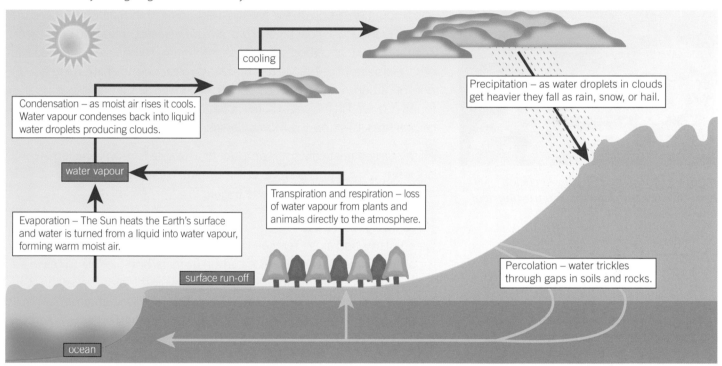

cooling

Precipitation – as water droplets in clouds get heavier they fall as rain, snow, or hail.

Condensation – as moist air rises it cools. Water vapour condenses back into liquid water droplets producing clouds.

water vapour

Transpiration and respiration – loss of water vapour from plants and animals directly to the atmosphere.

Evaporation – The Sun heats the Earth's surface and water is turned from a liquid into water vapour, forming warm moist air.

surface run-off

Percolation – water trickles through gaps in soils and rocks.

ocean

Figure 3 *The water cycle in nature*

1 What is a decomposer? [1 mark]
2 a State the main events in the water cycle. [5 marks]
  b Describe what happens in each of the main events of the water cycle. [5 marks]
3 Materials in an ecosystem are described as being recycled. Explain what this means. [4 marks]
4 Explain the importance of water to living organisms. [4 marks]
5 Explain why decomposers are so important in a stable ecosystem [3 marks]
6 Carbon and nitrogen are two of the key elements that are cycled in the environment. Discuss the importance of these two elements to living organisms and why it is so important that they are part of the decay cycle. ⊘ [6 marks]

Key points

● Material in the living world is recycled to provide building blocks for future organisms.
● Decay of dead animals and plants by microorganisms returns carbon to the atmosphere as carbon dioxide and mineral ions to the soil.
● Carbon dioxide in the atmosphere is used by plants in photosynthesis.
● The water cycle provides fresh water for plants and animals on land before draining into the seas. Water is continuously evaporated, condensed, and precipitated.

B16.3 The carbon cycle

Learning objectives

After this topic, you should know:

- what the carbon cycle is
- the processes that remove carbon dioxide from the atmosphere and return it again.

Figure 1 *Carbon is constantly cycled between living organisms and the physical environment*

Go further

Human actions are putting more and more carbon dioxide into the atmosphere. Scientists are working to develop mathematical models to help them predict how the carbon cycle will react to the measured changes in carbon dioxide levels. They are trying to model both abiotic changes and responses in living organisms.

Imagine a stable community of plants and animals. The processes that remove materials from the environment are balanced by processes that return materials back to the environment. Materials are constantly cycled through the environment. One of the most important of these is carbon. Every year about 166 gigatonnes of carbon are cycled through the living world. That's 166 000 000 000, which is 166×10^9 tonnes (or in standard form 1.66×10^{11} tonnes) – an awful lot of carbon!

All of the main molecules that make up our bodies (carbohydrates, proteins, fats, and DNA) are based on carbon atoms combined with other elements. The amount of carbon on the Earth is fixed. Some of the carbon is 'locked up' in fossil fuels such as coal, oil, and gas. It is only released when we burn them. They are known as carbon sinks. Huge amounts of carbon are combined with other elements in carbonate rocks such as limestone and chalk. There is a pool of carbon in the form of carbon dioxide in the air. Carbon dioxide is also found dissolved in the water of rivers, lakes, and oceans. These things all act as carbon sinks. This stored carbon is described as 'locked-up'.

All the time, a relatively small amount of available carbon is cycled between living things and the environment. This constant cycling of carbon is called the carbon cycle (Figure 2).

Photosynthesis

Green plants and algae remove carbon dioxide from the atmosphere for photosynthesis. They use the carbon from carbon dioxide to make carbohydrates, proteins, and fats. These make up the biomass of the plants and algae. The carbon is passed on through food chains to animals including primary, secondary, and tertiary consumers. This is how carbon is taken out of the environment. But how is it returned?

Respiration

Living organisms respire all the time. They use oxygen to break down glucose, providing energy for their cells. Carbon dioxide, as well as water, is produced as a waste product. This is how carbon is returned to the atmosphere.

When plants, algae, and animals die, their bodies are broken down by the decomposers including blowflies, moulds, and bacteria that feed on the dead bodies. Carbon is released into the atmosphere as carbon dioxide as the decomposers respire. All of the carbon (in the form of carbon dioxide) released by the various living organisms is then available again. It is ready to be taken up by plants and algae in photosynthesis.

Combustion

Wood from trees contains lots of carbon, locked into the molecules of the plant during photosynthesis over many years. Fossil fuels also contain lots of carbon, which was locked away by photosynthesising organisms millions of years ago.

When wood or fossil fuels are burnt, carbon dioxide is produced, so we release some of that carbon back into the atmosphere. Huge quantities of fossil fuels are burnt worldwide to power our vehicles and generate electricity. Wood is burnt to heat homes and (in many countries) to cook food.

Photosynthesis: carbon dioxide + water → glucose + oxygen

Respiration: glucose + oxygen → carbon dioxide + water

Combustion: fossil fuel or wood + oxygen → carbon dioxide + water

Figure 3 *Burning wood and fossil fuels to keep us warm, power our cars, or make our electricity, all releases 'locked-up' carbon in the form of carbon dioxide*

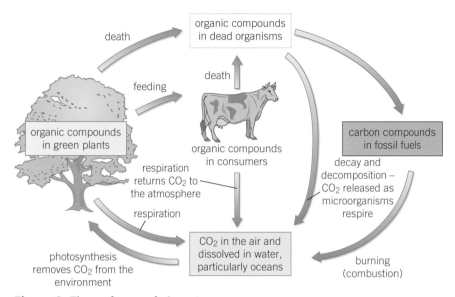

Figure 2 *The carbon cycle in nature*

The future

For millions of years, the carbon cycle has regulated itself. However, as we burn more fossil fuels we are pouring increasing amounts of carbon dioxide into the atmosphere. As the levels of carbon dioxide in our atmosphere increase, we are experiencing global warming. Scientists fear that the carbon cycle may not be able to maintain atmospheric carbon dioxide at a level that will support long-term human survival.

Synoptic links

For more on the role of carbon dioxide in possible climate change, see Topics B17.1, B17.4, and B17.5.

1 **a** What is the carbon cycle? [1 mark]
   b State the three main processes involved in the carbon cycle. [3 marks]
   c Explain why the carbon cycle is so important for life on Earth. [2 marks]

2 State three sources of carbon dixoide for photosynthesis in plants. [3 marks]

3 Describe the role of the processes of photosynthesis, respiration, and combustion in the carbon cycle. 🖊 [6 marks]

Key points

- The carbon cycle returns carbon from organisms to the atmosphere as carbon dioxide to be used by plants in photosynthesis.
- The decay of dead plants and animals by microorganisms returns carbon to the atmosphere as carbon dioxide.

B16 Organising an ecosystem

Summary questions

1 **a** Define the following terms used to describe feeding relationships:
  i producer [1 mark]
  ii primary consumer [1 mark]
  iii secondary consumer. [1 mark]
 b Give examples of:
  i **two** producers [2 marks]
  ii **two** primary consumers [2 marks]
  iii **two** secondary consumers. [2 marks]

2 **a** Explain the difference between predators and prey animals. [2 marks]
 b The graph in Figure 1 shows a typical predator-prey relationship. Describe what is happening at the points labelled A–D on the graph. [4 marks]

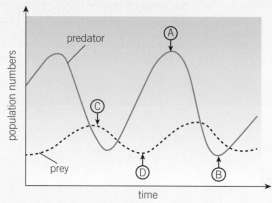

Figure 1 *Model of the relationship between a predator and its prey*

 c Figure 1 shows a theoretical model of the interaction between a predator and its prey. Suggest two reasons why observations from real communities might not look the same as this graph. [4 marks]

3 **a** Name the processes A–F which are involved in the different stages of the carbon cycle. [6 marks]

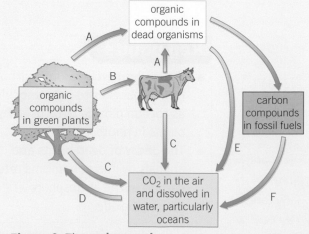

Figure 2 *The carbon cycle*

 b Describe how carbon dioxide is removed from the atmosphere in the carbon cycle. [2 marks]
 c Describe the three main processes by which carbon dioxide gets into and out of the atmosphere. [6 marks]
 d Give the four carbon stores where most carbon is stored. [4 marks]
 e Explain why the carbon cycle is so important. [6 marks]

4 **a** Give three reasons why water is important to living things? [3 marks]

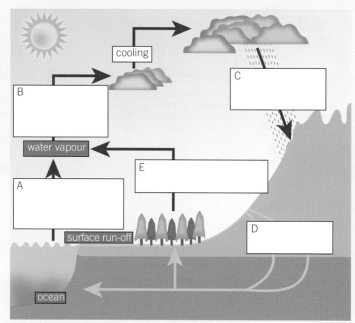

Figure 3 *The water cycle*

 b Describe the processes of the water cycle that are represented by the following letters:
  i A [2 marks]
  ii B [2 marks]
  iii C [2 marks]
  iv D [2 marks]
  v E [3 marks]

Practice questions

01 **Figure 1** shows a food chain.

Figure 1

Rose bush → Greenfly → Ladybirds → Spiders

01.1 Name the producer in this food chain. [1 mark]

01.2 Name the secondary consumer in this food chain. [1 mark]

01.3 Give the name of **one** predator in the food chain **and** the name of its prey. [2 marks]

01.4 What is the source of energy for this food chain? [1 mark]

02 **Figure 2** shows how the population of rabbits and foxes change over time.

Figure 2

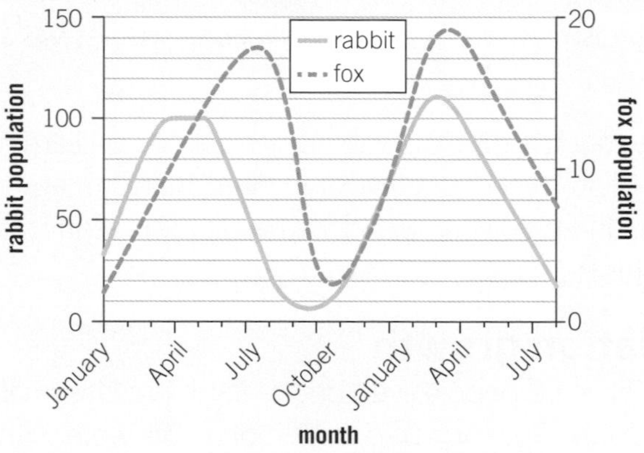

02.1 Draw a food chain that includes rabbits and foxes. [1 mark]

02.2 During which month is the rabbit population at its lowest? Suggest **one** reason for this. [2 marks]

02.3 What was the largest population of rabbits recorded in this study? [1 mark]

02.4 **H** Explain the changes in rabbit and fox populations during the year. [6 marks]

03 Many gardeners put horse manure on their plants to help them grow.

Horse manure contains carbon compounds.

Figure 3

Describe the processes taking place which enable plants to use carbon from manure. [5 marks]

04 Water is essential for the survival of plants and animals.

04.1 Give **one** function of water in animals. [1 mark]

04.2 Give **two** different functions of water in plants. [2 marks]

04.3 Give **three** ways that water is lost from the human body. [3 marks]

04.4 What is the name of the process by which water is lost from a plant? [1 mark]

17 Biodiversity and ecosystems

17.1 The human population explosion

Humans have been on Earth for less than a million years. Yet our activity has changed the balance of nature on the planet enormously. Several of the changes we have made seem to be driving many other species to extinction. Some people worry that we may even be threatening our own survival.

Biodiversity and why it matters

Biodiversity is a measure of the variety of all the different species of organisms on Earth, or within a particular ecosystem. In general, a high biodiversity ensures the stability of ecosystems. It reduces the dependence of one species on another for food, shelter, and the maintenance of the physical environment. As you have already seen, if there are relatively few species in an ecosystem and one is removed in some way, it has a dramatic effect on other species. If there are lots of different types of organisms, the loss or removal of one species has a limited effect because others take its place.

Scientists now realise that the future of the human species on Earth relies on us maintaining a good level of biodiversity. Unfortunately, many of our activities are reducing biodiversity, and we have only recently started attempts at halting this reduction.

Synoptic link

For more information on communities see Topic B15.1.

Human population growth

For many thousands of years, people lived on the Earth in quite small numbers. There were only a few hundred million of us. We were scattered all over the world, and the effects of our activities were usually small and local. Any changes could easily be absorbed by the environment where we lived. However, in the past 200 years or so, the human population has grown very quickly. In 2015, the human population passed 7 billion people, and it is still growing.

If the population of any other species of animal or plant suddenly increased like the human population has, nature would tend to restore the balance. Predators, lack of food, build-up of waste products, or diseases would reduce the population again. Yet we have discovered how to grow more food than we could ever gather from the wild. We can cure or prevent many killer diseases. We have no natural predators. This helps to explain why the human population has grown so fast.

People use the resources of the Earth – we use fossil fuels to generate electricity, for transport, and to make materials such as plastics, minerals from the rocks, and soil to grow food. The more people there are, the more resources we use.

Figure 1 *The Earth – as the human population grows, our impact on the planet increases daily*

The effect on land and resources

The increase in the human population has had an enormous effect on our environment. All of these people need land to live on.

- More and more land is used for building houses, shops, industrial sites, and roads on. This destroys the habitats of other living organisms and reduces biodiversity.

- We use billions of acres of land around the world for farming. Wherever people farm, the natural animal and plant populations are destroyed.

- We dig up vast areas of land for quarries to obtain rocks and metal ores, reducing the land available for other organisms.

- The waste produced by humans pollutes the environment and processing it takes up land, affecting biodiversity.

The huge human population drains the resources of the Earth. People are rapidly using up our finite reserves of metal ores and non-renewable energy resources such as crude oil and natural gas that cannot be replaced.

Managing waste

Rapid growth in the human population, along with improvements in the standards of living in many places, means that increasingly large amounts of waste are produced. This includes human bodily waste and the rubbish from packaging, uneaten food, and disposable goods. The dumping of this waste is another way in which we reduce the amount of land available for any other life apart from scavengers. There has also been an increase in manufacturing and industry to produce the goods we want. This in turn has led to increased industrial waste.

The waste we produce presents us with some very difficult problems. If it is not handled properly, it can cause serious pollution. Our water may be polluted by sewage, by fertilisers from farms, and by toxic chemicals from industry. The air we breathe may be polluted with smoke and poisonous gases, such as sulfur dioxide. The land itself can be polluted with toxic chemicals from farming, such as pesticides and herbicides. It can also be contaminated with industrial waste, such as heavy metals. These chemicals can be washed from the land into waterways. If our ever-growing population continues to affect the ecology of the Earth, reducing biodiversity, everyone will suffer the effects.

1 **a** Give three reasons why the human population has increased so rapidly over the past couple of hundred years. [3 marks]
  b Describe how people reduce the amount of land available for other animals and plants. [4 marks]

2 **a** Give three examples of resources that humans are using up. [3 marks]
  b List five examples of how the standard of living has improved over the past 100 years. [5 marks]

3 Explain in detail the different ways that the ever-increasing human population is causing pollution. 🖊 [6 marks]

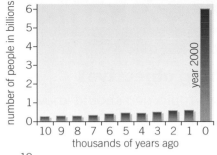

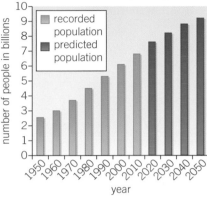

Figure 2 *Human population growth. Current UN predictions suggest that the world population will soar to 244 billion by 2150*

Key points

- Biodiversity is the variety of all the different species of organisms on Earth, or within an ecosystem.
- High biodiversity helps ensure the stability of ecosystems by reducing the dependence of one species on another for food, shelter, and the maintenance of the physical environment.
- Humans reduce the amount of land available for other animals and plants by building, quarrying, farming, and dumping waste.
- The future of the human species on Earth relies on us maintaining a good level of biodiversity. Many human activities are reducing biodiversity and only recently have measures been taken to address the problem.
- Rapid growth in the human population and an increase in the standard of living mean that increasingly more resources are used and more waste is produced.

B17.2 Land and water pollution

Learning objectives

After this topic, you should know:

- how human activities pollute the land
- how human activities pollute the water.

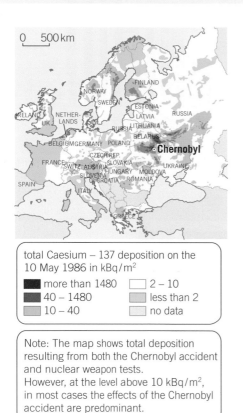

total Caesium – 137 deposition on the 10 May 1986 in kBq/m²

- ■ more than 1480
- ■ 40 – 1480
- ■ 10 – 40
- □ 2 – 10
- ■ less than 2
- □ no data

Note: The map shows total deposition resulting from both the Chernobyl accident and nuclear weapon tests.
However, at the level above 10 kBq/m², in most cases the effects of the Chernobyl accident are predominant.

Figure 1 *The accident at Chernobyl nuclear power plant polluted the land a long way away – including areas of the UK*

As the human population grows, more waste is produced. If it is not handled properly, this waste may pollute the land, the water, or the air. Increased pollution kills plants and animals and reduces biodiversity in affected habitats.

Polluting the land

People pollute the land in many different ways. The more people there are, the more bodily waste and waste water from our homes (sewage) is produced. If human waste is not treated properly, the soil becomes polluted with unpleasant chemicals and gut parasites.

In the developed world, people produce huge amounts of household waste and hazardous (dangerous) industrial waste. The household waste goes into landfill sites, which take up a lot of room and destroy natural habitats. Toxic chemicals can spread from the waste into the soil.

Toxic chemicals are also a problem in industrial waste. They can poison the soil for miles around. For example, after the Chernobyl nuclear accident in 1986, the soil was contaminated thousands of miles away from the original accident. For 26 years, until 2012, the lamb from some farms in North Wales could not be sold for food because the radioactivity levels in the soil and the grass were too high.

Land can also be polluted as a side effect of farming. Weeds compete with crop plants for light, water, and mineral ions. Animal and fungal pests attack crops and eat them. Farmers increasingly use chemicals to protect their crops. Weedkillers (or herbicides) kill weeds but leave the crop unharmed. Pesticides kill the insects that might attack and destroy the crop.

The problem is that these chemicals are poisons. When they are sprayed onto crops, they also get into the soil. From there, they can be washed out into streams and rivers (see later). They can also become part of food chains when the toxins get into organisms that feed on the plants or live in the soil. The level of toxins in the animals that first take in the affected plant material is small, but sometimes it cannot be broken down in their bodies. So, at each stage along the food chain, more and more toxins build up in the organisms. This is known as bioaccumulation, and eventually it can lead to dangerous levels of poisons building up in the top predators (Figure 2).

Polluting the water

A growing human population means a growing need for food. Farmers add fertilisers to the soil to make sure it stays fertile year after year. The minerals in these fertilisers, particularly the nitrates, are easily washed from the soil into local streams, ponds, and rivers. Untreated sewage that is washed into waterways or pumped out into the sea also causes high levels of nitrates in the water. The nitrates and other mineral ions stimulate the growth of algae and water plants, which grow rapidly. Some plants die naturally. Others die because there is so much competition for light that they are unable to photosynthesise. There is a big increase in microorganisms feeding on the dead plants. These microorganisms use up a lot of oxygen during respiration.

This increase in decomposers leads to a fall in the levels of dissolved oxygen in the water. This means there isn't enough oxygen to support some of the fish and other aerobic organisms living in it. They die – and are decomposed by yet more microorganisms. This uses up even more oxygen. Eventually, the oxygen levels in the water fall so low that all aerobic aquatic animals die, and the pond or stream becomes 'dead'.

Toxic chemicals such as pesticides and herbicides or poisonous chemicals from landfill sites can also be washed into waterways. These chemicals can have the same bioaccumulation effect on aquatic food webs as they do on life on land. The largest carnivores die or fail to breed because of the build-up of toxic chemicals in their bodies. In many countries, including the UK, there are now strict controls on the use of chemicals on farms. The same restrictions apply to the treatment of sewage and to landfill sites, to help avoid these problems arising.

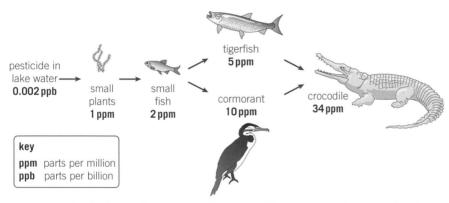

Figure 2 *The feeding relationships between different organisms can lead to dangerous levels of toxins building up in the top predators*

Pollution levels in water can be measured in many different ways. Oxygen and pH levels are measured using instruments. The water can be analysed to show the levels of polluting chemicals such as pesticides or industrial waste. Bioindicators – species such as salmon and bloodworms that can only be found in very clean or very polluted water – are also used to monitor pollution levels in waterways.

Figure 3 *This pond may look green and healthy, but all the animal life it once supported is dead as a result of increased competition for light and oxygen*

1   **a**   What is sewage?   [1 mark]
    b   Explain why is it important to dispose of sewage carefully.   [5 marks]
    c   What are bioindicators used for?   [1 marks]

2   **a**   Farming can cause pollution of the land. Describe the polluting effects farming can have on **i** land, and **ii** water.   [6 marks]
    b   In the UK, a chemical called DDT was used up until the 1980s to kill insects. Large birds of prey and herons began to die and their bodies were found to have very high levels of DDT in them. Discuss how this might have happened and suggest why it took a long time for any link to be made.   [6 marks]

Key points

- Unless waste and chemical materials are properly handled, more pollution will be caused.
- Pollution can occur on land, from landfill and from toxic chemicals such as pesticides and herbicides, which may also be washed from land to water.
- Pollution can occur in water from sewage, fertilisers, or toxic chemicals.
- Pollution kills plants and animals, which can reduce biodiversity.

B17.3 Air pollution

Learning objectives

After this topic, you should know:

- how acid rain is formed
- how acid rain affects living organisms
- how air pollution causes global dimming and smog.

Go further

In 2016 Chinese schools, factories, and building sites were closed regularly when air pollution levels reached dangerous levels. It is difficult to balance rapid industrialisation with environmental issues. Scientists and politicians are working together to try and find ways of improving air quality while maintaining economic progress.

A major source of air pollution is burning fossil fuels. As the human population grows and living standards improve, people are using more non-renewable fossil fuels including oil, coal, and natural gas.

The formation of acid rain

When fossil fuels are burnt in vehicles and factories, acid gases are formed. Carbon dioxide is released into the atmosphere. Fossil fuels often contain sulfur impurities, which react with oxygen when they burn to form sulfur dioxide gas. Sulfur dioxide can cause serious breathing problems for people if the concentrations get too high. Acidic sulfur dioxide and nitrogen oxides also dissolve in rainwater and react with oxygen in the air to form dilute sulfuric acid and nitric acid. This produces **acid rain**, which has been measured with a pH of up to 2.0 – more acidic than vinegar!

The effects of acid rain

Acid rain directly damages the environment. If it falls onto trees, it can kill the leaves, buds, flowers, and fruit. As it soaks into the soil, it can also destroy the roots. Whole ecosystems can be destroyed. As acid rain falls into lakes, rivers, and streams, the water in them becomes slightly acidic. If the concentration of acid gets too high, plants and animals can no longer survive. Many lakes and streams have become 'dead' – no longer able to support life – as a result of this. In some countries, such as Finland, the acid rain falls as 'acid snow'. This can be even more damaging as all the acid is released in the first meltwater of spring. This causes an 'acid flush' that magnifies the effect of the acid rain, producing water with a very low pH.

The worst effects of acid rain are often not felt by the country that produced the pollution (Figure 1). The sulfur dioxide and nitrogen oxides are carried high in the air by the winds. As a result, relatively 'clean' countries can receive acid rain from their 'dirtier' neighbours. Their own clean air moves on to benefit somewhere else.

The UK and other countries have worked hard to stop their vehicles, factories, and power stations producing acidic gases. They have introduced measures to reduce the levels of sulfur dioxide in the air. Low-sulfur petrol and diesel are now used in vehicles, catalytic converters are fitted to remove polluting gases from the exhaust fumes, and strict emission levels are set.

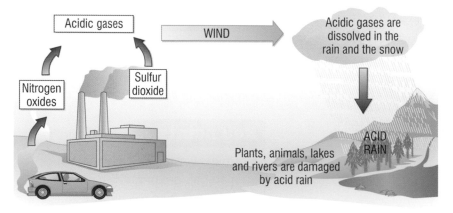

Figure 1 *Air pollution in one place can cause acid rain – and serious pollution problems – somewhere else entirely, even in another country*

In the UK and much of Europe, cleaner, low-sulfur fuels have been introduced such as gas, rather than coal, in power stations and an increase in electricity generated from nuclear power. Systems have been put in place in power station chimneys to clean the flue gases before they are released into the atmosphere. These processes also produce sulfuric acid as a useful by-product, which can be used in industry. As a result, the levels of sulfur dioxide in the air, and of acid rain, have fallen steadily over the past 40 years. However, there are still many countries around the world that do not have such controls in place (Figure 2). There is also an increasing interest in the use of biofuels, which only produce carbon dioxide and water as they burn.

Smoke pollution

Smoke pollution causes an increase in the number of tiny solid particles, called particulates, in the air. The sulfur products obtained from burning fossil fuels are part of this problem, as is smoke from any type of burning. These particles reflect sunlight so less light hits the surface of the Earth. This causes a dimming effect that could lead to a cooling of the temperatures at the surface of the Earth. Smoke pollution also affects human health directly. The particles are breathed in and can damage the lungs and the cardiovascular system.

Smog

Both smoke and chemicals such as sulfur dioxide and nitrogen oxides also add to another form of air pollution – smog. Smog forms a haze of small particles and acidic gases that can be seen in the air over major cities around the world. When China hosted the Olympics in 2008, it was decided to lower pollution so that the air was clean enough for the athletes to compete. The government introduced measures to halve the number of cars on Beijing's roads and close down factories.

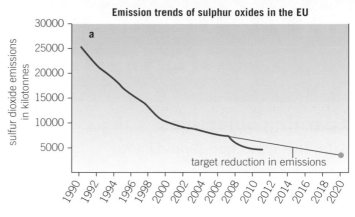

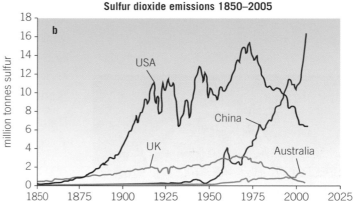

Figure 2 *The reduction of sulfur dioxide emissions and the resulting fall in levels of acid rain has been a great success in some countries – but there is still a lot of work to be done in countries that are still developing, and increasing their sulfur dioxide emissions rapidly*

1 Explain how acidic gases from cars and factories burning fossil fuels can pollute:
   a air [2 marks] b water [2 marks] c land. [2 marks]

2 a Complete a flow chart to show how acid rain is produced. [5 marks]
   b Explain why some countries that have strict controls on sulfur emissions still suffer acid rain damage to their buildings and ecosystems. [3 marks]

3 a Look at Figure 2a. What was the percentage reduction in sulfur dioxide emissions in the EU between 1990 and 2010? [3 marks]
   b Suggest two reasons for the observed reductions. [2 marks]
   c Explain how you would expect these observations to affect the levels of acid rain in Europe. [3 marks]

4 a Using Figure 2b, calculate the change in sulfur dioxide emissions for:
   i China, between 1850 and 1975 [2 marks]
   ii the USA, between 1975 and 2000. [2 marks]
   b Discuss the implications for global biodiversity of the trends seen in Figure 2b. [6 marks]

Key points

- Unless waste and chemical materials are properly handled, more pollution will be caused.
- Pollution can occur in the air from smoke and from acidic gases.
- Air pollution kills plants and animals, which can reduce biodiversity.

B17.4 Deforestation and peat destruction

Learning objectives

After this topic, you should know:

- what is meant by deforestation
- why loss of biodiversity matters
- the environmental effects of destroying peat bogs.

Figure 1 *Deforestation and the changed use of the land show up clearly here*

Figure 2 *The loss of biodiversity, from large mammals such as the orangutan, to the smallest mosses or fungi, will potentially have far-reaching effects in the local ecosystems and for human beings*

Study tip

Remember that trees, plants in peat bogs, and algae in the sea all use carbon dioxide for photosynthesis. Carbon compounds are then 'locked up' in these plants.

As the world population grows, humans need more land, more food, and more fuel. One way to deal with these increased demands has been to cut down huge areas of forests. The loss of our forests may have many long-term effects on the environment and ecology of the Earth.

The effects of deforestation

All around the world, especially in tropical areas, large-scale **deforestation** is taking place to obtain timber and to clear the land for farming. When the land is to be used for farming, the trees are often felled and burnt. This is known as 'slash-and-burn' clearance where the wood is not used, it is just burnt. No trees are planted to replace those that are cut down.

There are three main reasons for deforestation:

1 To grow staple foods such as rice, or ingredients for making cheap food in the developed world, such as palm oil from oil palms.

2 To rear more cattle, particularly for the beef market.

3 To grow crops that can be used to make biofuels based on ethanol. These include sugarcane and maize, which are readily fermented.

The destruction of large areas of trees, whether in tropical areas or in cooler climates, has a number of negative effects.

It increases the amount of carbon dioxide released into the atmosphere in two ways. Burning the trees leads to an increase in carbon dioxide levels from combustion. After deforestation, the dead vegetation decomposes and the microorganisms use up oxygen and release more carbon dioxide as they respire. Deforestation also reduces the rate at which carbon dioxide is removed from the atmosphere. Normally, trees and other plants use carbon dioxide in photosynthesis. They take it from the air and it gets locked up for years (sometimes for hundreds of years) in plant material such as wood. When we destroy trees, we lose a vital carbon dioxide 'sink'. Dead trees don't take carbon dioxide out of the atmosphere. In fact, they add to the carbon dioxide levels when they are burnt or decay.

Loss of biodiversity

Tropical rainforests contain more diversity of living organisms than any other land environment. When we lose these forests, we reduce **biodiversity** as many species of animals and plants may become extinct. Many of these species have not yet been identified or studied. We could be destroying sources of new medicines or food for the future.

For an animal such as the orang-utan, which eats around 300 different plant species, losing the forest habitat is driving the species to extinction. This is just one of hundreds if not thousands of species of living organisms of all different types that are endangered by the loss of their rainforest habitat.

Deforestation is taking place at a tremendous rate. In Brazil alone, an area about a quarter of the size of England is lost each year. When the forests are cleared, they are often replaced by a monoculture (single species) such as oil palms. This process also greatly reduces biodiversity.

Peat bog destruction

Peat bogs are another resource that is being widely destroyed. Peat forms over thousands of years, originally in peat bogs. Over time the bogs may dry out to form peatlands. Peat is made of plant material that cannot decay completely because the conditions are very acidic and lack oxygen. Peatlands and peat bogs act as a massive carbon store. They are also unique ecosystems, home to a wide range of plants, animals, and microorganisms that have evolved to grow and survive in the acidic conditions of a peat bog. These species include a number of carnivorous plants such as sundews and Venus fly traps.

Peat is burnt as a fuel. It is also widely used by gardeners and horticulturists to improve the properties of the soil and provide an ideal environment for seed germination, helping to increase food production. When peat is burnt or used in gardens, carbon dioxide is released into the atmosphere and the carbon store is lost.

Peat is formed very slowly so it is now being destroyed faster than it is made. The destruction of peat bogs also means the destruction of the organisms that depend on them, and more loss of biodiversity.

In the UK, the government is trying to persuade gardeners to use alternative 'peat-free' composts. This will reduce carbon dioxide emissions and conserve peat bogs and peatlands as habitats for biodiversity. Compost can be made from bark, garden waste, coconut husks, and other sources – the problem is persuading gardeners to use them.

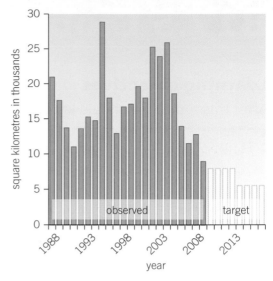

Figure 3 *The rate of deforestation is devastating. Even in the high profile Brazilian Amazon, where deforestation rates are dropping, around 8–10 000 km² of tropical rainforest is being lost each year*

Figure 4 *These pitcher plants have evolved to grow on the acid soil of a peat bog*

1 **a** Define deforestation. [2 marks]
 b Explain how deforestation affects biodiversity – and why it matters. [4 marks]

2 Give three reasons why deforestation increases the amount of carbon dioxide in the atmosphere. [3 marks]

3 **a** Explain why the numbers of peat bogs and peatlands in the world are decreasing. [2 marks]
 b Discuss why this is cause for concern. [4 marks]

4 Discuss the conflict between the need for cheap available compost and the need to conserve peat bogs and peatlands. [6 marks]

Key points

- Large-scale deforestation in tropical areas has occurred to provide land for cattle and for rice fields and to grow crops for biofuels.
- The destruction of peat bogs and other areas of peat to produce garden compost reduces the area of this habitat and thus the biodiversity associated with it.
- The decay or burning of peat releases carbon dioxide into the atmosphere.

B17.5 Global warming

Learning objectives

After this topic, you should know:

- what is meant by global warming
- how global warming could affect life on Earth.

Many scientists are very worried that the average temperature of the Earth is getting warmer. This is commonly called global warming.

Changing conditions

For millions of years, the carbon dioxide released by living things into the atmosphere from respiration has been matched by the amount removed. Carbon dioxide is removed from the atmosphere all the time by plants for photosynthesis. What is more, huge amounts of carbon dioxide are dissolved in the oceans and lakes. Plants and water act as carbon dioxide sinks.

As a result, carbon dioxide levels in the air have stayed about the same for a long period. Now, partly as a result of human activities, the levels of carbon dioxide are currently increasing at the same time as the numbers of plants available to absorb the carbon dioxide are decreasing. The speed of these changes means that the natural sinks cannot cope. The levels of carbon dioxide in the atmosphere are building up (Figure 1). At the same time, the levels of methane gas are also increasing. The land produced by deforestation is often used to grow rice (which releases methane as it grows) or cattle (which produce methane and release it regularly as they digest grass).

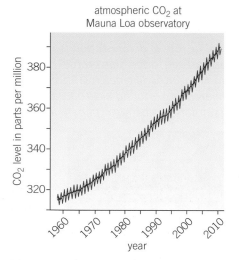

Figure 1 *The atmospheric carbon dioxide readings for this graph are taken monthly on a mountain top in Hawaii. There is a clear upward trend that shows no signs of slowing down*

The greenhouse effect

Energy from the Sun reaches the Earth, warming it up, and much of it is radiated back out into space. However, gases such as carbon dioxide and methane in the atmosphere absorb some of the energy transferred as the Earth cools down (Figure 2). As a result, the Earth and its surrounding atmosphere are kept warm and ideal for life. Because carbon dioxide and methane act like a greenhouse around the Earth, they are known as greenhouse gases. The way they keep the surface of the Earth warm is known as the greenhouse effect, and it is vital for life on Earth.

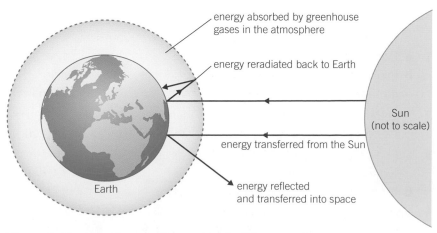

Figure 2 *The greenhouse effect – vital for life on Earth*

Global warming

The greenhouse effect is necessary to keep the Earth's surface at a suitable temperature for life. However, as the levels of carbon dioxide and methane in the atmosphere go up, the greenhouse effect increases. There are more greenhouse gases in the atmosphere and the average temperature at the Earth's surface is going up. The change is very small – only about 0.55 °C from the 1970s to the present day. This is not much – but an increase of only a few degrees Celsius may cause climate change and rising sea levels.

Many scientists think that a change in climate due to global warming will give a rise in severe and unpredictable weather events. Some people think that the very high winds and huge floods seen globally in the early years of the 21st century are early examples of these effects. If the Earth warms up, the ice caps at the North and South Poles will melt and so will many glaciers. As a result sea levels will rise and parts of countries or even whole countries may disappear beneath the sea. As temperatures rise and the water gets warmer, less carbon dioxide gas can dissolve in it. This in turn makes the problem worse – global warming will affect people all over the the world. Both climate change and a rise in sea levels will have a number of biological consequences:

- Loss of habitat – when low-lying areas are flooded by rising sea levels, habitats will be lost and so will the biodiversity of the area.

- Changes in distribution – as temperatures rise or fall, and rainfall patterns change, climate change may make conditions more favourable for some animals and they may be able to extend their range. Others may find that their range shrinks. Some will disappear completely from an area or a country.

- Changes in migration patterns – as climates become colder or hotter, and the seasons change, the migration patterns of birds, insects, and mammals may change.

- Reduced biodiversity – as the climate changes, many organisms will be unable to survive and will become extinct, for example, the potential loss of polar bears as Arctic ice melts.

Figure 3 *Puffin populations in northern Scotland are failing to rear their chicks because a rise in sea temperatures reduces the numbers of small fish that puffins feed on. They may need to move to new breeding sites if they are to survive*

Synoptic links

For information on how plants use carbon dioxide to make food, look back to Topic B8.1.

Synoptic links

To find out more about changes in the distribution of organisms, look at Topic B15.8.

Key points

- Levels of carbon dioxide and methane in the atmosphere are increasing, and contribute to global warming.
- Biological consequences of global warming include loss of habitat when low-lying areas are flooded by rising sea levels, changes in the distribution patterns of species in areas where temperature or rainfall has changed, and changes to the migration patterns of animals.

1 **a** Use the data in Figure 1 to produce a bar chart showing the maximum recorded level of carbon dioxide in the atmosphere every tenth year from 1960 to 2010. [3 marks]
   b Explain the trend you can see on your chart. [3 marks]
   c Describe the greenhouse effect. [5 marks]
2 **a** Explain why global warming is occuring. [4 marks]
   b Describe two of the the biological consequences of global warming. [6 marks]
3 Give an example of an organism that has been, or might be, affected by global warming. Explain the effect of global warming on the survival of this organism. [6 marks]

B17.6 Maintaining biodiversity

Learning objectives

After this topic, you should know:

- how waste, deforestation, and global warming all have an impact on biodiversity
- some of the ways people are trying to reduce the impact of human activities on ecosystems and maintain biodiversity.

Synoptic links

You learnt about some of the problems that can result from inbreeding in Topic B13.3.

Figure 1 *Sand lizards are just one of six types of reptiles found on UK lowland heaths*

Figure 2 *Losing a single tree reduces biodiversity, for example, around 1000 different species live on an English oak, and 19 trees studied in Panama had 1200 species of beetles alone*

People are becoming more aware of the importance of biodiversity and of the many human threats to biodiversity around the world. Increasingly, scientists and concerned citizens are putting programmes in place to try and reduce these negative effects on ecosystems and biodiversity. People can help maintain biodiversity in many ways, including the following.

Breeding programmes for endangered species

Breeding programmes can restore an endangered species to a sustainable population, but it is difficult. Many rare animals and plants do not reproduce easily or fast and artificial breeding programmes must avoid inbreeding. Often the habitat that the organisms need to survive has also been lost. There are some high-profile international breeding programmes of animals such as the panda and Przewalski horses. They have been bred in zoos with the genetics of each animal carefully recorded. Herds of these horses now live wild again in a Mongolian national park. Many rare snails and amphibians now only exist in captive breeding programmes. They will only be released when their natural habitats are safe again.

Protection and regeneration of rare habitats

Many habitats have become increasingly rare, and so the species of animals and plants adapted to living in them are increasingly under threat. People are protecting some of these rare habitats, and sometimes enabling them to regenerate. This protects the biodiversity, which may even increase again. Coral reefs, mangroves, and heathlands are all rare habitats that are threatened and are now becoming protected. To protect coral reefs, carbon dioxide emissions and global warming must be tackled as raised temperatures and decreased pH levels are the major threats to these most biodiverse marine habitats. Mangroves are vital sites for young fish to develop and are easily destroyed by too much or too little water or changes in salinity (salt content).

Around 20% of all the lowland heaths in the world are in the UK but this rare habitat has been disappearing fast. Now people are managing it to maintain its unique features and protect it from developers. They are even re-establishing lost heathland by removing trees, reversing drainage, and allowing ponies and cows to graze wild.

Reintroduction of field margins and hedgerows

In many agricultural areas farmers removed the hedgerows to produce huge fields in which to grow a single crop. This removed a wide variety of plants, birds, insects, and mammals – and in some places led to soil erosion and a reduction in soil fertility. Gradually farmers are replanting hedgerows and leaving wildflower margins round the edges of their fields, and the biodiversity of the countryside is increasing again.

Reduction of deforestation and carbon dioxide emissions

Some governments are recognising the damage deforestation and increasing carbon dioxide emissions are doing to the environment and biodiversity, and are working hard to reduce the effect. At one point the rainforests in Costa Rica were being felled rapidly until the government recognised what was happening. Now the Costa Rican rainforests and cloud forests are largely protected. Farm land originally produced by felling forests is being bought and allowed to return to forest over time. Tourists pay to visit the amazing habitats and see the biodiversity for themselves.

Many governments are working with the transport and electricity generation industries to reduce carbon dioxide emissions. For example, the carbon dioxide emissions of new cars are falling steadily as a result of more efficient engines.

Recycling resources

Waste placed in landfill sites affects biodiversity by using land and producing pollution. Globally, many countries are working to recycle as much waste as possible – including paper, glass, plastics, and metal – rather than dumping it in landfill. There is also a drive to recycle organic waste as compost or in methane generators. It isn't just households that are recycling waste – companies from car manufacturers to brewers are doing the same thing. In 2012, General Motors recycled or reused 84% of their manufacturing waste. This saved 2.2 million tonnes of landfill and prevented 11 million tonnes of CO_2 emissions. Many governments have introduced taxes on putting material in landfill – and as the tax has gone up, the amount of material put in landfill has gone down!

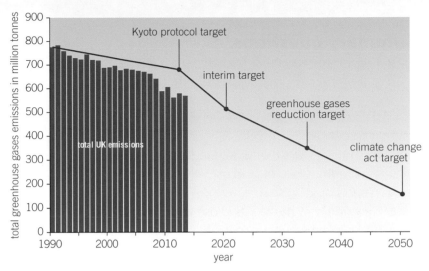

Figure 3 *Carbon dioxide emissions in the UK are falling as a result of the Kyoto agreement between many governments, the UK Climate Change Act 2008, and other legislation*

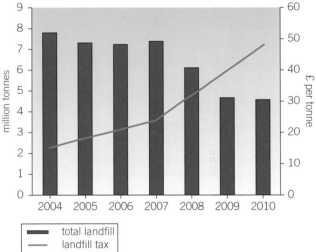

Figure 4 *Financial penalties can help reduce landfill and protect biodiversity*

1 a Explain why it is important to maintain biodiversity. [3 marks]
   b Summarise the main ways people can help to maintain biodiversity. [5 marks]
   c Suggest an example of where there might be a conflict between maintaining biodiversity and human needs. [4 marks]

2 Using the data in Figure 3:
   a Describe the trend in carbon dioxide emissions in the UK since 1990. [1 mark]
   b Suggest how this data demonstrates the effect of governments on carbon emissions. [3 marks]
   c Discuss three examples of how a fall in carbon dioxide emissions globally might help maintain biodiversity. [6 marks]

3 Using the data in Figure 4, suggest how taxes can be used to help reduce human damage to ecosystems and biodiversity. [4 marks]

> ### Key points
>
> - Scientists and concerned citizens have put programmes in place to reduce the negative effects of humans on ecosystems and biodiversity.

B17 Biodiversity and ecosystems

Summary questions

1 a i List three ways that human beings reduce the amount of land available for other living things. [3 marks]

ii Explain why each of these land uses is necessary to people. [3 marks]

iii Suggest ways that would reduce two of these different types of land use. [4 marks]

b What is biodiversity? [1 mark]

c Describe how human land use affects biodiversity and explain why this is a problem. 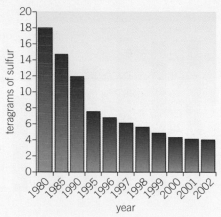 [6 marks]

2 a Draw a flow chart showing acid rain formation. [5 marks]

b Figure 1 shows the sulfur emissions made by European countries from 1980 to 2002. Use this graph to help you answer the following questions:

Figure 1 *Sulfur dioxide emissions made by European countries*

i What was the level of sulfur emissions in 1980? [1 mark]

ii What was the approximate level of sulfur in the air in the year that you were born? [1 mark]

iii What was the level of sulfur emissions in 2002? [1 mark]

c What does this data tell you about trends in the levels of sulfur emissions since 1980? [2 marks]

d Sketch the bar chart and show how you would expect it to have continued to the present day. Discuss your predictions. [6 marks]

3 In Figure 2 you can see clearly annual fluctuations in the levels of carbon dioxide recorded each year. These fluctuations are thought to be due to seasonal changes in the way plants are growing and photosynthesising throughout the year.

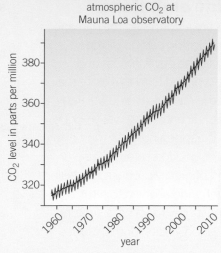

atmospheric CO$_2$ at Mauna Loa observatory

Figure 2 *The atmospheric carbon dioxide readings for this graph are taken monthly on a mountain top in Hawaii. There is a clear upward trend that shows no signs of slowing down*

a Explain how changes in plant growth and the rate of photosynthesis might affect carbon dioxide levels. [5 marks]

b How could you use this data as evidence to support the arguments against deforestation? [6 marks]

c Describe how the ever-increasing human population is affecting the build-up of greenhouse gases in the air. [6 marks]

d Explain how, in theory, this build-up of greenhouse gases will affect the climate of the Earth. [6 marks]

e Give two examples of evidence that the climate of the Earth is changing. [4 marks]

4 a Suggest how each of the following can reduce levels of biodiversity:

i Water pollution caused by sewage and fertilisers [5 marks]

ii Acid rain [3 marks]

iii Deforestation [3 marks]

b Explain how each of the following can help to maintain biodiversity:

i Breeding programmes for endangered species. [4 marks]

ii Protection of rare and endangered habitats. [4 marks]

iii Reduction of deforestation and carbon dioxide emissions. [4 marks]

Practice questions

01 Match each pollutant to its possible effect.

Pollutant	**Possible effect**
	acid rain
methane	blackening of buildings
sewage	increase risk of diseases
smoke	global warming

[3 marks]

02 There are areas in deep seas where there are hot water springs called hydrothermal vents.

Figure 1

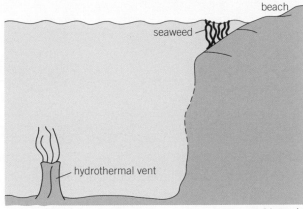

not to scale

Volcanic activity creates an environment of very hot, acidic water. No light reaches the deep seabed. The environment is very extreme but many different organisms can live there.

02.1 Complete the following sentence.
Organisms that can survive in extreme conditions are called .... [1 mark]

02.2 The food chains show some organisms that live near a hydrothermal vent and some organisms that live near the surface of the sea.
A food chain found near a hydrothermal vent:

bacteria → limpet → crab

A food chain found near the surface of the sea:

green seaweed → limpet → crab

- Give a difference between the food chain found near a hydrothermal vent and the food chain found near the surface of the sea.
- Use information given in this question and your own knowledge to suggest reasons for the difference. [6 marks]

AQA, 2013

03 Figure 2 shows how the world human population has increased.

Figure 2

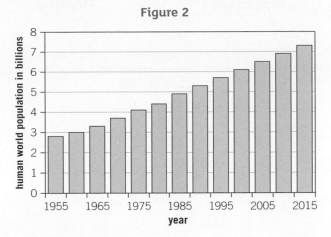

03.1 By how much has the world population increased between 1955 and 2015? [1 mark]

03.2 Use **Figure 2** to predict what the population will be in 2050. [1 mark]

03.3 The rapid growth in the human population means more food needs to be produced.
Suggest **one** other problem associated with an increasing human population. [1 mark]

03.4 To increase food production farmers started to grow one type of crop in a very large field. This led to an increased use of fertilisers and pesticides and caused environmental problems.
Today many farmers are reintroducing field margins and hedgerows.
Explain **one** way that this can help to improve the environment. [2 marks]

04 Deforestation is the clearing of large areas of forest.

04.1 Suggest **two** reasons why large areas of forest might be cleared. [2 marks]

04.2 Explain **two** ways deforestation can lead to global warming. [4 marks]

04.3 Explain **one** biological consequence of global warming. [2 marks]
Deforestation can lead to a reduction in biodiversity.

04.4 What is biodiversity? [1 mark]

04.5 Explain how deforestation can lead to a reduction in biodiversity. [2 marks]

Paper 1 questions

01.1 Match each type of cell to the correct drawing.

Type of cell	Drawing

a

animal cell

b

plant cell

c

bacterial cell

[3 marks]

01.2 Use the correct word from the box to match each description about cell structure.

| chloroplasts | nucleus | cell membrane |
| mitochondria | ribosomes | wall |

A .... contains chromosomes.

A .... controls what enters and leaves a cell.

Aerobic respiration takes place in ....

Photosynthesis takes place in the

.... [4 marks]

02 Some types of bacteria multiply as often as once every 20 minutes if given the optimum conditions.

02.1 What type of cell division happens in bacterial cells? [1 mark]

02.2 Give **two** factors that would affect the rate of cell division in bacterial cells. [2 marks]

02.3 If the mean division time was 20 minutes, how many divisions would there be in 7 hours? [1 mark]

02.4 Use the following equation to calculate the number of bacteria there would be after 7 hours if the population at the start was 1 bacterium. [2 marks]

$$\begin{array}{c}\text{number of}\\\text{bacteria at end of}\\\text{growth period}\end{array} = \begin{array}{c}\text{number of}\\\text{bacteria at start}\\\text{of growth period}\end{array} \times 2^{\text{number of divisions}}$$

03 Root hair cells are specialised cells.
Figure 1 shows a root hair cell. The symbols represent water molecules and nitrate and magnesium ions.

Figure 1

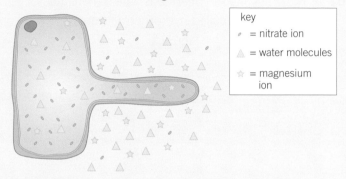

key
- = nitrate ion
△ = water molecules
☆ = magnesium ion

03.1 Describe how a root hair cell is adapted to absorb substances from the soil. [1 mark]

03.2 Substances can move into and out of cells by active transport, diffusion, or osmosis.
State how each of the following substances would move into the root hair cell shown in **Figure 1**.
water molecules ....
magnesium ions ....
nitrate ions ....
[3 marks]

04 Gonorrhoea is a sexually transmitted disease (STD). Some strains of the pathogen that cause gonorrhoea are resistant to antibiotics. This makes the disease difficult to treat.

04.1 What type of pathogen causes gonorrhoea?
A virus
B protist
C fungus
D bacterium [1 mark]

04.2 Give one other way to reduce the spread of gonorrhoea. [1 mark]

04.3 Suggest one way we could reduce the rate of development of antibiotic resistant strains. [1 mark]

05 A student investigated the effect of protease on egg white suspension.
He set up four tubes, as shown in **Table 1**, and put them in a water bath at 37 °C.

Table 1

Test tube A	Test tube B	Test tube C	Test tube D
• egg white • protease • hydrochloric acid	• egg white • protease • water	• egg white • boiled protease • hydrochloric acid	• egg white • water • hydrochloric acid

05.1 Give **two** variables the student should have controlled in the investigation. [2 marks]

05.2 At the start of the investigation the contents of the tubes looked cloudy due to the egg white suspension.
The student observed the tubes every minute and recorded whether the contents of each tube looked cloudy or clear. When the contents went clear this indicated that the protein in the egg white had been digested.
Explain why the contents of the tubes turned from cloudy to clear when the protein had been digested. [3 marks]
The results of the investigation are shown in **Table 2**.

Table 2

Time in minutes	Tube A	Tube B	Tube C	Tube D
0	cloudy	cloudy	cloudy	cloudy
1	cloudy	cloudy	cloudy	cloudy
2	clear	cloudy	cloudy	cloudy
3	clear	cloudy	cloudy	cloudy
4	clear	cloudy	cloudy	cloudy
5	clear	cloudy	cloudy	cloudy
6	clear	cloudy	cloudy	cloudy
7	clear	cloudy	cloudy	cloudy
8	clear	cloudy	cloudy	cloudy
9	clear	cloudy	cloudy	cloudy
10	clear	cloudy	cloudy	cloudy
11	clear	cloudy	cloudy	cloudy
12	clear	cloudy	cloudy	cloudy
13	clear	clear	cloudy	cloudy
14	clear	clear	cloudy	cloudy
15	clear	clear	cloudy	cloudy

05.3 Analyse the results in **Table 2**.
You should include in your answer:
● an explanation of the results for each tube
● conclusions [6 marks]

05.4 Suggest how the investigation could be improved to measure the digestion of the egg white protein? [3 marks]

06 Drugs must be trialled before the drugs can be used on patients.
Before the clinical trials, drugs are tested in the laboratory.
The laboratory trials are not trials on people.

06.1 What is the drug tested on in these laboratory trials? [1 mark]

06.2 Drugs must be trialled before the drugs can be used on patients.
Give **three** reasons why. [3 marks]

06.3 Read the information about cholesterol and ways of treating high cholesterol levels.
Diet and inherited factors affect the level of cholesterol in a person's blood.
Too much cholesterol may cause deposits of fat to build up in blood vessels and reduce the flow of blood. This may cause the person to have a heart attack.
Some drugs can lower the amount of cholesterol in the blood.
The body needs cholesterol. Cells use cholesterol to make new cell membranes and some hormones. The liver makes cholesterol for the body.
Some drugs can help people with high cholesterol levels.
Statins block the enzyme in the liver that is used to produce cholesterol.
People will normally have to take statins for the rest of their lives. Statins can lead to muscle damage and kidney problems. Using some statins for a long time has caused high numbers of deaths.
Cholesterol blockers reduce the absorption of cholesterol from the intestine into the blood.
Cholesterol blockers can sometimes cause problems if the person is using other drugs.
Evaluate the use of the two types of drug for a person with high cholesterol levels. [6 marks]

AQA, 2012

07 An athlete is running on a treadmill. The athlete's heart rate is being measured.
The athlete's heart rate at rest was 67 beats per minute.
The athlete jogged at a steady pace for 10 minutes and did not get out of breath.
His heart rate increased to 138 beats per minute.

07.1 Explain why heart rate needs to increase when jogging. [4 marks]

07.2 **H** The athlete then ran at a fast pace for a further 10 minutes.
When he stopped he was out of breath for several minutes.
Explain why the athlete was out of breath after he had stopped running. [4 marks]

08.1 Complete the equation for photosynthesis by selecting the correct answer from each box.

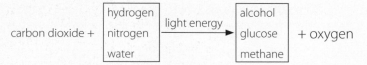

[2 marks]

Some students investigated the effect of light intensity on the rate of photosynthesis in pondweed.
Figure 2 shows the apparatus the students used.

Figure 2

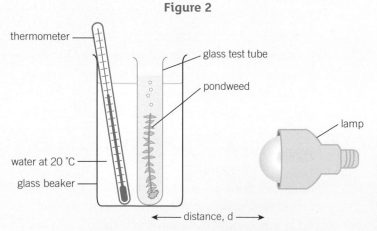

The closer the lamp is to the pondweed, the more light the pondweed receives.
The students placed the lamp at different distances, d, from the pondweed.
They counted the number of bubbles of gas released from the pondweed in one minute for each distance.
A thermometer was placed in the glass beaker.

08.2 Why was it important to use a thermometer in this investigation? [3 marks]

08.3 The students counted the bubbles four times at each distance and calculated the correct mean value of their results.
Table 3 shows the students' results.

Table 3

Distance d in cm	Number of bubbles per minute				
	1	2	3	4	Mean
10	52	52	54	54	53
20	49	51	48	52	50
30	32	30	27	31	30
40	30	10	9	11	

Calculate the mean number of bubbles released per minute when the lamp was 40 cm from the pondweed. [2 marks]

08.4 Draw a graph to show the students' results:
- label the axes
- plot the mean values of the number of bubbles
- draw a line of best fit. [4 marks]

08.5 **H** One student concluded that the rate of photosynthesis was inversely proportional to the distance of the lamp from the plant.
Does the data support this conclusion?
Explain your answer. [2 marks]

08.6 **H** Light intensity, temperature and concentration of carbon dioxide are factors that affect the rate of photosynthesis.
Scientists investigated the effects of these three factors on the rate of photosynthesis in tomato plants growing in a greenhouse.
Figure 3 shows the scientists' results.

Figure 3

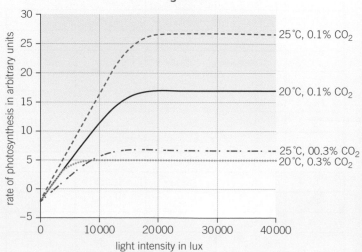

A farmer in the UK wants to grow tomatoes commercially in a greenhouse.

The farmer read about the scientists' investigation. During the growing season for tomatoes in the UK, natural daylight has an intensity higher than 30 000 lux.

The farmer therefore decided to use the following conditions in his greenhouse during the day:

- 20 °C
- 0.1% CO_2
- no extra lighting.

Suggest why the farmer decided to use these conditions for growing the tomatoes.

You should use information from the scientists' graph in your answer.   [4 marks]

AQA, 2014

09   Ⓗ A student wanted to find the concentration of sugar solution inside potato cells.

The student did the following:

- cut five pieces of potato
- blotted each piece of potato on paper and weighed each piece
- put each piece into a different concentration of sugar solution
- left the pieces in the solution for 1 hour
- blotted and weighed each piece of potato again.

Table 4 shows the student's results.

Table 4

Concentration of sugar solution in moles per dm³	Mass of potato at start in grams	Mass of potato after one hour in grams	Percentage change in mass of potato
0.2	8.3	8.5	+ 2.4
0.4	8.3	8.2	
0.6	8.2	8.0	− 2.4
0.8	8.1	7.8	− 3.7
1.0	8.3	7.9	− 4.8

09.1 Why did the student blot each piece of potato before weighing it?   [1 mark]

09.2 Calculate the percentage change in mass of potato for the 0.4 moles per dm³ sugar solution.   [2 marks]

09.3 Plot a graph to show the percentage change in mass of potato against the concentration of sugar solution.

You should:

- consider where to draw the axes so positive and negative values can be plotted
- draw a line of best fit through the points.   [5 marks]

09.4 Use your graph to determine the concentration of sugar in the potato cells.   [1 mark]

09.5 Explain why you have chosen this concentration.   [3 marks]

Paper 2 questions

01 The nervous system enables humans to react to their surroundings and to coordinate their behaviour.

01.1 Use the correct word from the box to complete the sentence.

| coordinator | response | sense | stimulus |

A change in the environment is called a

.... [1 mark]

01.2 Match each structure in the nervous system with its correct description. [5 marks]

Structure	Description
central nervous system	A muscle or a gland.
	Brain and spinal cord.
effector	Creates a stimulus.
	Carries electrical impulses to a muscle.
motor neurone	Detects changes in the environment.
receptor	Gap between neurones.
synapse	Sense organ.

[5 marks]

02 This question is about coordination in animals. Nerve impulses travel at high speeds through the nervous system.

Figure 1 shows the reflex pathway of an impulse from a receptor in the hand to the muscle that contracts to pull the hand away from the sharp pin.

Figure 1

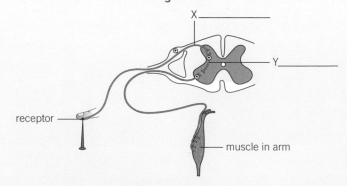

02.1 Label neurones **X** and **Y** on the diagram. [2 marks]

02.2 Describe how information is transmitted from neurone **X** to neurone **Y**. [3 marks]

02.3 The average speed an impulse travels through this pathway is 65.4 m/s.
The length of the pathway, from the receptor in the hand to the muscle in the arm, is 1.38 m.

$$\text{speed} = \frac{\text{distance}}{\text{reaction time}}$$

Use the equation to calculate the reaction time for this person.
Give the unit. [3 marks]

02.4 Reflex actions have a shorter reaction time than voluntary actions.
Suggest **two** reasons why. [2 marks]

02.5 Many processes in the body are coordinated by chemical substances called hormones.
A frightened mouse is trapped in the corner of a barn by a cat.
As the cat prepares to attack, the mouse suddenly runs away and escapes.
Name the hormone that is released in the mouse when it is frightened. [1 mark]

02.6 Explain how this hormone brings about changes in the body of the mouse to help it to escape.

[4 marks]

03 Human body cells contain 23 pairs of chromosomes.

03.1 Describe the similarities and differences in the chromosomes of a male and a female human.

[3 marks]

03.2 Copy and complete the Punnett square to show the inheritance of sex.

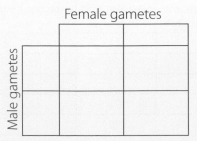

[3 marks]

03.3 A couple have three children. They are all boys.
They would like to have a daughter to complete their family.
Use your Punnett square to predict the probability that their fourth child would be a girl. [1 mark]

03.4 🕭 It is possible to produce a female baby using IVF treatment. However, it is illegal to choose the sex of your baby in the UK unless there is a risk of passing on a serious genetic disorder to your child. Describe the steps involved in IVF treatment to produce a female baby. [6 marks]

03.5 🕭 Suggest the pros and cons of the UK law that only allows parents to choose the sex of their baby if there is a risk of passing on a serious genetic disorder. [4 marks]

04 **Figure 2** shows an Arctic fox. The Arctic fox is a predator.

Figure 2

04.1 Describe **three** features of the Arctic fox that help it to be a successful predator. [3 marks]

04.2 The number of prey organisms in its habitat is one biotic factor that can affect the size of the Arctic fox population.
Give **two** other biotic factors that could affect the population of Arctic fox. [2 marks]

05 Variation in organisms can be caused by genes, the environment, or a mixture of both factors.
A student investigated the variation in the height of daffodils growing in the school grounds.

05.1 Suggest **one** cause of variation in the height of daffodils.
Give a reason for your answer. [2 marks]

05.2 **Table 1** shows the student's results.

Table 1

Height in cm	Tally
15.0–19.9	
20.0–24.9	卌 IIII
25.0–29.9	卌 卌 卌
30.0–34.9	卌 卌 II
35.0–39.9	III
40.0–44.9	I
45.0–49.9	

Suggest why the student recorded the height of the daffodils in groups. [1 mark]

05.3 Plot the student's results on graph paper. [4 marks]

05.4 What is the modal height range? [1 mark]

05.5 What is the range in the height of daffodils the student measured? [1 mark]

06 Some students were carrying out a fieldwork investigation on the school field.
One pair of students decided to estimate the total number of dandelion plants on the field.
The students:
- measured the size of the field
- placed a 0.5 m by 0.5 m sized quadrat in five different parts of the field where there were dandelions growing
- counted how many dandelions were in the quadrat
- used their results to estimate the total population of dandelions on the whole field.

The students' results are shown in **Table 2**.

Table 2

Quadrat number	Number of dandelions
1	9
2	12
3	15
4	14
5	10

06.1 What was the mean number of dandelions per quadrat. [1 mark]

06.2 The field measured 190m by 130m. Estimate how many dandelions there were on the field using the students' results. [4 marks]

06.3 The students' estimate was not accurate. Suggest **one** improvement to the students' investigation.

06.4 Explain how your answer to question **06.3** would produce a more accurate estimate. [2 marks]

Another pair of students decided to investigate how the distribution of two types of plantain varied across a path that students walked on.
Figure 3 shows the two types of plantain.

Figure 3

ribwort plantain   greater plantain

The students used a quadrat that was divided into 25 squares to estimate the percentage cover of each type of plantain.
Figure 4 shows one of the quadrats.

Figure 4

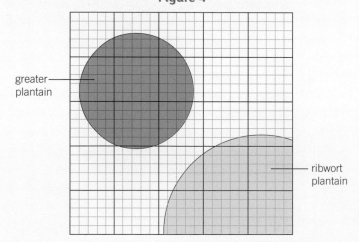

greater plantain

ribwort plantain

06.5 The percentage cover of the Greater plantain in **Figure 4** is 16%.
Estimate the percentage cover of Ribwort plantain. [2 marks]

06.6 The students placed quadrats across the path as shown in **Figure 5**.

Figure 5

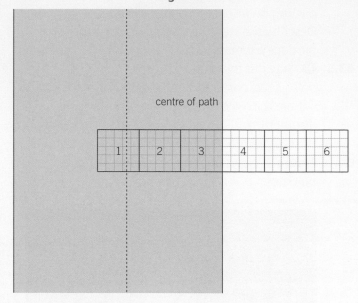

centre of path

1   2   3   4   5   6

06.7 What is this type of sampling, which measures changes across a line, called? [1 mark]
The students' results are shown in **Table 3**.

Table 3

	Percentage cover					
Quadrat number	1	2	3	4	5	6
Ribwort plantain	2	3	14	20	29	36
Greater palntain	30	28	18	5	3	2

06.8 Describe the distribution of Ribwort plantain and of Greater plantain across the path.
Use information from **Figure 3** to suggest reasons for this. [4 marks]

07 Ⓗ In a healthy individual the blood glucose concentration is maintained within a narrow range. Explain how the pancreas, liver, and muscles coordinate in order to maintain an optimum blood sugar concentration. [6 marks]

08 Humans can use different methods to produce plants and animals with desired characteristics. Selective breeding is one method that has been used for centuries by farmers.

08.1 Describe how a farmer could produce a cow that gives a high yield of creamy milk. [4 marks]

Another method to produce organisms with desired characteristics is genetic engineering. *Bacillus thuringiensis* (Bt) is a soil bacterium. This bacterium produces a poison that kills several different species of insect that feed on cotton plants.

Scientists have genetically modified cotton plants so that they produce the Bt poison.

08.2 Describe how cotton plants can be genetically modified to produce the Bt poison. [3 marks]

08.3 Explain the advantages of growing Bt cotton. [3 marks]

08.4 Some people are opposed to growing Bt cotton. Suggest **two** reasons why. [2 marks]

09 **Figure 6** shows the organs in the endocrine system of a woman.

Figure 6

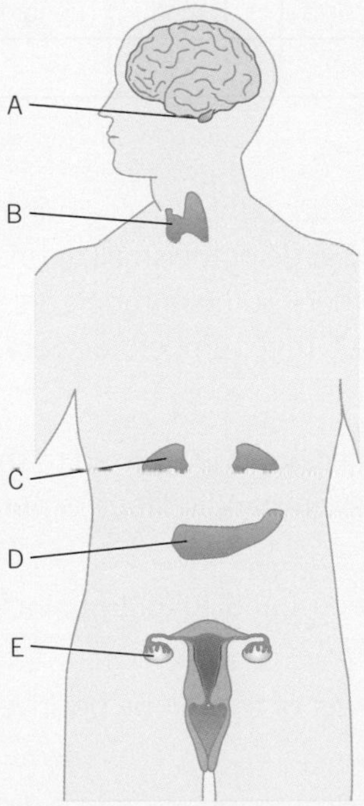

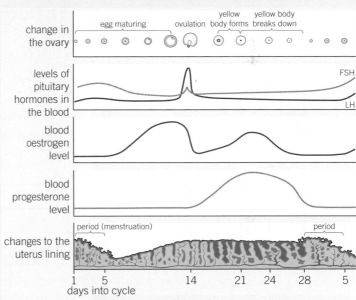

09.1 What is an endocrine gland? [2 marks]

09.2 Give the letter of each of the following organs shown in Figure 6.
Adrenal gland
Pituitary gland
Thyroid gland [3 marks]

09.3 The pituitary gland is often called the master gland. Explain why. [2 marks]

Several hormones are involved in control of the menstrual cycle in women.

Figure 7 shows how the concentrations of these hormones in the blood change during the menstrual cycle.

Figure 7

09.4 🄷 Explain the interactions between FSH, oestrogen, LH, and progesterone in the control of the menstrual cycle. [6 marks].

Maths skills for Biology
MS1 Arithmetic and numerical computation

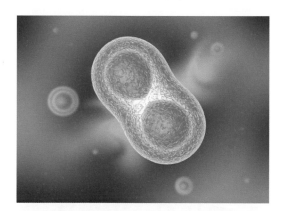

Figure 1 *Cells divide at different rates*

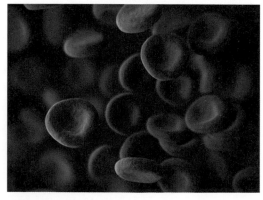

Figure 2 *These red blood cells have a diameter of approximately 7.0×10^{-6} m*

Synoptic link

To see how biologists use numbers in standard form, see Maths skills MS1b.

How big is a human cell? If one bacterium starts dividing, how many will there be in 24 hours? What is the probability that a child will inherit dimples from its parents?

Biologists use maths all the time – when collecting data, looking for patterns, and making conclusions. This chapter will support you in developing the key maths skills you need during your GCSE biology course. The rest of the book gives you many opportunities to practise using maths when it is needed as you learn about biology.

1a Decimal form

There will always be a whole number of bees in a swarm, and a whole number of eggs in a nest.

When you make measurements in biology the numbers may *not* be whole numbers but numbers *in between* whole numbers. These are numbers in decimal form, for example, the height of a boy could be 177.8 cm, or the mass of a soil sample could be 25.5 g.

The value of each digit in a number is called its place value.

thousands	hundreds	tens	units	·	tenths	hundredths	thousandths
4	5	1	2	·	3	4	5

1b Standard form

Place value can help you to understand the size of a number, however some numbers in biology are too large or too small to understand when they are written as ordinary numbers. Some numbers are very large, like the number of cells in your body. Other numbers are very small, such as the size of a bacterium or a virus.

Standard form (also called scientific notation) is used to show very large or very small numbers more easily. In standard form, a number is written as $A \times 10^n$.

- A is a decimal number between 1 and 10 (but not including 10), for example 7.0.
- n is a whole number. The power of ten can be positive or negative, for example 10^{-6}.

Remember to add the unit if applicable, for example, a distance measurement could have the unit m (for metres).

This gives you a number in standard form, for example, 7.0×10^{-6} m. This is the diameter of a single human red blood cell.

Table 1 explains how you convert numbers to standard form.

Table 1 *Converting numbers into standard form*

The number	The number in standard form	What you did to get to the decimal number	...so the power of ten is...	What the *sign* of the power of ten tells you
1000 m	1.0×10^3 m	You moved the decimal point 3 places to the *left* to get the decimal number	+3	The positive power shows the number is *greater* than one.
0.01 s	1.0×10^{-2} s	You moved the decimal point 2 places to the *right* to get the decimal number	−2	The negative power shows the number is *less* than one.

100 000 000 bacteria per cm³ of culture solution = 1.0×10^8 bacteria per cm³ of culture solution.

Multiplying numbers in standard form

You can use a scientific calculator for calculations using numbers written in standard form. You should work out which button you need to use on your own calculator (it could be **EE**, **EXP**, **10ˣ**, or **×10ˣ**).

Worked example: Standard form

1 mm³ of blood contains around 5 000 000 red blood cells.

How many red blood cells would there be in a standard blood donation of 470 cm³?

Solution

Step 1: 470 cm³ = 470 000 mm³

Convert the numbers to standard form.

$5\,000\,000 = 5 \times 10^6$ and $470\,000 = 4.7 \times 10^5$

Step 2: Calculate the total number of red blood cells in a blood donation.

Total number of red blood cells = number of red blood cells × mm³ blood

$= (5 \times 10^6 \text{ rbcs}) \times (4.7 \times 10^5 \text{ mm}^3 \text{ blood})$

$= (5 \times 4.7) \times (10^6 \times 10^5)$ pages

$= 23.5 \times 10^{11}$

In standard form $= 2.35 \times 10^{12}$ red blood cells per blood donation

---

Figure 3 *You can use a scientific calculator to do calculations involving standard form*

1c Ratios, fractions, and percentages

Ratios

A ratio compares two quantities. A ratio of $2:20$ of foxes to chickens means that for every two foxes, there are 20 chickens.

You can compare ratios by changing them to the form $1:n$ or $n:1$.

$1:n$ Divide both numbers by the *first* number.
For every one fox there are ten chickens.

$$\div 2 \left(\begin{array}{c} 2:20 \\ 1:10 \end{array} \right) \div 2$$

$n:1$ Divide both numbers by the *second* number.
For every tenth of a fox there is one chicken (even though you can't really get 'a tenth of a fox').

$$\div 20 \left(\begin{array}{c} 2:20 \\ 0.1:1 \end{array} \right) \div 20$$

You can describe the foxes in relation to the number of chickens using three different ratios $2:20$, $1:10$ and $0.1:1$. All of the ratios are equivalent – they mean the same thing.

You can simplify a ratio so that both numbers are the lowest whole numbers possible.

Worked example: Simplifying ratios

In an investigation to look at the effect of pH on the rate of enzyme action, a student mixed $15\,cm^3$ of acid with $90\,cm^3$ of water. Calculate the simplest ratio of the volume of acid to the volume of water.

Solution

Step 1: Write down the ratio of *acid : water*

Step 2: Both 15 and 90 have a common factor, 5.

Divide both numbers by 5.

Step 3: Both 3 and 18 have a common factor, 3.

Divide both numbers by 3.

$$\div 5 \left(\begin{array}{c} 15:90 \\ 3:18 \end{array} \right) \div 5$$
$$\div 3 \left(\begin{array}{c} \\ 1:6 \end{array} \right) \div 3$$

To get the simplest form of the ratio, you have divided by 15 (i.e., 3×5), which is the highest common factor of 15 and 90.

Synoptic link

To see examples of how biologists find ratios useful, see look at Topic B1.10, Topic B13.7, and Topic B13.8.

Worked example: Calculating the fraction of a quantity

A student has a 25 g sample of glucose. Calculate the mass of $\frac{2}{5}$ of this sample.

Solution

$\frac{2}{5}$ of $25 = \frac{2}{5} \times 25$

Step 1: Divide the total mass of the sample by the denominator.

$$25\,g \div 5 = 5\,g$$

Step 2: Multiply by the numerator.

$$5\,g \times 2 = 10\,g$$

Fractions

A fraction is a part of a whole.

$\frac{1}{3}$ The numerator tells you how many parts of the whole you have.

    The denominator tells you how many equal parts the whole has been divided into.

To convert a fraction into a decimal, divide the numerator by the denominator.

$\frac{1}{3} = 1 \div 3 = 0.33333\ldots = 0.\dot{3}$ (the dot shows that the number 3 recurs, or repeats over and over again).

To convert a decimal to a fraction, use the place value of the digits, then simplify. $0.045 = \frac{45}{1000} = \frac{9}{200}$

Percentages

A percentage is a number expressed as a fraction of 100.

$$77\% = \frac{77}{100} = 0.77$$

Worked example: Calculating a percentage

In an area of 75 cm² on a rocky shore a student found that 27.3 cm² was covered by mussels. Calculate the percentage of the rock covered by mussels on this site.

Solution

$$\text{percentage of rock covered in mussels} = \left(\frac{\text{area covered in mussels}}{\text{area of sample}}\right) \times 100\%$$

$$= \frac{27.3}{75} \times 100 = 36.4\%$$

Figure 4 *What is the percentage coverage of these mussels on the rocky shore?*

You may need to calculate a percentage of a quantity.

Worked example: Using a percentage to calculate a quantity

The biomass of the primary consumers in a food chain is 1500 kg. Only 11% of this biomass forms new material in the secondary consumers. Calculate the mass of material forming new secondary consumer.

Solution

Step 1: Convert the percentage to a decimal.

$$11\% = \frac{11}{100} = 0.11$$

Step 2: Multiply the answer to Step 1 by the total biomass in the primary consumers.

$$0.11 \times 1500 = 165 \text{ kg}$$

You may need to calculate a percentage increase or decrease in a quantity from its original value.

Worked example: A percentage change

A student heats a 4.75 g sample of soil and finds its mass decreases to 3.04 g. Calculate the percentage change in mass.

Solution

Step 1: Calculate the decrease in mass.

$$4.75 \text{ g} - 3.04 \text{ g} = 1.71 \text{ g}$$

Step 2: Divide the decrease in mass by the original mass.

$$\frac{1.71}{4.75} = 0.36$$

Step 3: Convert the decimal to a percentage.

$$0.36 \times 100\% = 36\%$$

Remember that in this case the answer is a percentage *decrease*.

1d Estimating the result of a calculation

When you use your calculator to work out the answer to a calculation you can sometimes press the wrong button and get the wrong answer. The best way to make sure that your answer is correct is to estimate it in your head first.

Worked example: Estimating an answer

A photosynthesising plant produced gas at a constant rate of $0.34\,mm^3/min$. How much gas is produced after eight minutes?

Solution

Find $0.34\,mm^3/min \times 8\,min$. Estimate the answer and then calculate it.

Step 1: Round each number up or down to get a whole number multiple of 10.

$0.34\,mm^3/min$ is about $0.3\,mm^3/min$ $8\,min$ is about $10\,min$

Step 2: Multiply the numbers in your head.

$0.3\,mm^3/min \times 10\,min = 3\,mm^3$

Step 3: Do the calculation and check it is close to your estimate.

Distance $= 0.34\,mm^3/min \times 8\,min = 2.72\,cm^3$

This is quite close to 3 so it is probably correct.

Notice that you could do other things with the numbers:

$0.34 + 8 = 8.34$ $\dfrac{0.34}{8} = 0.0425$ $8 - 0.34 = 7.66$

Not one of these numbers is close to 3. If you got any of these numbers you would know that you needed to repeat the calculation.

Study tip

When carrying out multiplications or divisions using standard form, you should add or subtract the powers of ten to work out roughly what you expect the answer to be. This will help you to avoid mistakes.

Sometimes the calculations involve more complicated equations, or standard form.

1 If the diameter of cell is $1.25 \times 10^7\,\mu m$ when viewed under an electron microscope at a magnification of $\times 500\,000$, what is the actual diameter of the cell? [1 mark]

2 a The concentration of a weak glucose solution was given as $0.000038\,mol/dm^3$. Express this concentration in standard form. [1 mark]

   b Bacteria in a fermenter produce human insulin at a rate of about 5×10^2 tonnes every year. If this rate is maintained, how many years will it take the bacteria to produce 1.25×10^4 tonnes? [1 mark]

3 a Students carried out several crosses between the same pair of black mice. Adding up all of the offspring they found they had 17 brown mice and 51 black mice. What is the simplest ratio of these offspring? [1 mark]

   b In a sample of 350 wheat plants, students discovered that only 24% of them were healthy. The rest had been attacked by wheat rust or other pests. How many plants were diseased? [1 mark]

   c The approximate percentage oxygen gas in the air is 20%. Express this percentage as a fraction. [1 mark]

MS2 Handling data

2a Significant figures

Numbers are rounded when it is not appropriate to give an answer that is too precise.

When rounding to significant figures, count from the first non-zero digit.

These masses each have three significant figures (s.f.). The s.f. are underlined in each case.

153 g, 0.153 g, 0.00153 g

Table 1 shows some more examples of measurements given to different numbers of s.f.

In general, you should give your answer to the same number of s.f. as the data in the question that has the lowest number of s.f.

Remember that rounding to s.f. is *not* the same as decimal places. When rounding to decimal places, count the number of digits that follow the decimal point.

> **Worked example: Significant figures**
> Calculate the rate of an enzyme controlled reaction that gives off 25 cm³ of gas in 7.85 s.
>
> **Solution**
> **Step 1:** Write down what you know.
>
> Volume of gas given off = 25 cm³ (2 s.f.)   Time = 7.85 s (3 s.f.)
>
> You should give your answer to 2 s.f.
>
> **Step 2:** Write down the equation that links the quantities you know and the quantity you want to find.
> $$\text{mean rate of reaction (cm}^3\text{/s)} = \frac{\text{amount of product formed (cm}^3)}{\text{time } t \text{ (s)}}$$
> **Step 3:** Substitute values into the equation.
> $$\text{speed} = \frac{25\ \text{cm}^3}{7.85\ \text{s}}$$
> $$= 3.184\,713\,375\ \text{cm}^3\text{/s}$$
> $$= 3.2\ \text{cm}^3\text{/s to 2 s.f.}$$

2b Arithmetic means

To calculate the **mean** (or average) of a series of values:

1   add together all the values in the series to get a total

2   divide the total by the number of values in the data series.

You will often need to do this with your sets of repeat readings when conducting investigations. This helps you to obtain more accurate data from sets of repeat readings where you have some random measurement errors.

Learning objectives

After this topic, you should know how to:

- use an appropriate number of significant figure
- find arithmetic means
- construct and interpret frequency tables and bar charts
- make order of magnitude calculations.

Table 1 *The number of significant figures – the significant figures in each case are underlined*

Number	0.05 s	5.1 nm	0.775 g/s	23.50 cm³
Number of significant figures	1	2	3	4

Synoptic link

There are many more examples of applying significant figures throughout your biology course.

Figure 1 *What is the mean height of this group?*

Table 2 *A frequency table for the blood group of 25 people*

Blood group	Frequency
A	10
B	3
AB	1
O	11

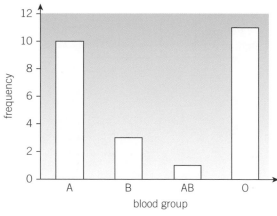

Figure 2 *Blood group O is the most common*

Table 3 *The pulse rate of 30 people*

Pulse rate, beats/minute	Frequency
60–64	1
65–69	4
70–74	12
75–79	8
80–84	5
85–89	1

Study tip

Remember to leave a gap between each bar.

Worked example: Calculating a mean

A student wanted to measure the average height of a group of five students.

Their results were as follows:

162 cm   159 cm   169 cm   157 cm   163 cm

Calculate the mean height of the group.

Solution

Step 1: Add together the recorded values.

162 cm + 159 cm + 169 cm + 157 cm + 163 cm = 810 cm

Step 2: Then divide by the number of recorded values (in this case, there 5 students).

$$\frac{810}{5} = 162 \text{ cm (3 s.f.)}$$

The mean height of the group was 162 cm (3 s.f.)

2c Frequency tables, bar charts, and histograms
Frequency tables and bar charts

The word data describes observations and measurements which are made during experiments or research.

Qualitative data is non-numerical data, such blood groups.

The frequency table (Table 2) shows the blood group of 25 people.

The height of the bars in the bar chart represent the frequency of each category.

Quantitative data is numerical measurements.

Discrete data can only take exact values (usually collected by counting).

The frequency table (Table 3) the pulse rate of 30 people. The measurements are grouped into intervals.

- make sure the values in each class do not overlap
- aim for a sensible number of classes – usually no more than six.

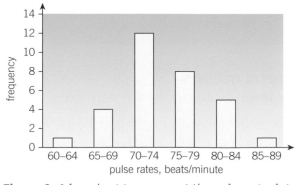

Figure 3 *A bar chart to represent the pulse rate data*

You can also use bar charts to compare two or more independent variables.

Histograms

Continuous data can take any value (usually collected by measuring), such as mass, volume, or density.

Continuous data can be displayed on a histogram. A histogram is similar to a bar chart, except that its classes and bars may have unequal widths. The key differences between a bar chart and a histogram are:

● The frequency of each class is represented by the area of each bar, not the height of each bar.

● The vertical axis is labelled as the frequency *density*, not the frequency.

● The horizontal axis is shows the value, not the name of the category.

● There are no gaps between the bars on a histogram.

Worked example: Calculating frequency density

A student investigated the heights of a type of plant in a small field.

Copy and complete the table to show the frequency density for each range of heights, and plot a histogram to represent this data.

Height of plants, h / cm	Frequency	Class width	Frequency density
$0 \leq h < 20$	40		
$20 \leq h < 30$	15		
$30 \leq h < 50$	20		
$50 \leq h < 60$	30		
$60 \leq h < 100$	20		

Solution

Step 1: Calculate each class width.

Step 2: Use the formula frequency density $= \dfrac{\text{frequency}}{\text{class width}}$ to calculate the frequency density

Height of plants, h / cm	Frequency	Class width	Frequency density
$0 \leq h < 20$	40	$20 - 0 = 20$	$\dfrac{40}{20} = 2.0$
$20 \leq h < 30$	15	$30 - 20 = 10$	$\dfrac{15}{10} = 1.5$
$30 \leq h < 50$	20	$50 - 30 = 20$	$\dfrac{20}{20} = 1.0$
$50 \leq h < 60$	30	$60 - 50 = 10$	$\dfrac{30}{10} = 3.0$
$60 \leq h < 100$	20	$100 - 60 = 40$	$\dfrac{20}{40} = 0.5$

Step 3: Plot a histogram of height against frequency density to illustrate the data (Figure 4).

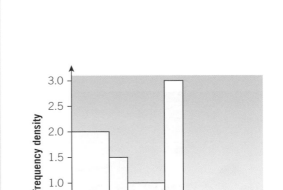

Figure 4 *A histogram to represent the plant height data*

Synoptic link

There are many examples of how biologists produce and use bar charts and histograms in Chapter B7 and throughout your biology book.

Worked example: Estimating animal population size

Scientists used the capture-recapture technique to estimate the grey seal population in an area off the coast of Wales.

First sample: 50 seals

Second sample: 40 unmarked seals, and 5 marked seals

Estimate the population size.

Solution

Step 1: Write out the formula.

estimated population size =

$$\frac{\text{first sample size} \times \text{second sample size}}{\text{number of recaptured marked individuals}}$$

Step 2: Enter the data.

estimated population
$$\text{size} = \frac{50 \times 45}{5}$$

Step 3: Work out the answer.

Estimated population
size = 450 seals

Figure 5 *Scientists use the capture-recapture method to estimate the population of grey seals*

Synoptic link

You can find examples of how biologists use sampling techniques in Topic B15.3.

2d Sampling

Sampling means taking a smaller number of observations or measurements from a larger population or area. You can scale up the sample to make estimates about the larger area.

You may want to estimate the abundance (number) of selected organisms on your school field. The population of animals in an area can be estimated using the capture-recapture technique. This technique scales up the results from a small sample area to estimate a population. To do this:

1 Capture organisms from a sample area.

2 Mark individual organisms, then release back into the community.

3 At a later date, recapture organisms in the original sample area.

4 Record the number of marked and unmarked individuals.

5 Estimate the population size using the formula:

estimated population size = $\dfrac{\text{first sample size} \times \text{second sample size}}{\text{number of recaptured marked individuals}}$

Estimating plant populations from a sample

To work out the plant population in an area:

1 Mark a small area of land – 1 m² is often ideal. Record the type and number of organisms in the area.

2 Take a number of samples of the area, and calculate the mean population for each organism present. The larger the number of samples taken, the more repeatable and reproducible your results.

3 To work out the total population of an organism use the formula,
Estimated population size = mean population per unit area × total area

Worked example: Estimating plant population size

Students looked at five 1 m² square areas of the school garden. They found that each area had two buttercup plants. Estimate the population of buttercup plants in their 60 m² garden.

Solution

Step 1: Write out the formula.

estimated population size = mean population per unit area (/m²) × total area (m²)

Step 2: Enter the data.

estimated population size = 2 /m² × 60 m²

Step 3: Work out the answer.

estimated population size = 120 buttercup plants

2e Probability

The probability of an event occurring tells you how likely it is for the event to happen. A probability can be written as a percentage, a decimal, or a fraction.

Table 4 shows some probabilities you might come across when looking at genetic crosses.

Worked example: Calculating probability

A student did a survey of students with straight and curved thumbs in his class. 15 students had curved thumbs, and 10 students had straight thumbs. Calculate the probability that a student has curved thumbs.

Solution

Step 1: Calculate the total number of students sampled.

number of students sampled = 15 + 10 = 25

Step 2: Calculate the percentage of curved thumbed students

Percentage of students with curved thumbs

$$= \frac{\text{number of students with curved thumbs} \times 100}{\text{total number of students}}$$

$$= \frac{15 \times 100}{25} = 60\%$$

The probability that a student in this class has curved thumbs is 60%.

Table 4 *Probabilities in genetic crosses*

Probability	Fraction	Percentage
4 out of 4	$\frac{4}{4} = 1$	100%
3 out of 4	$\frac{3}{4}$	75%
2 out of 4	$\frac{2}{4} = \frac{1}{2}$	50%
1 out of 4	$\frac{1}{4}$	25%
0 out of 4	$\frac{0}{4} = 0$	0%

Synoptic link

You can find examples of these types of calculations in Topics B12.4, B12.5 and B12.6.

2f Averages

When you collect data, it is sometimes useful to calculate an average. There are three ways you can calculate an average – the mean, the mode, and the median.

You saw how to calculate a mean earlier in this topic (section 2b).

How to calculate a median

When you put the values of a series in order from smallest to biggest the middle value is called the **median**.

Figure 6 *Whether your thumb is curved or straight is controlled by a single gene*

Worked example: Calculating a median (odd number of values)

The thicknesses of seven oak leaves are shown below.

    0.8 mm 1.2 mm 0.9 mm 0.9 mm 0.8 mm 1.2 mm 1.0 mm

Calculate the median thickness of the leaves.

Solution

Step 1: Place the values in order from smallest to largest:

    0.8 mm 0.8 mm 0.9 mm 0.9 mm 1.0 mm 1.2 mm 1.2 mm

Step 2: Select the middle value – this is the median value.

    median value = 0.9 mm

If you have an even number of values, you select the middle pair of values and calculate a mean. That then becomes your median value.

How to calculate a mode

The mode is the value that occurs most often in a series of results. If there are two values that are equally common, then the data is bimodal.

Figure 7 *What is the median height of these plants?*

Synoptic link

You can find examples of how biologists use these calculations in Topic B15.3.

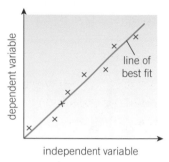

Figure 8 *A scatter graph*

Study tip

Use a transparent ruler to help you draw the line of best fit so you make sure that there are the same number of points on either side of the line.

Worked example: Calculating a mode

The masses of some invertebrates that fell into a pitfall trap are given below.

3.6g 4.2g 8.3g 6.5g 4.1g 4.2g 3.6g 4.2g 5.2g 3.2g 5.9g 3.2g

Calculate the modal mass of the invertebrates collected.

Solution

Step 1: Place the values in order from smallest to largest.

3.2g 3.2g 3.6g 3.6g 4.1g 4.2g 4.2g 4.2g 5.2g 5.9g 6.5g 8.3g

Step 2: Select the value which occurs the most often.

mode = 4.2g

2g Scatter diagrams and correlations

You may collect data and plot a scatter graph (Topic MS4). You can add a line to show the trend of the data, called a line of best fit. The line of best fit is a line that goes through as many points as possible and has the same number of points above and below it.

If the gradient of the line of best fit is:

positive it means as the independent variable *increases* the dependent variable *increases*

negative it means as the independent variable *increases* the dependent variable *decreases*

zero it means changing the independent variable has no effect on the dependent variable.

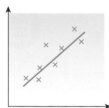 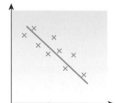 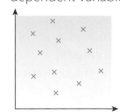

A relationship where there happens to be a link is called a correlational relationship, or **correlation**. You say that the relationship between the variables is positive, negative, or that there is no relationship.

For example:

- As you increase the light intensity, the rate of photosynthesis the increases – a positive correlation.

- As the number of predators in a habitat increases, the number of prey will decrease – a negative correlation.

- The colour of your hair has no effect on the likelihood that you will develop cancer – no relationship.

The presence of a relationship does not always mean that changing the independent variable *causes* the change in the dependent variable. In order to claim a causal relationship, you must use science to predict or explain *why* changing one variable affects the other.

Often there is a third factor that is common to both so it looks as if they are related. You could collect data for shark attacks and ice cream sales. A graph shows a positive correlation but shark attacks do not make people buy ice cream. Both are more likely to happen in the summer.

2h Estimates and order of magnitude

Being able to make a rough estimate is helpful. It can help you to check that a calculation is correct by knowing roughly what you expect the answer to be. A simple estimate is an order of magnitude estimate, which is an estimate to the nearest power of 10.

For example, to the nearest power of 10, you are probably 1 m tall and can run 10 m/s.

You, your desk, and your chair are all of the order of 1 m tall. The diameter of a molecule is of the order of 1×10^{-9} m, or 1 nanometre.

1 How many significant numbers are the following numbers quoted to?

a	13.0	[1 mark]	**b**	0.07	[1 mark]
c	560	[1 mark]	**d**	10.13	[1 mark]
e	0.999	[1 mark]	**f**	7×10^8	[1 mark]
g	2.605×10^{-3}	[1 mark]	**h**	1.01×10^{14}	[1 mark]

2 A student was testing how rapidly the carbohdrase enzyme amylase breaks down a solution of starch, using the disappearance of the blue-black colour of an iodine indicator as the end point. They repeated the experiment three times and obtained the following results:
1st test 180 s   2nd test 203 s   3rd test 175 s
Calculate the mean time taken for the starch to be digested, giving your answer to the appropriate number of significant figures.   [2 marks]

3 A group of students tried some of Mendel's breeding experiments with peas. After crossing two pea plants that both had round peas, they found that they collected 275 round peas and 89 wrinkled peas.
 a What percentage of the peas were wrinkled?   [2 marks]
 b If this cross is repeated, what is the probability that one of the offspring plants will produce round peas?   [1 mark]

4 Students surveyed the number of snails in the school greenhouse using a capture–recapture technique. They captured 25 snails and marked their shells. A week later they captured 30 snails, and four of them had marks on their shells. Estimate the snail population of the greenhouse.   [2 marks]

5 The heights of a class of students were recorded as follows:
1.80   1.55   1.49   1.76   1.64   1.50   1.52   1.90   1.52   1.76
1.58   1.58   1.70   1.76   1.64   1.80   1.64   1.70   1.50
Calculate the mean, mode, and median of this data.   [3 marks]

Worked example: Comparing orders of magnitude

If the size of a prokaryotic cell is of the order 1.2×10^{-6} m and eukaryotic cells are from around 1×10^{-5} m to 1×10^{-4}, estimate how the size of a prokaryotic cell compares to the size of:

a the smallest eukaryotic cell

b the largest eukaryotic cell

c to the nearest order of magnitude.

Solution

a Divide the size of the smallest eukaryotic cell by the size of the prokaryotic cells:

$1 \times 10^{-5} \div 1.2 \times 10^{-6} =$ approximately 10 (10^1)

So you can say that the prokaryotic cell is one order of magnitude (10^1) smaller than the smallest eukaryotic cells

b Divide the size of the smallest eukaryotic cell by the size of the prokaryotic cells:

$1 \times 10^{-4} \div 1.2 \times 10^{-6} =$ approximately 10^2

c So you can say that the prokaryotic cell is two orders of magnitude (10^2) smaller than the largest eukaryotic cells.

Synoptic link

You can practise making orders of magnitude calculations in Topic B1.3.

MS3 Algebra

Learning objectives

After this topic, you should know how to:

- understand and use the symbols: =, <, <<, >>, >, ∝, ~
- solve simple algebraic equations.

3a Mathematical symbols

You have used lots of different symbols in maths, such as +, −, ×, ÷. There are other symbols that you might meet in science. These are:

Symbol	Meaning	Example
=	equal to	$2\,cm^3/s \times 2\,s = 4\,cm^3$
<	is less than	The mean height of a child in a family < the mean of an adult in a family
<<	is very much less than	The diameter of a cell << the diameter of an apple
>>	is very much bigger than	The number of cells in an adult human being >> the number of cells in a two day embryo
>	is greater than	The number of producers in a food chain > the number of primary consumers
∝	is proportional to	rate of an enzyme controlled reaction ∝ the temperature
~	is approximately equal to	$272\,m \sim 300\,m$

3b Changing the subject of an equation

To solve an equation you often need to change the subject of the equation by rearranging the formula. Remember you must carry out the same operation on both sides of the equation (see Maths skill 3d).

3d Solving simple equations

Solving equations is an important maths skill. You will often have to use data from an experiment to calculate a variable, for example, if you know the length of an image of a cell, and the magnification of the image, you can calculate the actual length of the cell.

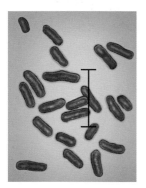

Figure 1 *Salmonella is a bacteria that is often a cause of food poisoning*

1 How would you read the following expressions as a sentence?
 a The enzymes of the stomach work best in pH < 7 [1 mark]
 b rate of reaction ∝ the concentration of reactant A [1 mark]
 c $22\,cm^3 \sim 24\,cm^3$ [1 mark]

2 A student looked at a plant cell under a light microscope. The length of the cell was 20.0 mm. The magnification was ×100. What is the actual length of the plant cell? [1 mark]

Worked example: Solving equations

Figure 1 shows a light microscope of *salmonella* bacteria. The magnification is ×1000. In the image, one of the bacteria has a length of 15 mm. What is the actual size of the bacterium?

Solution

Step 1: Substitute the values you have into the equation.

$$\text{magnificat ion} = \frac{\text{size of image}}{\text{size of real object}}$$

$$1000 = \frac{15}{x}$$

Step 2: Calculate the missing value.

$$x = \frac{15}{1000} = 0.015\,mm$$

MS4 Data and graphs

During your GCSE course you will collect data in different types of experiment or investigation. The data will either be:

- from an experiment where you have changed *one* independent variable (or allowed time to change) and measured the effect on a dependent variable
- from an investigation where you have collected data about *two* independent variables to see if they are related.

4a Collecting data by changing a variable

In many investigations you change one variable (the independent variable) and measure the effect on another variable (the dependent variable). In a fair test, the other variables are kept constant.

For example, you can vary the concentration of carbon dioxide (independent variable) and measure the effect on the rate of photosynthesis (dependent variable).

A scatter diagram lets you show the relationship between two numerical values.

- The independent variable is plotted on the horizontal axis.
- The dependent variable is plotted on the vertical axis.

The line of best fit is a line that goes roughly through the middle of all of the points on the scatter graph. The line of best fit is drawn so that the points are evenly distributed on either side of the line.

4b Graphs and equations

If you are changing one variable and measuring another you are trying to find out about the relationship between them. A straight line graph tells you about the mathematical relationship between variables but there are other things that you can calculate from a graph.

Straight line graphs

The equation of a straight line is $y = mx + c$, where m is the gradient and c is the point on the *y*-axis where the graph intercepts, called the *y*-intercept.

Straight line graphs that go through the origin (0,0) are special. For these graphs, *y* is directly proportional to *x*, and $y = mx$. If two quantities are directly proportional, as one quantity increases, the other quantity increases by the same proportion.

In science, plotting a graph usually means plotting the points then drawing a line of best fit.

When you describe the relationship between two *physical* quantities, you should think about the reason why the graph might (or might not) go through (0,0).

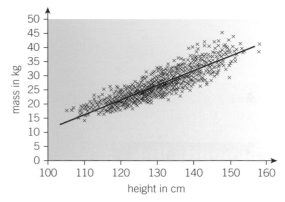

Figure 1 *A scatter graph with line of best fit showing the relationship between height and body mass*

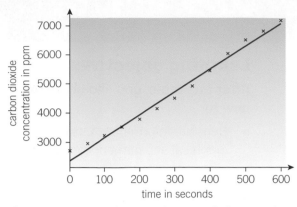

Figure 2 *A graph showing the rate of change of the volume of carbon dioxide produced by yeast during fermentation can tell you the reaction rate*

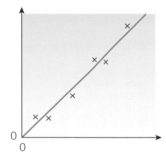

Figure 3 *A line of best fit that passes through the origin*

Synoptic link

You can learn more about using graphs to measure the rate of biological reactions in Topic B3.5.

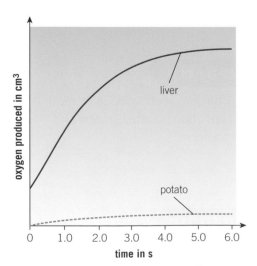

Figure 4 *Rate of reaction of catalase*

For example, if you are measuring the volume of a gas produced in a reaction over time, when the time = 0 s (at the start of the reaction), the volume of gas produced at that point will be obviously be 0 cm³.

However, if you are measuring the mass of reactants over time, in a reaction that gives off a gas, at time = 0 s, the *y*-intercept will not be zero but the starting mass of the reactants.

4c Plotting data

When you draw a graph you choose a scale for each axis.

● The scale on the *x*-axis should be *the same* all the way along the *x*-axis but it can be *different* to the scale on the *y*-axis.

● Similarly, the scale on the *y*-axis should be *the same* all the way along the *y*-axis but it can be *different* to the scale on the *x*-axis.

● Each axis should have a label and a unit, such as time in s.

4d Determining the gradient of a graph

The gradient of a straight line is calculated using the equation:

$$\text{gradient} = \frac{\text{change in } y}{\text{change in } x}$$

For all graphs where the quantity on the *x*-axis is time, the gradient will tell you the rate of change of the quantity on the *y*-axis with time.

When you are studying rates of reaction you might need to calculate a gradient from a graph of either:

● the amount of reactant as it decreases with time

● the amount of product as it increases with time.

1 Sketch a line graph, labelling the *x* and *y* axes, that shows:
   a A positive, constant gradient that passes through the origin (0,0).
   [1 mark]

   b A negative gradient – the gradient decreases as *x* increases.
   [1 mark]

2 a Write the general equation that describes a straight line on a graph, using the letters *y*, *x*, *m* and *c*.
   [1 mark]
   b State what letters *m* and *c* represent on the straight line graph.
   [1 mark]

   c Write the general equation that describes a straight line on a graph that passes through the origin (0,0).
   [1 mark]

3 Calculate the rate of the reaction of the enzyme catalase with hydrogen peroxide to produce oxygen by measuring the gradient of Figure 4 five seconds after the reaction starts in:
   a liver
   [1 mark]
   b potato.
   [1 mark]

MS5 Geometry and trigonometry

5c Shapes and structures

An important part of biology is visualising surface area to volume ratios in living organisms, and using simple models to explain the need for adaptations in exchange surfaces. It is helpful to be able to use 3D models of geometric shapes to demonstrate how these relationships work.

For biologists, the concept of surface area to volume ratio (SA:V) is very important when explaining, for example, factors that affect the rate of diffusion, the adaptations of exchange surfaces, and the adaptations of organisms for conserving water, warming up, and cooling down.

Surface area

You should remember the formulae for the area of rectangles and triangles.

- area of a rectangle = length × width (this also works for a square)
- area of a triangle = $\frac{1}{2}$ × base × height (this works for any triangle)

You can estimate the surface area of irregular shapes by counting squares on graph paper. This method is useful for estimating the area of a leaf, for example, if you trace the leaf onto graph paper.

The surface area of a 3D object is equal to the total surface area of all its faces. In a cuboid, the areas of any two opposite faces are equal. This allows you to calculate the surface area of the cuboid without having to draw a net.

> **Worked example: Calculating the surface area of cell**
> A plant cell in the shape of a cuboid, measuring 15 µm × 20 µm × 80 µm. Calculate its surface area.
>
> **Solution**
> **Step 1:** Calculate the area of each face.
>
> area of face 1 = 15 µm × 20 µm = 300 µm²
>
> area of face 2 = 15 µm × 80 µm = 1200 µm²
>
> area of face 3 = 20 µm× 80 µm = 1600 µm²
>
> **Step 2:** Calculate the total area of the three different faces.
>
> area = 300 µm² + 1200 µm² + 1600 µm² = 3100 µm²
>
> **Step 3:** Multiply the answer to **Step 2** by 2 because the opposite sides of a cuboid have equal areas.
>
> total surface area = 2 × 3100 µm² = 6200 µm²

After this topic, you should know how to:

- calculate areas of rectangles, and surface areas and volumes of cubes.

Synoptic link

Examples of the use of surface area to volume ratio can be found in in Chapter B1.

Figure 1 *You can estimate the surface area of irregular shapes using graph paper*

You need three steps to calculate the surface area of a cylinder:

Step 1: Calculate the total area of the two circular ends using the equation

$$\text{area} = 2 \times \pi \times \text{radius}^2$$

Step 2: Calculate the area of the curved surface using the equation

$$\text{area} = 2 \times \pi \times \text{radius} \times \text{length}$$

Step 3: Add the answers to Steps 1 and 2 together.

Volumes

Use this expression to calculate the volume of a cuboid:

$$\text{volume of cuboid} = \text{length} \times \text{width} \times \text{height}$$

You can calculate the volume in different units depending on the units of length, width, and height.

Worked example: Calculating volume

Calculate the volume of a piece of wood of length 15 cm, width of 6 cm, and depth of 1.5 cm, expressing your answer in cm^3 and m^3.

Solution

Step 1: Calculate the volume using the equation.

$$\text{volume} = \text{length} \times \text{width} \times \text{height}$$
$$= 15\,\text{cm} \times 6\,\text{cm} \times 1.5\,\text{cm}$$
$$= 135\,\text{cm}^3$$
$$= 140\,\text{cm}^3 \ (2\ \text{s.f.})$$

Step 2: Convert the measurements to metres.

$$\text{length} = 0.15\,\text{m}$$
$$\text{width} = 0.06\,\text{m}$$
$$\text{height} = 0.015\,\text{m}$$

Step 3: Use the equation to calculate the volume.

$$\text{volume} = \text{length} \times \text{width} \times \text{height}$$
$$= 0.15\,\text{m} \times 0.06\,\text{m} \times 0.015\,\text{m}$$
$$= 0.000\,135\,\text{m}^3$$
$$= 0.000\,14\,\text{m}^3 \ (2\ \text{s.f.})$$

1 You are given two cuboids, one with sides of 1 cm and the other with sides of 2 cm. Calculate the surface areas and volumes of these cuboids and compare their surface area : volume ratios. [1 mark]

Working Scientifically

WS1 Development of scientific thinking

Science works for us all day, every day. Working as a scientist you will have knowledge of the world around you, particularly about the subject you are working with. You will observe the world around you. An enquiring mind will then lead you to start asking questions about what you have observed.

Science usually moves forward by slow steady steps. Each small step is important in its own way. It builds on the body of knowledge that we already have. In this book you can find out about:

- how scientific methods and theories change over time (Topics B1.1, B5.3, and Chapter B17)

- the models that help us to understand theories (Topics B1.6, B1.7, B1.9, B5.3, B7.1, B7.2, B17.5, and Chapter B14)

- the limitations of science, and the personal, social, economic, ethical and environmental issues that arise (Topics B2.3, B2.4, B5.1, B6.3, B13.4, B13.5, B14.4, and Chapter B7)

- the importance of peer review in publishing scientific results (Topic B6.4, and Chapters B15 and B18)

- evaluating risks in practical work and in technological applications (Topics B6.1, B15.3, and Chapter 17).

The rest of this section will help you to 'work scientifically' when planning, carrying out, analysing and evaluating your own investigations.

Figure 1 *All around you, everyday, there are many observations you can make. Studying science can give you the understanding to explain and make predictions about some of what you observe*

WS2 Experimental skills and strategies

Deciding on what to measure

Variables are quantities that change or can be changed. It helps to know about the following two types of variable when investigating many scientific questions:

A *categoric* variable is one that is best described by a label, usually a word. For example, the fur colour of a mouse used in an experiment is a categoric variable.

A *continuous* variable is one that you measure, so its value could be any number. For example, temperature, as measured by a thermometer or temperature sensor, is a continuous variable. Continuous variables have values (called quantities). These are found by taking measurements, and S.I. units such as grams (g), metres (m), and joules (J) should be used.

Making your data repeatable and reproducible

When you are designing an investigation you must make sure that you, and others, can trust the data you plan to collect. You should ensure that each measurement is repeatable. You can do this by getting consistent sets of repeat measurements and taking their mean. You can also have more confidence in your data if similar results are obtained by different investigators using different equipment, making your measurements reproducible.

You must also make sure you are measuring the actual thing you want to measure. If you don't, your data can't be used to answer your original question. This seems very obvious, but it is not always easy to set up. You need to make sure that you have controlled as many other variables as you can. Then no-one can say that your investigation, and hence the data you collect and any conclusions drawn from the data, is not **valid**.

How might an independent variable be linked to a dependent variable?

- The *independent* variable is the one you choose to vary in your investigation.
- The *dependent* variable is used to judge the effect of varying the independent variable.

These variables may be linked together. If there is a pattern to be seen (e.g., as one thing gets bigger the other also gets bigger), it may be that:

- changing one has caused the other to change
- the two are related (there is a correlation between them), but one is not necessarily the cause of the other.

Starting an investigation

Scientists use observations to ask questions. You can only ask useful questions if you know something about the observed event. You will not have all of the answers, but you will know enough to start asking the correct questions.

When you are designing an investigation you have to observe carefully which variables are likely to have an effect.

An investigation starts with a question and is followed by a prediction, and backed up by scientific reasoning. This forms a hypothesis that can be tested against the results of your investigation. You, as the scientist, predict that there is a relationship between two variables.

You should think about carrying out a preliminary investigation to find the most suitable range and interval for the independent variable.

Making your investigation safe

Remember that when you design your investigation, you must:

- look for any potential hazards
- decide how you will reduce any risk.

You will need to write these down in your plan:

- write down your plan
- make a risk assessment
- make a prediction and hypothesis
- draw a blank table ready for the results.

Figure 2 *Safety precautions should be appropriate for the risk. Biuret reagent is corrosive but you do not need to wear a full face mask. Instead, chemical and splash-proof eye protection should be worn, and care should be taken.*

Different types of investigation

A **fair test** is one in which only the independent variable affects the dependent variable. All other variables are controlled and kept constant.

This is easy to set up in the laboratory, as long as no organisms are involved. Fair tests involving living things or fieldwork are almost impossible. Investigations in the environment and using living organisms involve complex variables that are changing constantly.

So how can we set up the fieldwork investigations? The best you can do is to make sure that all of the many variables change in much the same way, except for the one you are investigating. For example, if you are monitoring the effects of pollution on plants, they should all be experiencing the same weather, together – even if it is constantly changing.

If you are investigating two variables in a large population then you will need to do a survey. Again, it is impossible to control all of the variables. For example, imagine scientists investigating the effect of a new drug on diabetes. They would have to choose people of the same age and same family history to test. Remember that the larger the sample size tested, the more valid the results will be.

Control groups are used in these investigations to try to make sure that you are measuring the variable that you intend to measure. When investigating the effects of a new drug, the control group will be given a placebo. The control group think they are taking a drug but the placebo does not contain the drug. This way you can control the variable of 'thinking that the drug is working', and separate out the effect of the actual drug.

Designing an investigation
Accuracy

Your investigation must provide accurate data. Accurate data is essential if your results are going to have any meaning.

How do you know if you have accurate data?

It is very difficult to be certain. **Accurate results are very close to the true value.** However, it is not always possible to know what the true value is.

Sometimes you can calculate a theoretical value and check it against the experimental evidence. Close agreement between these two values could indicate accurate data.

You can draw a graph of your results and see how close each result is to the line of best fit.

Try repeating your measurements and check the spread or range within sets of repeat data. Large differences in a repeated measurement suggest inaccuracy. Or try again with a different measuring instrument and see if you get the same readings.

Precision

Your investigation must provide data with sufficient precision (i.e., **close agreement within sets of repeat measurements**). If it doesn't then you will not be able to make a valid conclusion.

Figure 3 *Imagine you wanted to investigate the effect pollution from a chemical factory has on nearby plants. You should choose a control group that is far away enough from the chemical plant to not be affected by the pollution, but close enough to be still experiencing similar environmental conditions*

Study tip

Trial runs will tell you a lot about how your investigation might work out. They should get you to ask yourself:

- do I have the correct conditions?
- have I chosen a sensible range?
- have I got sufficient readings that are close enough together? The minimum number of points to draw a line graph is generally taken as five.
- will I need to repeat my readings?

Study tip

Just because your results show precision it does not mean your results are accurate.

Imagine you carry out an investigation into the energy value of a type of food. You get readings of the amount of energy transferred from the burning food to the surroundings that are all about the same. This means that your data will have precision, but it doesn't mean that they are necessarily accurate.

Figure 4 *The green line shows the true value and the pink lines show the readings two different groups of students measured. Precise results are not necessarily accurate results*

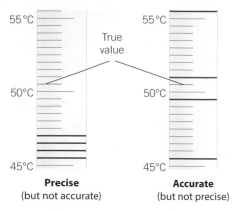

Figure 5 *Despite the fact that a stopwatch has a high resolution, it is not always the most appropriate instrument to use for measuring time*

Precision versus accuracy

Imagine measuring the temperature after a set time when a fuel is used to heat a fixed volume of water. Two students repeated this experiment, four times each. Their results are marked on the thermometer scales in Figure 4:

● A precise set of results is grouped closely together.

● An accurate set of results will have a mean (average) close to the true value.

How do you get precise, repeatable data?

● You have to repeat your tests as often as necessary to improve repeatability.

● You have to repeat your tests in exactly the same way each time.

● You should use measuring instruments that have the appropriate scale divisions needed for a particular investigation. Smaller scale divisions have better resolution.

Making measurements
Using measuring instruments

There will always be some degree of uncertainty in any measurements made (WS3). You cannot expect perfect results. When you choose an instrument you need to know that it will give you the accuracy that you want (i.e., it will give you a true reading). You also need to know how to use an instrument properly.

Some instruments have smaller scale divisions than others. Instruments that measure the same thing, such as mass, can have different resolutions. The resolution of an instrument refers to the smallest change in a value that can be detected (e.g., a ruler with centimetre increments compared to a ruler with millimetre increments). Choosing an instrument with an inappropriate resolution can cause you to miss important data or make silly conclusions.

But selecting measuring instruments with high resolution might not be appropriate in some cases where the degree of uncertainty in a measurement is high, for example, measuring reaction time using hand squeezes. (see Topic B10.2). In this case, a stopwatch measuring to one hundredths of a second is not going to improve the accuracy of the data collected.

WS3 Analysis and evaluation

Errors

Even when an instrument is used correctly, the results can still show differences. Results will differ because of a **random error**. This can be a result of poor measurements being made. It could also be due to not carrying out the method consistently in each test. Random errors are minimised by taking the mean of precise repeat readings, looking out for any outliers (measurements that differ significantly from the others within a set of repeats) to check again, or omit from calculations of the mean.

The error may be a systematic error. This means that the method or measurement was carried out consistently incorrectly so that an error was being repeated. An example could be a balance that is not set at zero correctly. Systematic errors will be consistently above, or below, the accurate value.

Presenting data
Tables
Tables are really good for recording your results quickly and clearly as you are carrying out an investigation. You should design your table before you start your investigation.

The range of the data
Pick out the maximum and the minimum values and you have the **range**. You should always quote these two numbers when asked for a range. For example, the range is between the lowest value in a data set, and the highest value. **Don't forget to include the units**.

The mean of the data
Add up all of the measurements and divide by how many there are. As seen in Topic WS2, you can ignore outliers in a set of repeat readings when calculating the mean, if found to be the result of poor measurement.

Bar charts
If you have a categoric independent variable and a continuous dependent variable then you should use a **bar chart**.

Line graphs
If you have a continuous independent and a continuous dependent variable then use a **line graph**.

Scatter graphs
These are used in much the same way as a line graph, but you might not expect to be able to draw such a clear line of best fit. For example, to find out if the size of mussels is related to their distance from the low tide line you might draw a scatter graph of your results.

Using data to draw conclusions
Identifying patterns and relationships
Now you have a bar chart or a line graph of your results you can begin looking for patterns. You must have an open mind at this point.

Firstly, there could still be some anomalous results. You might not have picked these out earlier. How do you spot an anomaly? It must be a significant distance away from the pattern, not just within normal variation.

A line of best fit will help to identify any anomalies at this stage. Ask yourself – 'do the anomalies represent something important or were they just a mistake?'

Secondly, remember a line of best fit can be a straight line or it can be a curve – you have to decide from your results.

The line of best fit will also lead you into thinking what the relationship is between your two variables. You need to consider whether the points you have plotted show a linear relationship. If so, you can draw a straight

Figure 6 *How you record your results will depend upon the type of measurements you are taking*

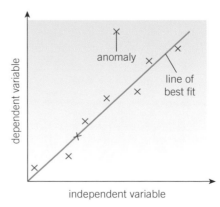

Figure 7 *A line of best fit can help to identify anomalies*

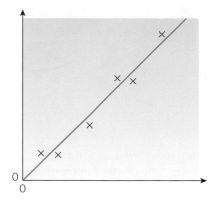

Figure 8 *When a straight line of best fit goes through the origin (0, 0) the relationship between the variables is directly proportional*

line of best fit on your graph (with as many points above the line as below it, producing a 'mean' line). Then consider if this line has a positive or negative gradient.

A **directly proportional** relationship is shown by a positive straight line that goes through the origin (0, 0).

Your results might also show a curved line of best fit. These can be predictable, complex or very complex. Carrying out more tests with a smaller interval near the area where a line changes its gradient will help reduce the error in drawing the line (in this case a curve) of best fit.

Drawing conclusions

Your graphs are designed to show the relationship between your two chosen variables. You need to consider what that relationship means for your conclusion. You must also take into account the repeatability and the reproducibility of the data you are considering.

You will continue to have an open mind about your conclusion.

You will have made a prediction. This could be supported by your results, it might not be supported, or it could be partly supported. It might suggest some other hypothesis to you.

You must be willing to think carefully about your results. Remember it is quite rare for a set of results to completely support a prediction or be completely repeatable.

Look for possible links between variables, remembering that a positive relationship does not always mean a causal link between the two variables.

Your conclusion must go no further than the evidence that you have. Any patterns you spot are only strictly valid in the range of values you tested. Further tests are needed to check whether the pattern continues beyond this range.

The purpose of the prediction was to test a hypothesis. The hypothesis can:

- be supported
- be refuted
- lead to another hypothesis.

You have to decide which it is on the evidence available.

Making estimates of uncertainty

You can use the range of a set of repeat measurements about their mean to estimate the degree of uncertainty in the data collected.

For example, in a test that looked at the effect of concentration on the rate of reaction between the enzyme catalase and hydrogen peroxide solution, a student got these results:

A given volume and concentration of hydrogen peroxide gave off 20 cm^3 of oxygen in 45 s (1st attempt), 49 s (2nd attempt), 44 s (3rd attempt) and 48 s (4th attempt).

The mean result = (45 + 49 + 44 + 48) ÷ 4 = 46.5 s

The range of the repeats is 44 s to 49 s = 5 s

So, a reasonable estimate of the uncertainty in the mean value would be half of the range.

In this case, we could say the time taken was 46.5 s plus or minus ± 2.5 s.

You can include a final column in your table of results to record the 'estimated uncertainty' in your mean measurements.

The level of uncertainty can also be shown when plotting your results on a graph (Figure 9).

As well as this, there will be some uncertainty associated with readings from any measuring instrument. You can usually take this as:

● half the smallest scale division. For example, 0.5 cm on a centimetre ruler, or

● on a digital instrument, half the last figure shown on its display. For example, on a balance reading to 0.01 g the uncertainty would be ± 0.005 g.

Anomalous results

Anomalies (or outliers) are results that are clearly out of line compared with others. They are not those that are due to the natural variation that you get from any measurement. Anomalous results should be looked at carefully. There might be a very interesting reason why they are so different.

If anomalies can be identified while you are doing an investigation, then it is best to repeat that part of the investigation. If you find that an anomaly is due to poor measurement, then it should be ignored.

Evaluation

If you are still uncertain about a conclusion, it might be down to the repeatability, reproducibility and uncertainty in your measurements. You could check reproducibility by: looking for other similar work on the Internet or from others in your class, or getting somebody else, using different equipment, to redo your investigation (this occurs in peer review of data presented in articles in scientific journals). You could also try an alternative method to see if it results in you reaching the same conclusion.

When suggesting improvement that could be made in your investigation, always give your reasoning.

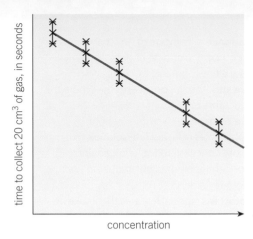

Figure 9 *Indicating levels of uncertainty. These are all the results of a group that chose five different concentrations to test and repeated each test three times.*

Glossary

abundance a measure of how common or rare a particular type of organism is in a given environment

active site the site on an enzyme where the reactants bind

active transport the movement of substances from a dilute solution to a more concentrated solution against a concentration gradient, requiring energy from respiration

adaptations special features that make an organism particularly well suited to the environment where it lives

ADH anti-diuretic hormone helps control the water balance of the body and affects the amount of urine produced by the kidney

adrenaline hormone that prepares the body for flight or fight

adult stem cells stem cells that are found in adults that can differentiate and form a limited number of cells

aerobic respiration an exothermic reaction in which glucose is broken down using oxygen to produce carbon dioxide and water and release energy for the cells

algae simple aquatic organisms (protista) that make their own food by photosynthesis

alleles different forms of the same gene sometimes referred to as variants

alveoli tiny air sacs in the lungs that increase the surface area for gaseous exchange

amino acids molecules made up of carbon, hydrogen, oxygen, and nitrogen that are the building blocks of proteins

amylase enzyme that speeds up the digestion of starch into sugars

anaerobic respiration an exothermic reaction in which glucose is broken down in the absence of oxygen to produce lactic acid in animals and ethanol and carbon dioxide in plants and yeast. A small amount of energy is transferred for the cells

aorta the artery that leaves the heart from the left ventricle and carries oxygenated blood to the body

aphids insects that penetrate the plant phloem and feed on the dissolved food. They act as plant pathogens and are also vectors that carry pathogenic viruses, bacteria, and fungi into healthy plant tissue

archaea one of the three domains, containing primitive forms of bacteria that can live in many of the extreme environments of the world.

arteries blood vessels that carry blood away from the heart. They usually carry oxygenated blood and have a pulse

asexual reproduction involves only one individual and the offspring is identical to the parent. There is no fusion of gametes or mixing of genetic information

atria the upper chambers of the heart

bacteria single-celled prokaryotic organisms

benign tumours growths of abnormal cells that are contained in one area, usually within a membrane, and do not invade other tissues

bile neutralises stomach acid to give a high pH for the enzymes from the pancreas and small intestine to work well. It is not an enzyme

biodiversity a measure of the variety of all the different species of organisms on earth

binary fission reproduction by simple cell division, for example in bacteria

biomass the amount of biological material in an organism

cancer the common name for a malignant tumour, formed as a result of changes in cells that lead to uncontrolled growth and division

capillaries the smallest blood vessels. They run between individual cells and have a wall that is only one cell thick

carbohydrases enzymes that speed up the breakdown of carbohydrates into simple sugars

carbohydrates molecules that contain only carbon, hydrogen, and oxygen. They provide the energy for the metabolism and are found in foods such as rice, potatoes, and bread

carbon cycle the cycling of carbon through the living and non-living world

carcinogens agents that cause cancer or significantly increase the risk of developing cancer

catalyst a substance that speeds up the rate of another reaction but is not used up or changed itself

causal mechanism something that explains how one factor influences another

cell cycle the three-stage process of cell division in a body cell that involves mitosis and results in the formation of two identical daughter cells.

cell membrane the membrane around the contents of a cell that controls what moves in and out of the cell

cell wall the rigid structure around plant and algal cells. It is made of cellulose and strengthens the cell

cellulose the complex carbohydrate that makes up plant and algal cell walls and gives them strength.

central nervous system (CNS) the part of the nervous system where information is processed. It is made up of the brain and spinal cord

chlorophyll the green pigment contained in the chloroplasts

chloroplasts the organelles in which photosynthesis takes place

clinical trials test potential new drugs on healthy and patient volunteers

classification the organisation of living organisms into groups according to their similarities

cloning the production of identical offspring by asexual reproduction

communicable disease disease caused by pathogens that can be passed from one organism to another

community group of interdependent living organisms in an ecosystem

competition the process by which living organisms compete with each other for limited resources such as food, light, or reproductive partners

coordination centres areas that receive and process information from receptors

coronary arteries the blood vessels that supply oxygenated blood to the heart muscle

correlation an apparent link or relationship between two factors

cystic fibrosis an inherited disorder that affects the lungs, digestive, and reproductive system and is inherited through a recessive allele

cytoplasm the water-based gel in which the organelles of all living cells are suspended and most of the chemical reactions of life take place

decomposers microorganisms that break down waste products and dead bodies

denatured the breakdown of the molecular structure of a protein so it no longer functions

differentiate the process where cells become specialised for a particular function

diffusion the spreading out of the particles of any substance in a solution, or particles in a gas, resulting in a net movement of particles from an area of higher concentration to an area of lower concentration down a concentration gradient

digestive system organ system where food is digested and absorbed

distribution where particular types of organisms are found within an environment

dominant allele the phenotype will be apparent in the offspring even if only one of the alleles is inherited

double circulatory system the circulation of blood from the heart to the lungs is separate from the circulation of blood from the heart to the rest of the body

effectors areas (usually muscles or glands) that bring about responses in the body

efficacy a measure of how effective a drug is

embryonic stem cells stem cells from an early embryo that can differentiate to form the specialised cells of the body

endocrine system the glands that produce the hormones that control many aspects of the development and metabolism of the body, and the hormones they produce.

endothermic reaction a reaction that requires a transfer of energy from the environment

enzymes biological catalysts, usually proteins

epidermal the name given to cells that make up the epidermis or outer layer of an organism

eukaryotic cells cells from eukaryotes that have a cell membrane, cytoplasm, and genetic material enclosed in a nucleus

evolutionary trees models used to explain the evolutionary links between groups of organisms

exothermic reaction a reaction that transfers energy to the environment

extinction the permanent loss of all members of a species from an area or from the world

extremophile an organism that can survive and reproduce in extreme conditions

fatty acids part of the structure of a lipid molecule

follicle stimulating hormone (FSH) causes the eggs to mature in the ovary

genetic engineering the process by which scientists can manipulate and change the genotype of an organism

genotype the genetic makeup of an individual for a particular characteristic, for example hair or eye colour

glucagon hormone involved in the control of blood sugar levels

glucose a simple sugar

glycerol part of the structure of a lipid molecule

glycogen carbohydrate store in animals

guard cells surround the stomata in the leaves of plants and control their opening and closing

haemoglobin the red pigment that carries oxygen around the body in the red blood cells

homeostasis the regulation of the internal conditions of a cell or organism to maintain optimum conditions for function, in response to internal and external changes

homozygote individual with two identical alleles for a characteristic

hormones chemicals produced in one area of the body of an organism that have an effect on the functioning of another area of the body. In animals hormones are produced in glands.

hypertonic (osmosis) a solution that is more concentrated than the cell contents

hypotonic (osmosis) a solution that is less concentrated than the cell contents

ionising radiation has enough energy to cause ionisation in the materials it passes through, which in turn can make them biologically active and may result in mutation and cancer

inoculate introducing microorganisms to a culture medium, or introducing modified microorganisms into an individual to protect them against disease

insulin hormone involved in the control of blood sugar levels

interdependence the network of relationships between different organisms within a community, for example each species depends on other species for food, shelter, pollination, seed dispersal, etc.

isotonic (osmosis) a solution that is the same concentration as the cell contents

lactic acid the end product of anaerobic respiration in animal cells

limiting factors limit the rate of a reaction, for example photosynthesis

lipase enzymes that speed up the breakdown of lipids into fatty acids and glycerol

lipids include fats and oils and are found in foods such as butter, olive oil, and crisps. They are made of carbon, hydrogen and oxygen

malignant tumours invade neighbouring tissues and spread to different parts of the body in the blood where they form secondary tumours. They are also known as cancers

mean the arithmetical average of a series of numbers

median the middle value in a list of numbers

medulla region of the brain concerned with unconscious activities such as controlling the heart rate and breathing rate

meiosis two stage process of cell division that reduces the chromosome number of daughter cells. It is Involved in making gametes for sexual reproduction

metabolism the sum of all the reactions taking place in a cell or the body of an organism

mitochondria the site of aerobic cellular respiration in a cell

mitosis part of the cell cycle where one set of new chromosomes is pulled to each end of the cell forming two identical nuclei during cell division

mode the number which occurs most often in a set of data

motor neurones carry impulses from the central nervous system to the effector organs

mutation a change in the genetic material of an organism

natural selection the process by which evolution takes place. Organisms produce more offspring than the environment can support. Only those that are most suited to their environment will survive to breed and pass on their useful characteristics to their offspring.

nerve bundle of hundreds or even thousands of neurones

neurones basic cells of the nervous system that carry minute electrical impulses around the body

nitrates mineral ions needed by plants to form proteins

non-communicable diseases are not infectious and cannot be passed from one organism to another

nucleus organelle found in many living cells containing the genetic information surrounded by the nuclear membrane

oestrogen – female sex hormone that controls the development of secondary sexual characteristics in girls at puberty, and the build-up and maintenance of the uterus lining during the menstrual cycle

organ an aggregation (collection) of different tissues working together to carry out specific functions

organ system a group of organs that work together to carry out specific functions and and form organisms

osmosis the diffusion of water through a partially permeable membrane from a dilute solution (which has a high concentration of water) to a concentrated solution (with a low concentration of water|) down a concentration gradient

ova the female sex cells, eggs

ovaries female sex organs that produce eggs and sex hormones

ovulation the release of a mature egg (ovum) from the ovary

oxygen debt the extra oxygen that must be taken into the body after exercise has stopped to complete the aerobic respiration of lactic acid

palisade mesophyll the upper layer of the mesophyll tissue in plant leaves made up of closely packed cells that contain many chloroplasts for photosynthesis

partially permeable membrane a membrane that allows only certain substances to pass through

pathogens microorganisms that cause disease

Penicillium the mould from which the antibiotic penicillin is extracted

permanent vacuole space in the cytoplasm filled with cell sap

phenotype the physical appearance/ biochemistry of an individual for a particular characteristic

phloem the living transport tissue in plants that carries dissolved food (sugars) around the plant

photosynthesis the process by which plants make food using carbon dioxide, water, and light

pituitary gland endocrine 'master gland' found in the brain that secretes a number of different hormones into the blood in response to different conditions to control other endocrine glands in the body

placebo a medicine that does not contain the active drug being tested, used in clinical trials of new medicines

plasma the clear yellow-liquid part of the blood that carries dissolved substances and blood cells around the body

plasmolysis the state of plant cells when so much water is lost from the cell by osmosis that the vacuole and cytoplasm shrink and the cell membrane pulls away from the cell wall

platelets fragments of cells in the blood that play a vital role in the clotting mechanism of the blood

polydactyly a dominant inherited disorder that results in babies born with extra fingers and/or toes

pulmonary artery the large blood vessel that takes deoxygenated blood from the right ventricle of the heart to the lungs

pulmonary vein the large blood vessel that carries oxygenated blood from the lungs back to the left atrium of the heart

Punnett square diagram a way of modelling a genetic cross and predicting the outcome using probability

preclinical testing is carried out on a potential new medicine in a laboratory using cells, tissues, and live animals

primary consumer animals that eat producers

producers organisms such as plants and algae that can make food from raw materials such as carbon dioxide and water.

prokaryotic cells from prokaryotic organisms have a cytoplasm surrounded by a cell membrane, and a cell wall that does not contain cellulose. The genetic material is a DNA loop that is free in the cytoplasm and not enclosed by a nucleus. Sometimes there are one or more small rings of DNA called plasmids

proteases enzymes that speed up the breakdown of proteins into amino acids

proteins molecules that contain carbon, hydrogen, oxygen, and nitrogen and are made of long chains of amino acids. They are used for building the cells and tissues of the body and to form enzymes

quadrat a sample area used for measuring the abundance and distribution of organisms in the field

quantitative sampling records the numbers of organisms rather than just the type

range the maximum and minimum values for the independent or dependent variables – important in ensuring that any patterns are detected

recessive a phenotype that will only show up in the offspring if both of the alleles coding for that characteristic are inherited

receptors – cells that detect stimuli – changes in the internal or external environment.

red blood cells biconcave cells that contain the red pigment haemoglobin and carry oxygen around the body in the blood

reflexes rapid automatic responses of the nervous system that do not involve conscious thought

resolving power a measure of the ability to distinguish between two separate points that are very close together

ribosomes the site of protein synthesis in a cell

sample size the size of a sample in an investigation

secondary consumer animals that eat the primary consumers

selective breeding speeds up natural selection by selecting animals or plants for breeding that have a required characteristic

sensory neurone neurone that carries impulses from the sensory organs to the central nervous system

sex chromosomes carry the information that determines the sex of an individual

sexual reproduction involves the joining (fusion) of male and female gametes producing genetic variation in the offspring

sexually transmitted disease (STD) transmitted from an infected person to an uninfected person by unprotected sexual contact.

simple sugars – small carbohydrate units, for example glucose

species the smallest group of clearly identified organisms in Linnaeus's classification system, often described as a group of organisms that can breed together and produce fertile offspring

sperm the male sex cells or gametes that carry the genetic material from the male parent

spongy mesophyll the lower layer of mesophyll tissue in plant leaves that contains some chloroplasts and many large air spaces to give a big surface area for the exchange of gases

statins drugs used to lower blood cholesterol levels and improve the balance of HDLs to LDLs in the blood

stem cells undifferentiated cells with the potential to form a wide variety of different cell types

stent a metal mesh placed in a blocked or partially blocked artery. They are used to open up the blood vessel by the inflation of a tiny balloon

stimuli changes in the external or internal environment that can be detected by receptors

stomata openings in the leaves of plants, particularly on the underside and opened and closed by guard cells, allowing gases to enter and leave the leaf

testosterone the main male sex hormone that controls the male secondary sexual characteristics at puberty and the production of sperm

therapeutic cloning a process where an embryo is produced that is genetically identical to the patient so the cells can then be used in medical treatments

tissue a group of specialised cells with a similar structure and function

tissue culture a modern way of cloning plants that allows thousands of new plants to be created from one piece of plant tissue

toxicity a measure of how toxic (poisonous) a substance is

transect a measured line or area along which ecological measurements are made

translocation the movement of sugars from the leaves to the rest of the plant through the phloem

transpiration the loss of water vapour from the leaves of plants through the stomata when they are opened to allow gas exchange for photosynthesis. It involves evaporation from the surface of the cells and diffusion through the stomata

tumour a mass of abnormally growing cells that forms when the cells do not respond to the normal mechanisms that control growth and when control of the cell cycle is lost

turgor the pressure inside a plant cell exerted by the cell contents pressing on the cell wall

type 1 diabetes a disorder where the pancreas fails to produce sufficient insulin

type 2 diabetes a disorder where the body cells no longer respond to the insulin produced by the pancreas

urea the waste product formed by the breakdown of excess amino acids in the liver

vaccine dead or inactive pathogenic material used in vaccination to develop immunity to a disease in a healthy person

veins blood vessels that carry blood away from the heart. They usually carry deoxygenated blood and have valves to prevent the backflow of blood

vena cava the large vein that brings deoxygenated blood from the body into the heart

ventilated movement of air or water into and out of the gas exchange organ, for example lungs or gills

ventricles chambers of the heart that contract to force blood out of the heart

viruses pathogens that are much smaller than bacteria and can only reproduce inside the living cells of other organisms

white blood cells blood cells involved in the immune system of the body. They engulf pathogens and make antibodies and antitoxins

xylem the non-living transport tissue in plants that transports water from the roots to the leaves and shoots

zygote the single new cell formed by the fusion of gametes in sexual reproduction

Index